"The science is abundantly clear that glol[...] vironmental crisis of our age. However, tl[...] scientific management is yielding mostly [...] needed mitigation and adaptation action. [...] thinking, a 'bottom up' approach in which ac[...]ons are pragmatically achieved at the local level (adaptive governance). Successful outcomes from such local experiments will show what can work politically, and it is by 'scaling up' from these successes that policies will evolve that can respond globally to the very real threat to our habitable world posed by rapidly changing climate."

—VICTOR R. BAKER, Regents Professor, Department of Hydrology and Water Resources, The University of Arizona

"While science has made enormous contributions to understanding the nature of climate change, can it play a major role in finding and implementing solutions? Are we missing something important by focusing primarily on global-scale, science-driven solutions via international negotiations? This book provides some provocative suggestions on how to deal with the most complex threat to humanity in the 21st century."

—WILL STEFFEN, Executive Director, The Australian National University Climate Change Institute

"There is unfolding a global tragedy of our planetary commons: national leaders are risking the future of the earth for short-term benefits. Brunner and Lynch show how, by listening to the advice from scientists about the seriousness of climate warming risks and adopting adaptive governance, the global challenge we share can be turned into smaller, tractable problems with real solutions."

—A. HENDERSON-SELLERS, Former Executive Director of the UN's World Climate Research Programme and Australian Research Council Professorial Fellow in Climate Risk at Macquarie University

"As societies seek ways to avoid the unmanageable and manage the unavoidable from climate change, federal, state, and local governments need viable adaptation cases to inform their decisions but also effective policy framings so that 'bottom up' actions and 'top down' institutional support lead to adaptive solutions that can be communicated peer to peer."

—JOSH FOSTER, Climate Adaptation Manager, Urban Leaders Adaptation Initiative, Center for Clean Air Policy

Adaptive Governance and Climate Change

Ronald D. Brunner and Amanda H. Lynch

American Meteorological Society

Front cover photograph by Grace Redding.

Published by the American Meteorological Society
45 Beacon Street, Boston, Massachusetts 02108

For more AMS Books, see www.ametsoc.org/amsbookstore.
Order online, or call (617) 227-2426, extension 686.

Library of Congress Cataloging-in-Publication Data

Brunner, Ronald D.
  Adaptive governance and climate change / Ronald D. Brunner and Amanda H. Lynch.
    p. cm.
  Includes bibliographical references and index.
  ISBN 978-1-878220-97-4 (pbk.)
  1. Climatic changes—Political aspects. 2. Global warming—Political aspects. 3. Environmental policy—International cooperation. 4. Intergovernmental cooperation. I. Lynch, Amanda H. II. Title.
  QC903.B78 2009
  363.738'74—dc22
                    2009053569

Printed in the United States of America
at Capital Offset Company Inc.

Mixed Sources
Product group from well-managed forests, controlled sources and recycled wood or fiber
www.fsc.org   Cert no. BV-COC-080420
©1996 Forest Stewardship Council

CONTENTS

PREFACE

Major initiatives to mitigate global warming are pending as we write. One is the Waxman-Markey bill in Congress, officially called the American Clean Energy and Security Act. The goal is to reduce the 2005 level of greenhouse gas emissions in the United States by 17% by 2020, and 83% by 2050. The primary means is a national cap-and-trade system, which would create a market for trading a decreasing total of emission permits to be allocated free and auctioned by the federal government. Another is the Carbon Pollution Reduction Scheme, with similar goals and means, being debated in the Australian Parliament. A third is the United Nations Climate Conference scheduled for December 7–18, 2009, in Copenhagen, to clarify emissions reductions targets for industrialized nations and other issues that stand in the way of an ambitious successor to the Kyoto Protocol, which expires in 2012. The Conference is the 15th meeting of the Conference of the Parties to the U.N. Framework Convention on Climate Change and the 5th meeting of the Parties to the Kyoto Protocol.

Like others concerned about climate change, we hope such initiatives will succeed quickly in reducing the danger of increasing concentrations of greenhouse gases in the atmosphere. At the same time, we fear that these initiatives may be stymied by a continuing lack of political will. During the last two decades, a lack of political will in addressing the danger of climate change has been manifest in the substitution of scientific assessments and technology R & D for actions that deploy existing knowledge and technology; in the official proclamation of long-term goals that defer the costs of realizing them to future office holders and their constituents; in the negotiation of emissions-reduction targets and timetables lacking sanctions severe enough to be enforceable; in limited capabilities to measure compliance and otherwise enforce those regulations that do have some teeth; and in the neglect of appraisals to terminate policies that have not worked and to improve those that have. Progress in mitigating dangerous concentrations of greenhouse gases in the atmosphere has been disappointing so far, especially given the magnitude of the task ahead. Meanwhile, there is growing concern that we are running out of time; irreversible changes in climate may be imminent according to some scientists.

In this context it would be prudent to complement such major initiatives with adaptive governance. This approach to reducing net losses from climate change depends on factoring the global problem into thousands of local problems, each of which is more tractable scientifically and politically than the global one and somewhat different. From intensive case studies by ourselves and others it is clear that some local communities have developed and implemented innovative policies that effectively reduce their vulnerability to climate change or their greenhouse gas emissions while accommodating other community interests. Indeed, other community interests often have been enough to justify decisions and actions that mitigate or adapt to climate change; these are "no regrets" policies that make sense regardless of climate change projections. Through networks these communities have made the more successful adaptation and mitigation policies available for voluntary adaptation by local communities elsewhere, and have set aside policies that failed. They have also clarified shared needs for external resources from those central authorities who are interested in supporting what has worked on the ground.

In short, adaptive governance is an opportunity for field-testing in series and in parallel thousand of policies for adapting to those climate changes we cannot avoid, and for mitigating those we can. If our hopes for more effective national and international agreements on climate policy are realized, adaptive governance practiced on a broader scale can expedite and improve their implementation. If our fears turn out to be warranted, adaptive governance nevertheless can build on existing resources, including pockets of political will, to field-test policies and to diffuse and adapt on a broader scale those policies that succeed in reducing net losses from climate change. Under either contingency, there is more to be gained at the margin by encouraging adaptive governance than by relying *exclusively* on business as usual. This book provides an introduction to adaptive governance, as an opportunity for those concerned about climate change to expand the range of individual choices and collective decisions under consideration. Major initiatives like those pending have dominated attention for more than two decades, but they have not produced progress commensurate with the magnitude of the task ahead.

Collaboration on this book began at the University of Colorado at Boulder more than a decade ago, when Amanda Lynch and her colleagues invited Ron Brunner to participate in an assessment of climate change on the North Slope coast of Alaska. The invitation was prompted by Brunner's lecture on a paper in progress, "Science and the Climate Change Regime." It was accepted on assurances by Lynch and her colleagues that helping the people of the North Slope adapt to climate change was the project's top priority; contributing to the scientific literature was an important but secondary goal. The project proposal was informed by an exploratory grant and funded as "An Integrated Assessment of the Impacts of Climate Variability on the Alaskan North Slope Coastal Region" by the National Science Foundation. Our field research began in August 2000 in Barrow, the center of the North Slope Borough government, and concluded there in February 2008. During the field research it gradually became apparent that what we were learning about climate change adaptation, and reporting periodically to people in Barrow, could be significant to scientists and policy makers well beyond the North Slope.

Consequently, we began exploring beyond the project's boundaries to clarify the larger significance of the project's findings. In other case studies, in other climate-related literature, and in general theoretical literature, we found many corroborating observations and insights. But typically they were scattered, addressed exclusively to other scientists or to national and international policy makers, and marginalized by the mainstream quest to reduce greenhouse gas emissions through mandatory targets and timetables, legally binding on all nations with significant emissions. It seemed appropriate to pull together in a book what we learned on the ground in Barrow and in the literature. Early in 2007 we began to focus on writing this book to clarify adaptive governance and draw attention to it; the manuscript was substantially completed two years later. We conclude that adaptive governance is an emerging pattern of science, policy, and decision making, and so far a missed opportunity for reducing net losses from climate change on larger scales at all levels in the international system, from local to global.

Nonetheless, we expect adaptive governance to be controversial in some quarters (but not in others) because it is a departure from dominant ways of thinking about climate change. Moreover, specific steps toward adaptive governance are sometimes in conflict, sometimes complementary, with business as usual in various policy arenas. Any issues that arise are best engaged, we believe, through the comparative evaluation of specific action proposals—from business as usual, from adaptive governance, and from other promising approaches—in the light of explicit criteria and evidence that is detailed and comprehensive within practical constraints. This is nothing more than an affirmation of good practice in science, policy, and decision making.

ACKNOWLEDGMENTS

We would like to express our gratitude to the other principal investigators on the Barrow project: Jim Maslanik, who took over as lead investigator when Lynch returned to Australia; Judy Curry, who provided the initial inspiration as well as crucial ideas early in the design of the project; James Syvitski, who contributed to our understanding of coastal processes in the region; and Linda Mearns, an important connection to current thinking in regional climate change who also proved a doughty collaborator when faced with two hungry polar bears. Our hopes for the Barrow project would not have been realized without the contributions of Liz Cassano and Leanne Lestak in particular, for their research, their support, and their good humour. Other research contributors and project supporters based in the lower 48 states or in Australia were Ruth Doherty, Sheldon Drobot, Cinda Gillilan, Klaus Görgen, Melinda Koslow, Bill Manley, Christina McAlpin, Jon Pardikes, Scott Peckham, Matt Pocernich, Matthew Rothstein, Cove Sturtevant, Claudia Tebaldi, Casey Thornbrugh, Petteri Uotila, and Jason Vogel. Eric Parrish helped with graphics after conclusion of the project.

Of course the Barrow project would not have been possible without the openness, good will, and cooperation of the people of Barrow. In Barrow, project personnel included archeologists and long-time Barrow residents Anne Jensen, Senior Scientist at the Ukpeaġvik Iñupiat Corporation Real

Estate Science Division, and Glenn Sheehan, Executive Director of the Barrow Arctic Science Consortium (BASC). They guided us through the culture and the community, including its relevant past, and reviewed an earlier draft of Chapter 3 focused on Barrow and the North Slope. We received invaluable logistics support from Glenn and others at BASC, including Lewis Brower, Alice Brower Drake, Linda Fischel, Henry Gueco, Matt Irinaga, Lollie Hopson, Scott Oyagak, and Dave Ramey. We received important observations and insights on various occasions from Barrow residents Mike Aamodt, Tom Brower III, Rob Elkins, Dan Enders, Richard Glenn, Bob Harcharek, Edward Itta, Curt Thomas, Kenneth Toovak, and Chastity Olemaun.

Other residents of Barrow also contributed through meetings, interviews, and documentation, including Sheldon Adams, Bart Ahosagak, Johnny Aiken, Tom Albert, Dave Anderson, Summer Arnhart, Leslie Boen, Arnold Brower Sr., Eugene Brower, Fred Brower, Price Brower, Ronald Brower Sr., Frank Brown, Paul Bush, Marie Adams Carroll, Mike Donovan, George Edwardsen, Terri Ekalook, Earl Finkler, Ben Frantz, Craig George, John Gleason, Arlene Glenn, Allison Graves, George Leavitt, Doreen Lampe, Dave Logan, Andrew McLean, Ben Nageak, Charlie Neakok, Lars Nelson, Allen Nesteby, Joe Nicely, George Olemaun, Marvin Olson, David Ongley, Jerry Painter, Crawford Patkotak, Pat Patterson, Donovan Price, Delbert Rexford, Tim Russell, Dorothy Savok, Joe Stankowitz, Joe Stepak, Gina Sturm, Jim Vorderstrasse, Mary Ann Warden, and Laurie Wing. Additional insights on Barrow came from former residents or frequent visitors, including Karen Brewster, Forrest Brooks, Steve Chronic, Hajo Eichen, Owen Mason, Dave Norton, Grace Redding, Paul Schuitt, Dale Stotts, Jesse Walker, Charles Wohlforth, and Bernie Zak. Luci Eningowuk from Shishmaref, Earl Kignak from Point Hope, and Tom Brower III of Barrow reviewed initial ideas for organizing a network of Native villages vulnerable to coastal erosion and flooding in the Arctic.

Michael Ledbetter at the National Science Foundation gave us tremendous encouragement in getting the project on Barrow and the North Slope off the ground, first by providing an exploratory grant and then guiding us through the funding of a 5-year project (NSF OPP-0100120) as part of the Human Dimensions of the Arctic component of the Arctic Systems Science program. Neil Swanberg of NSF provided crucial continuing support. Amanda Lynch also received substantial support from the Australian Research Council (FF0348550). Rachelle Hollander supported Ron Brunner's earlier project on capitalizing the policy sciences for climate change research (NSF SBR-95-12026). Any opinions, findings, and conclusions or recommendations are those of the authors and do not necessarily reflect the views of the Australian Research Council or the National Science Foundation.

But this is a book on adaptive governance and climate change. As such it goes well beyond Barrow and the North Slope, the focus of just one of five chapters. For helpful comments or discussions on other parts of the manuscript as it evolved, we are especially grateful to Cheryl Anderson, Kate Auty, Josh Foster, Joe Friday, Mike Hamnett, Søren Hermansen, Susan Iott, Eileen Shea, John Thwaites, Will Steffen, Henk Tennekes, and Jason Vogel. Of course our gratitude does not imply that they agree with any of our arguments. We are also grateful for feedback from participants at various workshops, meetings, and conferences where either or both of us presented parts of the manuscript in progress. These included the workshop of WAS*IS (Weather & Society*Integrated Studies) at Mt. Macedon, Australia, January 2007; the International Workshop on Dialogue Between Social and Natural Sciences sponsored by a consortium of Japanese universities, IR3S (Integrated Research System for Sustainability Science), in Honolulu, February 2007; the workshop on Arctic Coastal Zones at Risk sponsored by four scientific organizations (LOICZ, AMAP, IASC, and IHDP) in Tromsø, Norway, October 2007; a symposium on extreme events held by Emergency Management Australia in Melbourne, November 2007; the 87th and 88th General Meetings of the American Meteorological Society in San Antonio, January 2007, and New Orleans, January 2008, respectively; the annual Science at the Shine Dome meeting of the Australian Academy of Sciences in Canberra, May 2008; the 27th Annual Institute of the Society of Policy Scientists and a lecture at the National Center for Atmospheric Research, both in Boulder, CO, in October 2008; and the Climate Change Congress in Copenhagen, February 2009. Sarah Jane Shangraw at the American Meteorological Society shepherded the manuscript supportively and efficiently though the review and publication process.

Finally, we would both like to express a deep and abiding gratitude to our spouses, for all that they have done in support of us and this book.

ACRONYMS

AAAS	American Association for the Advancement of Science
ACIA	Arctic Climate Impact Assessment
AEC	Atomic Energy Commission
AEWC	Alaska Eskimo Whaling Commission
AFN	Alaska Federation of Natives
AGGG	Advisory Group on Greenhouse Gases
APEC	Asia-Pacific Economic Cooperation
ASNA	Arctic Slope Native Association
ASRC	Arctic Slope Regional Corporation
AUD	Australian dollars
BIA	Bureau of Indian Affairs
BTS	Barrow Technical Services
CARB	California Air Resources Board
CCAP	Climate Change Action Plan
CCP	Cities for Climate Protection
CCSP	U.S. Climate Change Science Program
CCTP	U.S. Climate Change Technology Program
CEI	Competitive Enterprise Institute
CES	Committee on Earth Sciences
CIPM	Capital Improvements Project Management, NSB

CND	Canadian dollars
CMDL	Climate Monitoring and Diagnostics Laboratory
CO_2	carbon dioxide
DEW	Distant Early Warning
ENSO	El Niño–Southern Oscillation
EA	Environmental Assessment
EIS	Environmental Impact Statement
EPA	U.S. Environmental Protection Agency
ERDA	Energy Research and Development Administration
ESSP	Earth Systems Science Partnership
ETS	Emissions Trading System
EU	European Union
EUA	EU Allowance
FAR	First Assessment Report of the IPCC
FEMA	U.S. Federal Emergency Management Administration
GAO	U.S. Government Accountability Office (formerly General Accounting Office)
GHG	greenhouse gases
ICAS	Iñupiat Community of the Arctic Slope
ICC	Inuit Circumpolar Conference
ICLEI	International Council for Local Environmental Initiatives
ICSU	International Council of Scientific Unions
IGBP	International Geosphere–Biosphere Programme
IHDP	International Human Dimensions Program
IPCC	Intergovernmental Panel on Climate Change
IWC	International Whaling Commission
NAP	National Allocation Plan
NAPA	National Adaptation Programmes of Action
NARL	Naval Arctic Research Laboratory
NAST	National Assessment Synthesis Team
NFIP	National Flood Insurance Program
NOAA	National Oceanic and Atmospheric Administration
NSB	North Slope Borough
NWS	U.S. National Weather Service (formerly Weather Bureau)
OGP	Office of Global Programs
OSTP	Office of Science and Technology Policy in the White House
OTA	Office of Technology Assessment, formerly a research arm of the U.S. Congress
PEAC	Pacific ENSO Applications Center

RGGI   Regional Greenhouse Gas Initiative
RISA   Regional Integrated Sciences and Assessments program
RSWG   Response Strategies Working Group
SAPs   Synthesis and Applied Products
SRES   Special Report on Emission Scenarios
UIC   Ukpeaġvik Iñupiat Corporation
UNEP   United Nations Environment Programme
UNESCO   United Nations Educational, Scientific, and Cultural
            Organization
UNFCCC   United Nations Framework Convention on Climate Change
USD   U.S. dollars
USGCRP   U.S. Global Change Research Program
WCRP   World Climate Research Program
WERI   Weather and Energy Research Institute, University of Guam
WMO   World Meteorological Organization

BOXES AND FIGURES

"Knowing is not one thing that we do among many others, but a quality of any of our doings. . . . To say that we know is to say that we do rightly in the light of our purposes, and that our purposes are thereby fulfilled is not merely a sign of our knowledge, but its very substance."

Abraham Kaplan, *The Conduct of Inquiry*

CLARIFYING THE PROBLEM

In *An Inconvenient Truth*, an award-winning documentary film, Al Gore used startling graphs and dramatic photos to summarize the scientific consensus on global warming and its past and projected impacts on planet Earth and its inhabitants. Along with human population growth and more powerful technologies, he cited "our way of thinking" as a third major factor that has transformed humanity's relationship to the earth. Supporting our way of thinking are certain misconceptions the former vice president of the United States attributed to special interests. For example, he reported a leaked internal memo that advised lobbyists and public relations specialists for a group of companies including ExxonMobil to "reposition global warming as theory, rather than fact." And they succeeded to the extent that 53% of the 623 news stories in a random sample from influential newspapers did raise doubts about global warming. But in the end, Gore's message about meeting this planetary emergency, as he called it, was upbeat in view of the array of technologies available to curb global warming and its impacts: "We already know everything we need to know to effectively address this problem. We've got to do a lot of things, not just one." But he concluded with an important qualification: "We have everything we need, save perhaps *political will*."[1]

The film emphasized the more extreme climate changes and adverse impacts projected within the mainstream scientific consensus.[2] But the balance

1

of evidence supports enough of the inconvenient truth dramatized in the film to underscore the need for action to mitigate global warming and other climate changes and to adapt to the changes we cannot prevent. The evidence also supports the former vice president's emphasis on the influential role of political opponents in propagating misconceptions that inhibit if not block individual and collective action. But those of us who are dissatisfied with the outcomes of climate change mitigation and adaptation efforts to date, or who are concerned about the magnitude of the task ahead, might go beyond blaming political opponents to reconsider our own roles. Have we done all that we can to reduce past losses from climate change and future vulnerabilities? Will persistence in business as usual be sufficient to contain losses from climate change? In response to such questions a bit of introspection suggests another inconvenient truth: We need to open the established climate change regime to additional approaches to science, policy, and decision making.

An Appraisal

The United Nations Framework Convention on Climate Change (UNFCCC) was opened for signature at the Earth Summit in Rio de Janeiro in June 1992. The climate change regime was formally established in March 1994 when the 50th national government ratified the convention. Since then, at least 189 national governments have ratified it and joined the Conference of the Parties to the UNFCCC. The ultimate objective of the convention, as stated in Article 2, is "stabilization of greenhouse gas concentrations in the atmosphere at a level that would prevent dangerous anthropogenic interference in the climate system."[3] As a first step toward meeting the ultimate objective, the UNFCCC itself included Article 4(2), a nonbinding commitment of 36 industrialized countries specified in Annex I to the convention to reduce their emissions of greenhouse gases to 1990 levels by the year 2000. In December 1997, the Conference of the Parties to the convention negotiated a successor policy, the Kyoto Protocol. The protocol went into effect in February 2005, after ratification by the Russian Federation met the prescribed threshold, ratification by 55 parties to the convention, including Annex I countries accounting for at least 55% of the total carbon dioxide (CO_2) emissions of Annex I countries in 1990. Each of the parties to the protocol formally committed itself not to exceed a specified amount of emissions of six greenhouse gases, calculated as CO_2-equivalent emissions and averaged over the five-year period 2008–2012. In the aggregate, the parties overall would reduce their emissions by at least

5% below 1990 levels using whatever joint and national means they deemed appropriate.[4] Australia ratified the protocol late in 2007, leaving the United States as the only Annex I country *not* party to it. As early as 2004, attention turned to negotiating another emissions-reductions policy to follow termination of the Kyoto Protocol in 2012.[5]

The significance of these and other policies in light of the ultimate objective depends on the standards applied. To be sure, some progress has been made in reducing greenhouse gas emissions by the counter-factual standard—the level of emissions that would have occurred without the UNFCCC—but this standard is relatively difficult to estimate as a gauge of progress. A more widely used standard is the level of greenhouse gas emissions that occurred in 1990 when measurements became available for most industrialized countries and climate change was still a relatively new issue. Using this standard, the UNFCCC reported in 2009 that the country with the most emissions, the United States, *increased* its emissions of greenhouse gases by 14.0% from 1990 to 2006, the lastest year for which data are available.[6] The United States and 25 other industrialized countries *increased* their aggregate emissions by 9.1% from 1990 to 2006. However, when 14 industrialized countries in transition to market economies are folded in, aggregate emissions *decreased* by 4.5% during the same period. This reflects primarily the depth of economic decline in the former Soviet bloc after the end of the Cold War, not policies implemented under UNFCCC auspices. These percentage increases and decreases in emissions vary with statistical adjustments in the 1990 baseline and to a lesser extent in annual emissions reported since then.[7] Meanwhile, developing countries not included in the Kyoto Protocol or Annex I to the UNFCCC have significantly added to global greenhouse gas emissions, but their reporting capabilities are not sufficient to gauge aggregate emissions trends.[8] However, in November 2006 the International Energy Agency in Paris projected that China, with soaring coal consumption fueling rapid economic development, would surpass the United States in CO_2 emissions in 2009.[9] In June 2008 the Netherlands Environmental Assessment Agency reported that China's CO_2 emissions had in fact exceeded the United States's by 7% in 2006 and 14% in 2007.[10]

These outcomes are especially disappointing in light of the magnitude of the task implied by the ultimate objective of the UNFCCC. What stabilization level would prevent "dangerous" anthropogenic interference in the climate system? That depends on judgments of facts and values. European Union sources recommend limiting average global warming to no more than 2°C above our best estimate of the preindustrial "equilibrium" temperature.

The atmospheric greenhouse gas concentration that would allow this outcome is not well understood. Current scientific analyses confirm that *cumulative* emissions determine the likelihood of keeping warming below 2°C. Current estimates suggest that a stabilization level of no more than 450 parts per million in the concentration of CO_2-equivalent greenhouse gases in the atmosphere could achieve an even chance, at best, of constraining warming to this level. Even this level would require emissions to peak earlier than 2020, and then decline to 50% of 1990 emissions by 2050. This implies that the world will use less than half the current economically recoverable fossil fuel reserves prior to 2050.[11] Different assumptions lead to different projections, of course. But to put the magnitude of the task ahead in broader historical context, global greenhouse gas emissions are estimated to have increased 70% from 1970 to 2004, and they are expected to continue to increase *under current policies and practices*.[12] It is worth emphasizing that all policy-relevant climate change research is justified on the premise that current policies and practices are subject to change. Meanwhile, global emissions appear to be growing at a faster rate than any of the emissions scenarios considered in the Fourth Assessment Report of the Intergovernmental Panel on Climate Change (IPCC) in 2007.[13] And global mean temperature and sea level since 1990 are rising at a rate in the upper range or above that projected in the IPCC scenarios.[14]

Already, concentrations of greenhouse gases in the atmosphere are sufficient to force significant changes on natural and human systems.[15] And perhaps losses from climate change have already arrived through extreme weather events. Munich Re, a global reinsurance company, estimated that the number of great natural catastrophes worldwide increased nearly threefold from the 1960s to the 1990s, while economic losses from these same catastrophes increased ninefold during the same period.[16] Using more recent data from Munich Re, the Earth Policy Institute at Columbia University reports that economic losses from major hurricanes worldwide increased from $24 billion in the 1980s to $113 billion in the 1990s and $272 billion in this decade through 2005.[17] Included in the estimate for this decade are losses from Hurricane Katrina, which struck New Orleans and the Gulf Coast of the United States in late August 2005. From another source we calculate that nearly 12,000 lives were lost and over 153 million lives were impacted by wind storms, wildfires, floods, temperature extremes, and droughts worldwide in 2005 alone.[18] However, it should be emphasized that many other factors interact with extreme weather events to cause losses. These include the location of people, property, and other things of value with respect to extreme weather, as well as response capabilities before, during, and after extreme weather. Moreover, while changes in the frequency of extreme weather events have been observed, scientists still debate the physical and statistical linkages between the frequency and intensity of extreme weather events and global climate change.[19]

Underlying these disappointing outcomes and possibilities for improving upon them are two patterns of governance. One is scientific management, the foundation of the climate change regime established by the UNFCCC and a major reason why progress under the regime has been so modest. In *Seeing Like a State*, James Scott traced parts of this pattern back to the origins of modern states in Europe, when monarchs standardized measures and practices to extract more revenues, conscripts, and control more efficiently from diverse subject communities. In a more pungent critique, John Ralston Saul underscored the rise of *Voltaire's Bastards* who separated the Enlightenment emphasis on reason from common sense morality. Parallel to developments in Europe, Frederick Winslow Taylor formally introduced American engineers to principles of scientific management in 1895. From such origins, "Scientific Management has worked its way into the fabric of all modern industrial societies, where it is now so common as to go unnoticed by most people."[20] In the United States not long after the turn of the twentieth century, "Scientific Management aspired to rise above politics, relying on science as the foundation for efficient policies made through a single central authority—a bureaucratic structure with the appropriate mandate, jurisdiction, and expert personnel."[21] But during the last century it became increasingly clear, in the United States at least, that effective control is dispersed among multiple authorities and interest groups; that efficiency is only one of many interests to be reconciled in policy decision processes; and that science on important issues is politically contested. Amid such twenty-first-century realities, scientific management typically leads to policy gridlock.

Adaptive governance emerged more or less spontaneously here and there at the local level as a loosely coordinated array of pragmatic responses to manifest failures of scientific management. Not surprisingly, recognition of the pattern in the last decade or two and the term itself followed innovations in practice. Adaptive governance is characteristically more responsive to differences and changes on the ground; often but not always it proceeds from the bottom up rather than the top down.[22] From intensive case studies it is clear that some local communities working separately can integrate scientific and local knowledge into policies that advance their common interest on contested issues; in the process, politics and profound uncertainties are unavoidable. Similar communities working together can harvest their collective experience to make successful innovations anywhere in a network available for adaptation elsewhere on a voluntary basis. They can also clarify their common needs for national authorities, who typically control the resources necessary to move ahead, including knowledge and information, funding, and legal authority. This pattern of governance has been documented in natural resource policy in the American West, but it is *not* limited to that sector or region; it appears to be an adaptation to twenty-first-century realities more generally.[23] Adaptive governance suggests factoring the global climate change problem into thousands of local problems, each of which is more tractable scientifically and politically than the global problem. Insights from research on adaptive governance turned out to be constructive, if not essential in our project to assist one local community, Barrow, AK, in adapting to climate change. Since the project initiated field work in August 2000, we have found aspects of adaptive governance scattered through climate change and related literatures, corroborating our experience on the ground in Barrow.

This book integrates what we have learned about adaptive governance as a set of proposals for opening the established regime to accelerate progress on climate change problems. It is intended primarily for those scientists, environmentalists, administrators, policymakers, and other citizens of the world who are sufficiently dissatisfied with disappointing outcomes to date, or sufficiently concerned about the magnitude of the task ahead, to consider changes in business as usual. The next section provides a brief historical overview and comparison of scientific management and adaptive governance in climate change, deferring most of the substantive details to later chapters. Building on these constructs, the final section introduces normative considerations that also have a bearing on understanding disappointing outcomes to date and what might be done to improve them.

It should be understood at the outset that scientific management and adaptive governance are simplified constructs that frame inquiry on, and action in, a much more complex reality that lies beyond anyone's complete or completely objective understanding. But we can improve our understanding through inquiry and action even if no one is omniscient. Scientific management, the established frame, has for the most part restricted attention to only part of the relevant picture.[24] Adaptive governance is a means of directing attention to otherwise neglected parts that can help reduce our vulnerability to climate change.

Constructing the Context

The framing of climate change as a global problem recognizes that CO_2 and other greenhouse gases, regardless of their geographic origins, are dispersed more or less uniformly in the atmosphere through the global circulation. These gases absorb and reirradiate heat that would otherwise escape into space, warming the earth and making life as we know it possible. But increases in concentrations of greenhouse gases in the atmosphere force temperature increases and other climate changes. These in turn force changes on natural and human systems, mostly adverse changes because these systems evolved under different climate conditions. The basic physics of the greenhouse effect are neither new nor controversial. In 1896, building on the work of others, "the Swedish scientist Svante Arrhenius was the first to make a quantitative link between changes in CO_2 concentration and climate."[25] For a doubling of CO_2 concentrations in the atmosphere, he predicted a global temperature increase of 5° to 6°C.

Scientific interest in global warming increased in the late 1950s with direct measurements of atmospheric CO_2 at Mauna Loa in Hawaii, confirming the Revelle–Suess hypothesis that concentrations were on the rise (Fig. 1.1). In a famous framing of the hypothesis, Roger Revelle and Hans Suess wrote in 1957 that "human beings are now carrying out a large scale geophysical experiment of a kind that could not have happened in the past nor be reproduced in the future. Within a few centuries we are returning to the atmosphere and oceans the concentrated organic carbon stored in sedimentary rocks over hundreds of millions of years. This experiment, if adequately documented, may yield a far-reaching insight into the processes determining weather and climate."[26] Two years later, Swedish atmospheric scientists Bert Bolin and Erik Eriksson identified a bottleneck in the transfer of carbon dioxide from the atmosphere to oceans, and they suggested that the burning of fossil fuels would lead to an accumulation of carbon dioxide in the atmosphere. They linked the accumulation with the energy balance of the earth and global warming.[27]

Other scientific developments included improvements in climate records, recognition that gases in addition to CO_2 contribute to the greenhouse effect, and improvements in global circulation models of the atmosphere that increased confidence in predictions of global warming. In 1987, discovery of a depleted ozone layer in the atmosphere over the Antarctic (popularly known as "the ozone hole") underscored the unintended consequences of human activities for the atmosphere. Such developments added credibility

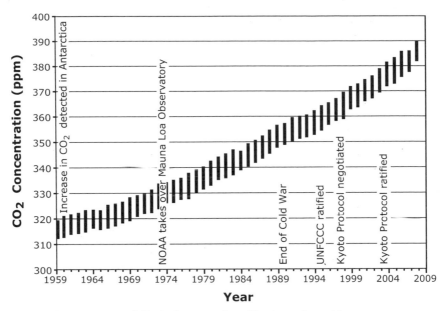

Fig. 1.1. Concentrations of CO₂ in the atmosphere. The range of monthly averages, 1959–2008. Source: Data from Mauna Loa Observatory, Hawaii.

to scientists' predictions of climate change and intensified demands for action to mitigate it. This led to establishment of the climate change regime. But the footprint of the regime in the Mauna Loa measurements (Fig. 1.1) is less evident than economic decline in the former Soviet bloc after the end of the Cold War.

Scientific Management

Scientists and their allies put climate change on the political agenda and took the lead in policy initiatives.[28] Daniel Bodansky reports that "a number of scientists and non-governmental organizations [NGOs] acted as entrepreneurs, promoting the climate change issue through conferences, reports and personal contacts" beginning in earnest early in the 1970s.[29] Among the most important initiatives was a conference held at Villach, Austria, in October 1985. It approved the establishment of a small group of independent scientists as the Advisory Group on Greenhouse Gases (AGGG) to advise the UN Environment Programme (UNEP), the World Meteorological Organization (WMO), and the International Council of Scientific Unions (ICSU). It was set up to ensure that "periodic assessments are undertaken of the state of scientific understanding and its practical applications and to initiate, if deemed necessary, consideration of a global convention, and to advise on further mechanisms and actions required at the national and international levels."[30] The AGGG "remained a major influence and organizing force behind the dissemination of the climate threat after 1985."[31] At the time of the Villach Conference, ICSU had just finalized its International Geosphere–Biosphere Programme (IGBP); an Ad Hoc Planning Group for this program of research met under ICSU auspices in Bern, Switzerland, in September 1986 and published Report No. 12 in 1990.[32] We summarize this major research plan in the next chapter.

Perhaps "the high-water mark of policy declarations on global warming" came from the World Conference on the Changing Atmosphere: Implications for Global Security, held in Toronto in June 1988.[33] The Toronto Conference Statement called for a 20% reduction in CO_2 emissions, a provocative demand that "caused considerable unease among governments and industry."[34] Also in June 1988, NASA scientist James Hansen testified before a U.S. Senate committee that he was 99% certain that global warming was under way.[35] Concurrently, an extreme heat wave and drought in eastern North America attracted attention to Hansen's testimony and the Toronto Conference Statement and amplified their influence. As climate scientist Stephen Schneider observed, "In 1988, nature did more for the notoriety of global warming in 15 weeks than any of us [scientists] or sympathetic journalists or politicians were able to do in the previous fifteen years."[36] Nature was assisted by Sen. Tim Wirth and colleagues who scheduled Hansen's testimony for the day with the maximum expected temperature for Washington in that period and who opened the windows of the hearing room the night before to make the atmosphere inside sweltering during the hearings.[37] With or without such political support, nature in this role penalizes inaction and sometimes forces action. This is an underutilized resource in compensating for a lack of political will—a critical limiting factor in Al Gore's appraisal, our appraisal, and others.[38]

Built on climate science foundations, such highly visible initiatives toward mitigation policy attracted the attention of officials and nonofficials who perceived their interests to be threatened and provoked political backlash. "By June 1987, the U.S. State Department had become unhappy about the AGGG as representing little more than 'free wheeling academics.' Governmental bodies therefore began to wrest the policy initiative from the AGGG network by replacing parts of it and extending others to include governmental research bodies, especially those close to WMO."[39] In 1988, the United States and allied governments requested that WMO and UNEP establish an organization which became the IPCC and effectively replaced the independent AGGG. "Through the WMO, governments gained the power to veto participants to the IPCC and influence its brief."[40] But "there was considerable bitterness in some quarters about the fate of the AGGG, which was disbanded under pressure from the U.S. State Department. . . ."[41] The Executive Council of the WMO, which had taken the lead, endorsed the IPCC and its brief in June 1988, just before the Toronto Conference and after many changes had been made.[42] The formal mandate of the IPCC was to "(i) assess available scientific information on climate change, (ii) assess the environmental and socio-economic impacts of climate change, and (iii) formulate response strategies."[43] The informal mandate was "in part to reassert governmental control and supervision over what was becoming an increasingly prominent political issue."[44] The IPCC's First Assessment Report was completed in 1990; we summarize the results of its Working Group III on response strategies in the next chapter. The report was updated in 1992 to provide the scientific foundation for negotiation of the UNFCCC by the International Negotiating Committee.[45]

Meanwhile, in the United States, threatened industries organized against taxes and other policies proposed to reduce greenhouse gas emissions. The Global Change Coalition, for example, spun off from the National Association of Manufacturers in 1989 to lobby on behalf of utility, oil, coal, and automobile interests. As established around that time, U.S. policy effectively substituted research for action to mitigate greenhouse gas emissions.[46] At an international conference in Noordwijk, the Netherlands, in 1989, the head of the U.S. Environmental Protection Agency (EPA), William Reilly, "spoke mainly of scientific uncertainties and declared that 'the United States is substantially increasing its budgets for scientific research into the causes and consequences of climate change.'"[47] The chief beneficiary was the U.S. Global Change Research Program (USGCRP), an initiative of the interagency Committee on Earth Sciences.[48] The committee's first major research strategy and research plan, both published in 1989, prompted substantial budget increases for global change research. In the next chapter we summarize the Fiscal Year 1990 Research Plan for the USGCRP. Participants in USGCRP and in the national research programs of other countries cooperated formally and informally through various scientific organizations, including the IGBP and IPCC.[49] In the division of labor, the IPCC assesses original research funded for the most part by national research programs, while the IGBP and the World Climate Research Program (WCRP), among others, organize research at the international level.

These science programs formalized an existing epistemic community focused on earth systems science and scientific assessments. According to Paul Edwards, an epistemic community is "a knowledge-based professional group which shares a set of beliefs about cause-and-effect relationships and a set of practices for testing and confirming them. Crucially, an epistemic community also shares a set of values and an interpretative framework; these guide the group in drawing policy conclusions from knowledge. Its ability to stake an authoritative claim to knowledge is what gives an epistemic community its power."[50] The epistemic community, participating most directly through the IPCC, became an integral part of the coalition supporting establishment of the climate change regime through the UNFCCC. But the coalition and the regime also included governments reluctant to do much more than fund research. Sonja Boehmer–Christiansen concluded that "[t]o protect its own interests, science had to respond to a new international context in which most governments needed time and therefore welcomed uncertainties. Governmental interest was in funding more research rather than enforcing changes in energy policy. Scientific institutions nationally and globally could not reject this offer. If forced to choose, the interests of science cannot but lie with research rather than policy change. Policy neutrality was therefore becoming increasingly attractive."[51] The outcome served the scientific interest in reducing uncertainty and political interests in deferring action.[52] But there is no necessary implication that scientists directly or indirectly involved in the emergence of the regime wittingly struck a deal. Indeed, working scientists removed from political arenas typically assume their work is objective and politically neutral. Other scientists closer to the action typically assume they can separate themselves and their work from politics.[53] But that assumption is mistaken when scientific research impinges on political interests, as indicated by the ongoing political backlash to science-based demands for emissions reductions.

The IPCC institutionalizes expert advice for national and international policymakers in the climate change regime. The scientific foundation of the regime, as stated in the IPPC's Second Assessment Report in 1995, is the expectation that climate change is an "irreducibly global problem." This implies that "effective protection of the climate system requires international cooperation in the context of wide variations in income levels, flexibility and expectations of the future. . . ."[54] International cooperation on behalf of the ultimate objective stated in Article 2 is manifest most directly in commitments by industrialized countries to reduce greenhouse gas emissions under Article 4(2), the Kyoto Protocol, and perhaps a successor that continues the quest begun more than two decades ago for mandatory targets and timetables to reduce emissions. The formal structure of decision making implies that the important decisions on climate change are made from the top down by national governments working together in the Conference of the Parties to the UNFCCC and in joint implementation schemes, and working separately to implement their respective policy commitments by national means. This global construction of climate change science, policy, and decision-making structures reflects the aspirations of scientific management. But the aspirations of scientific management in the political arenas of our time are seldom matched by achievements, and climate change is no exception.

Scientific assessments of dangerous anthropogenic interference in the climate system had little direct bearing on the targets and timetables for emissions reductions set in Article 4(2) or the Kyoto Protocol.[55] According to Jerry Mahlman, then director of the Geophysical Fluid Dynamics Laboratory at Princeton, "'it might take another 30 Kyotos over the next century' to cut global warming down to size."[56] The industrialized parties in Annex I made compliance with Article 4(2) policy commitments voluntary, withheld severe sanctions to effectively enforce compliance with policy commitments under the Kyoto Protocol, and continue to rely on self-reporting of compliance by the separate parties. Nevertheless, the industrial country parties still took more than seven years to ratify the Kyoto Protocol. The developing country parties made it clear in Article 4(7) of the UNFCCC that "economic and social development and poverty eradication are [their] first and overriding priorities" despite the ultimate objective in Article 2. A lack of political will is manifest in such reservations, the history of the regime's establishment, and disappointing outcomes to date.[57] The global framing of the problem leaves out of the decision-making structure the citizens of diverse local communities around the world whose support and cooperation are necessary for effective mandates from the top down. The IPCC's scientific assessments support the interests of national and international authorities exclusively, as if their interests override all others or are equivalent to them. We review the evolution of the established regime in more substantive detail in Chapter 2.

Toward Adaptive Governance

Although the climate change regime is still well established, the coalition of interests supporting its quest for mandatory targets and timetables to reduce greenhouse gas emissions has begun to break down in recent years.

Additional policy alternatives are gaining attention with growing concerns that emissions reductions have been miniscule and that time is running out, as James Hansen observed on the 20th anniversary of his June 1988 testimony.[58] Moreover, critiques of the regime's dominant way of thinking and doing, scientific management, began to emerge nearly two decades ago and have increased in the last decade. Here we single out three of the critiques that point toward opening the regime. They do not represent acceptance of the adaptive governance construct as a whole; they are independent insights that overlap with important parts of that construct.

An early challenge was provoked by a single-topic issue of *Scientific American* titled "Managing Planet Earth" published in September 1989. The introduction acknowledged that "[c]hanges in individual behavior are surely necessary but are not sufficient" responses to the problem of global environmental change. But it emphasized expansion of the issue: "It is as a global species that we are transforming the planet. It is only as a global species—pooling our knowledge, coordinating our actions and sharing what the planet has to offer—that we have any prospect for managing the planet's transformation along pathways of sustainable development. Self-conscious, intelligent management of the earth is one of the great challenges facing humanity as it approaches the 21st century."[59] In *Weather* in 1990, Director of Research for the Royal Netherlands Meteorological Institute Hendrik Tennekes responded with some passion to "Managing Planet Earth" and "Stabilize the Climate System," the name of a program launched by the U.S. EPA. "I am terrified by the hubris, the conceit, the arrogance implied by words like these," Tennekes wrote. "Who are we to claim that we can manage the planet? We can't even manage ourselves. Who are we to claim we can run the planetary ecosystem? In an ecosystem no one is boss, virtually by definition. Why are we, with our magnificent brains, so easily seduced by technocratic totalitarianism." Tennekes later explained that "technocratic totalitarianism" referred to a confluence of scientific and management pretenses that ignore the simple feedback suggested by another question, "But how does it play in Peoria?" (Here, the city of Peoria, Illinois, serves as a symbol of ordinary local communities worldwide, not just in America.) He identified and critiqued some of these pretenses in his 1990 article, including the linear logic in a sequence of rational arguments. In contrast, we are constrained by the planetary ecosystem to "sustained adjustment and permanent adaptation."[60]

In a commentary published in *Nature* in 1997, Steve Rayner and Elizabeth Malone invoked Zen Buddhism on behalf of "breaking through mental boundaries imposed by established ways of looking at the world." Noting disappointing outcomes in climate change mitigation and adaptation to date, they contended that "[t]he time is ripe for a fresh look at climate policy strategy." They challenged the "grip emissions reduction strategy has on policy" by drawing attention to pressing issues of human welfare. As a starting point they chose "assessing human vulnerability and social adaptation" over existing efforts to "predict the unpredictable": "The problem is that there is no way of knowing whether rapid social and technological change will prove a saving grace or another challenge to global environmental governance." The new starting point

> . . . may be more relevant to stakeholders as it allows a varied response to local conditions. For instance, a measure designed to protect a coastal community from sea-level rise may have nothing in common with measures to stem desertification. Adaptation is a bottom-up strategy that starts with changes and pressures experienced in people's daily lives. This contrasts with the top-down approach of national targets for emissions reductions. The connections between emissions targets and people's everyday behaviour and responsibilities seem less direct, even abstract. Designing adaptation strategies may be more sensitive to real trade-offs made by real people.

Rayner and Malone conceived adaptation strategies as a possible path to emissions reductions and urged policymakers to consider "whether there is anything to be learned from adaptation to assist the process of actual emissions reductions (as distinct from the formal process of agreeing national targets)." They offered five suggestions for policymakers, and later expanded them to ten in the concluding chapter of a four-volume assessment of social science research. In these suggestions we found many important aspects of adaptive governance as we understand it; we return to them in Chapter 4. It is worth noting that this social science assessment framed climate change science, policy, and decision making as a matter of human choice and welfare, not scientific predictions to reduce uncertainty.[61]

In a viewpoint published in *Global Environmental Change* in 2000, David Cash observed that nearly all international and regional treaties "require scientific assessment and monitoring to support decision-making. The institutional structure of these assessments has historically been top-down, centralized, and primarily focused on producing written reports." He cited the assessments of the IPCC as an example and acknowledged their benefits in understanding large-scale phenomena and informing international negotiations. However, "[i]n doing so, traditional centralized assessments have failed in assisting local decision-makers in taking actions to help prevent global environmental problems, or in implementing responses to adapt to local impacts of global change." Moreover, "[h]eterogeneity of local impacts and vulnerabilities, the interactions of multiple environmental stresses, and large geographic variance in costs and benefits highlight the potential pitfalls of centralized assessment systems which are poorly linked to decision-makers at multiple levels." More concretely, Cash emphasized that "[g]lobal mean temperature change, while perhaps spurring international action, is irrelevant to local emergency relief managers in Bangladesh or farmers in Nebraska." He recommended distributed information–decision support systems that characteristically provide "(1) multiple connections between researchers and decision-makers which cut across various levels (polycentric networks); and (2) sustained and adaptive organizations which allow for iterated interactions between scientists and decision-makers." Iterated interactions increase the relevancy and legitimacy of an assessment for decision makers. Polycentric networks encourage innovation and flexibility; through redundancy they also protect the overall system from the failure of any one of its parts. Cash cited as an example the Pacific ENSO Applications Center (PEAC).[62] We review PEAC's early work along with other exceptions to scientific management in Chapter 2.

Such critiques of the established global frame for climate change science, policy, and decision making were part of the milieu but not the primary impetus for our project in Barrow, AK. In collaboration with a number of colleagues, we sought primarily to help the people of Barrow adapt to climate change by drawing upon our diverse specializations; we also agreed at the outset that scientific publications were important but secondary outcomes of the project.[63] Barrow is a community of several thousand people, mostly Iñupiat Eskimos whose ancestors sustained themselves in that unforgiving environment for several thousand years.[64] It is located within the Arctic Circle on the North Slope coast of Alaska, at the northernmost point in the United States, and about 1,100 miles from the North Pole (Fig. 1.2). Climate change is not an issue there; the signs are obvious for all to see. Acceptance of it can be inferred from the headline of a commercial announcement circulated in Barrow in 2003: "Is Global Warming affecting your future?? You bet . . . if you're utilizing the barge services of Bowhead Transportation."[65] Thus Barrow is a microcosm of things to come as signs of climate change become more obvious at lower latitudes. Our fieldwork under a pilot project began in Barrow on August 20, 2000, with a week of conversations to explore the

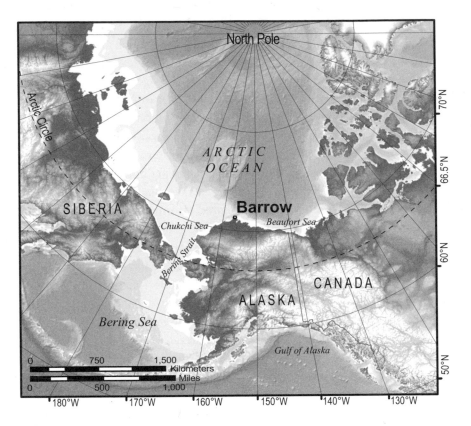

Fig. 1.2. Barrow in regional context. Source: Leanne Lestak and Eric Parrish.

climate-related problems experienced by a variety of Barrow residents. We concluded that the least tractable of these problems for the community as a whole was its vulnerability to coastal erosion and flooding from big storms— one of which had cost Barrow millions of dollars in damages earlier in the month.[66] Based on this initial finding, we secured support from community leaders for research on this problem as well as a five-year grant from the National Science Foundation.

As research proceeded in Barrow we continued to apply insights from concurrent research on cases of adaptive governance in the American West. We focused first on understanding the extreme weather events that had damaged the community beginning with the great storm of October 1963 and on what might be done to reduce the community's vulnerability in the future. We sought to serve the community's common interest by expanding the range of informed choices, leaving any policy decisions to be made by community members on the basis of their many interests including but not limited to adaptation to climate change. Both the shared policy purpose and the common focus of attention on extreme weather events were essential in pulling together the local knowledge of residents, who had experienced big storms and their impacts firsthand, and the expertise of colleagues from more than a dozen scientific and technical specializations. We reported our findings to community leaders and the general public in Barrow annually as a group and intermittently on separate field trips, through technical meetings, public lectures, classroom visits, and interviews one on one, in small groups, and on local public radio (KBRW). Concurrently we sought advice on further research from community members as contributors to what they dubbed "the big storms project." Since the official conclusion of the project early in 2006, and with the help of a no-cost extension, we have attempted to maintain contacts informally. Whether we satisfactorily served the community's common interest is a judgment best left to the people of Barrow.

However, we can report that our findings have been used in and for Barrow. The U.S. Army Corps of Engineers cited our photometric measurements of coastal erosion from 1948 to 2002 as reason for deleting the beach nourishment alternative from its Barrow Storm Damage Reduction Project in 2006. Our documentation of climate change in Barrow during the last half century, indicating profound uncertainties, may have influenced the Army Corps's decision to substitute a more robust scenario-based method for the customary cost-benefit analysis more dependent on projections. Our numerical-model simulations of historical extreme events addressed an empirical question of interest to subsistence hunters in Barrow: Do storms track the edge of the sea ice? Other simulations produced rules of thumb, to be field tested by local weather forecasters, to gauge the damage potential of approaching storms. Our reconstruction of the great storm of October 1963 was used in various distributed decisions: locating a site for the new hospital outside the area flooded in 1963; designing the new $62 million Barrow Arctic Research Center[67] on a pad and pilings high enough to prevent flooding by an equivalent storm; designing an emergency management exercise to prepare for another great storm; and perhaps reinforcing the need for an inland road to serve as an evacuation route when the coastal road is washed out. In addition, our policy history helped bring back to the agenda various nonstructural alternatives, especially relocation and planning and zoning, that were briefly considered and prematurely dropped in the planning and promotion of a beach nourishment program funded in 1992. Toward the end of the project we drew attention to critical vulnerabilities of the utility corridor (or "utilidor") that provides potable water and sewer services year-round. The utilidor, buried in permafrost, is justifiably a matter of pride in a community subject to winter temperatures as low as −50°C. In Chapter 3 we tell the story of Barrow's experience, including our integrated assessment of coastal erosion and flooding problems there.[68]

It should not be assumed that integrated assessments of climate change are actually used in policy decisions on the ground. A case in point is the "plain language synthesis of the key findings of the Arctic Climate Impact Assessment (ACIA), designed to make the scientific findings accessible to policy makers and the broader public." ACIA was commissioned by the Arctic Council, a "high-level intergovernmental forum that provides a mechanism to address the common concerns and challenges faced by arctic people and governments."[69] The synthesis was modeled on the IPCC's summaries for policymakers and published with excellent color graphics in November 2004. At the conclusion of mostly separate interviews in Barrow in late March 2006, one of us showed a copy of the ACIA synthesis to 10 people involved in ameliorating various parts of Barrow's coastal erosion and flooding problems, and asked if they knew of it. Only two were aware of the synthesis, and none had read it. In this and other instances, scientific excellence is no guarantee that an assessment of climate impacts will inform decisions on the ground. Conversely, a scientific assessment is not necessary for successful adaptations on the ground even though it can help.

Summary and Comparison

Box 1.1 distills what we have learned from our field research in Barrow and from various other cases and literature relevant to climate change science, policy, and decision making. Each column describes a hypothetical pure specimen of a distinguishable pattern of governance. It is properly used as a heuristic for detailed empirical inquiry into particular specimens, and not as a substitute for such inquiry or as a normative model. In short, these patterns are ideal types in the sense of Max Weber and the policy sciences: They are as "useful in calling attention to *deviations* from the type as in characterizing the few cases that exactly conform to it."[70] Further, only the ideal types are mutually exclusive. Actual cases of governance in diverse and dynamic societies are mixed, with aspects of emerging patterns complementing or competing with the established pattern. The established pattern of scientific management in climate change is clear enough to identify emerging exceptions to it. Throughout this book we use Box 1.1 as a guide to describe historical cases

Box 1.1. Ideal Types
Opening the Established Climate Change Regime

Established Regime
SCIENTIFIC MANAGEMENT

Proposed Openings
ADAPTIVE GOVERNANCE

Centralized Decision Making
Top-Down. Central authorities at the top of international and national hierarchies make the important policy decisions.
Bureaucracies. Policies are implemented uniformly and impersonally by subordinates accountable to central authorities.
Expertise. Disinterested experts develop technologies and integrated scientific assessments for central authorities.

Decentralized Decision Making
Bottom-Up. Needs shared in networks can help central authorities allocate resources to support what worked on the ground.
Networks. Case studies of local policies that worked can be diffused by networks for voluntary adaptation by other communities.
Experience. Local communities working in parallel can adapt and field test policies in their own contexts; diversity is an asset.

Technical Rationality
Planning. Policy process is discrete, relying on formal methods and metrics to evaluate planned alternatives and avoid failure.
Targets. Comprehensive policy depends on science-based technologies to realize a given target efficiently and above politics.
Linear. Unfettered basic research to reduce scientific uncertainty is a prerequisite for rational and cost-effective policy decisions.

Procedural Rationality
Appraisal. Policy processes are serial, relying on appraisals for terminating failed policies and building on successful ones.
Interests. Incremental policies integrate or balance interests in a community to advance its common interest; politics are necessary.
Cooperative. Scientists and policymakers work together toward overlapping practical aims, sharing differently informed insights.

Extensive Science
Generalized. Research generalizes across human or natural systems for results of broad national or international scope.
Predictive. Stable and standard parts are integrated into numerical models to derive falsifiable predictions to reduce uncertainty.
Reductive. Research selects from diverse systems separate parts relevant to a stable relationship or standard measure or method.

Intensive Science
Centered. Inquiry focuses on understanding and reducing losses from extreme events in single cases; context matters.
Integrative. Each factor is contingent on a working "model" of the whole case; gaps and inconsistencies in it prompt revisions.
Comprehensive. Inquiry strives to cover all the major interacting factors, human and natural, shaping outcomes in the single case.

and constructing their significance for climate change adaptation primarily and for climate change mitigation secondarily.

Various logical and empirical connections are implicit in each column. At the risk of oversimplification, consider a few connections for illustrative purposes. The first column lists characteristics of scientific management in the established frame. Centralized decision making is specialized to comprehensive policy decisions with major consequences of broad scope and complexity. The broad scope of each decision reinforces other tendencies to assert control from the top down. The major consequences attract the attention of many interest groups concerned about opportunity costs if nothing else, increasing the odds of indecision and delay. The major consequences also make the cost of policy failure prohibitive, justifying a reliance on experts to provide scientific validation and a requirement to get the policy "right" the first time through planning. Policy failures are difficult to acknowledge under these circumstances. This is technical rationality, an approach to policy closely approximated in geoengineering schemes, considered in Chapter 5, to deploy science-based technologies on a scale large enough to mitigate warming of the total earth system. In contrast, the second column lists proposals for opening the established frame to improve outcomes through adaptive governance. In this pattern, decentralized decision making is specialized to many concurrent policy decisions, each with relatively modest consequences of limited scope and complexity in a somewhat different context. Each decision tends to attract fewer interest groups, facilitates innovation and action on policies adapted to particular contexts, enables field testing of many policies in series and in parallel, and supports actions that terminate the failures and build on the successes. Adaptation to different contexts justifies intensive research for planning, which includes local knowledge, and for harvesting local experience to inform policy decisions elsewhere. This is procedural rationality, an approach to policy approximated in distributed policies and incremental learning by doing in the Barrow case in Chapter 3.

Comparison of the two columns row by row highlights not only complementarities between scientific management and adaptive governance but also differences that sometimes surface in politics. Consider for example the *Synthesis* of the IPCC's Second Assessment Report in 1995, which took the top-down perspective of scientific management, exemplified by *Seeing Like a State*, in the following passage: "Significant reductions in net greenhouse gas emissions are technically possible and can be economically feasible. . . . The degree to which [this] technical potential and cost-effectiveness are realized is dependent on initiatives to counter lack of information and overcome cultural, institutional, legal, financial and economic barriers which can hinder diffusion of technology or behavioral changes."[71] (Notice that "political" was omitted from the list of barriers; "counter" and "overcome" are about as far as the *Synthesis* went toward acknowledging the reality of political differences of any kind.) But from the perspectives of some people on the ground in Barrow, and no doubt in other communities, specific local examples of some items on the IPCC's generic list are not barriers to be overcome but values to be preserved, as important in themselves or instrumental to other community values. And a lack of information at higher levels about such realities on the ground is another barrier to realizing the potential for significant net reductions in greenhouse gas emissions. "Effective communication about climate change issues requires understanding of the frames of reference being used by all participants."[72] This understanding in turn depends on the simple feedback implied by Tennekes's question, "But how does in play in Peoria?"

The alternative to imposing the top-down perspective of the established regime (even if the Conference of the Parties had the necessary political power and will to use it) is opening the established frame to a larger picture of reality. This would finesse unresolved political difficulties diagnosed as early as 1989 by William Ruckelshaus: "The difficulty of converting scientific findings into political action is a function of the uncertainty of the science and the pain generated by the action. . . . It is hard for people—hard even for the groups of people who constitute governments—to change in response to dangers that may not arise for a long time or that just might not happen at all."[73] If this is so, then it would be pragmatic and politically expedient to reframe the climate change problem in part for people directly impacted by extreme weather events: The pain of loss from their perspectives is immediate and tangible, not remote or hypothetical. It also would be principled to bring people on the ground into the climate regime as active participants in making decisions within their reach; they need not be relegated to stakeholder status and treated as pawns in national and international political arenas effectively beyond their reach. The IPCC's 1995 *Synthesis* recognized that "[e]quity is an important element for legitimizing decisions and promoting cooperation."[74] From our perspective equity includes the distribution of power, or participation in making important decisions, on the broadest practical basis. Equity is not limited to the distribution of economic costs and benefits, knowledge and information, or any one value.

Of course it would be foolish to abandon the established climate regime, even if that were feasible. The regime may yet pay off in terms of the ultimate objective prescribed in Article 2, and in any case the national governments in the Conference of the Parties control many of the resources necessary to deal with climate change problems. But in view of disappointing outcomes after two decades of effort, including an expenditure of at least $38.8 billion on climate change research in the United States alone,[75] it would be equally foolish to bet the farm or the planet by relying exclusively on the established regime as it is and to ignore the potential for opening it to adaptive governance. To help realize that potential, we use Box 1.1 as a guide to bring more of the relevant past and possible futures into the picture in the chapters that follow. In doing so we also provide concrete examples that approximate the ideal types introduced in Box 1.1. Like lines, polygons, and other graphic devices used in a road map, the characteristics of the two patterns do not represent a map of science, policy, and decision making in any particular context. Rather, they are to be used for mapping and self-orientation in many particular contexts, each of which is ultimately unique.

The Common Interest

Our construction of the context makes it clear that a choice between defending the established climate change regime as it is and opening it up to adaptive governance is neither value free nor independent of social and political interests. Both alternatives answer to "the facts," although to somewhat different facts; hence, a choice cannot be made on an empirical basis alone under the guise or illusion of objectivity. One's own interests, including values, are necessarily implicated in the choice of any alternative. Moreover, any choice has social and political consequences for the many interests at stake in climate change adaptation and mitigation policies. Politically, one alternative is biased toward leaving people on the ground as mere stakeholders, if not pawns in an international regime; another is biased toward empowering them as active participants taking primary responsibility for addressing climate-related problems within their reach. The implication is that each of us might make our interests explicit to ourselves, at least, and consider the social and political consequences of a choice in order to act with our eyes open. An interest is a pattern of value demands on behalf of an individual or group identity, supported by expectations that the value demands are advantageous for that identity.[76] We recommend a value commitment to serve the common interest of the community at hand, whether that community is local, national, or global in scope. This commitment is a basis for reconciling our individual and group interests as scientists, for example, with our other interests as citizens. This commitment is easily justified and defended by those who recognize and respect the United Nation's Universal Declaration of Human Rights.[77]

"In the simplest terms, the common interest is composed of interests widely shared by members of a community. It would benefit the community as a whole and be supported by most community members, *if they can find it.*"[78] Operationally, the common interest is not to be assumed or taken for granted. It must be constructed in each community, with or without outside help, on the basis of the valid and appropriate interests of community members. (Invalid or inappropriate interests may be discounted.) An appropriate interest includes a value demand consistent with the more basic values of community members, such as equity, democracy, and sustainability. If the interest is also valid, the value demand is supported by expectations consistent with the evidence available. For example, the ultimate objective in Article 2—"stabilization of greenhouse gas concentrations in the atmosphere at a level that would prevent dangerous anthropogenic interference in the climate system"—is a value demand that can be justified by a larger value commitment to the sustainability of ecosystems and human communities. The value demand is supported by the preponderance of evidence available—including observations on CO_2, temperature, and the like—scientific theories such as the greenhouse effect, and the observed and expected impacts of climate change. Thus, the interest expressed in the ultimate objective is both valid and appropriate. In this example, expectations arising from scientific developments activated and supported interests in doing something about climate change. As in all things, interests are subject to change.

But many other interests are involved in finding the common interest in response to climate change. Article 2 itself continues with other interests of the parties that are overlooked when the level of greenhouse gas concentrations is taken as a single target: "Such a level should be achieved within a time frame sufficient to allow ecosystems to adapt naturally to climate change, to ensure that food production is not threatened and to enable economic development to proceed in a sustainable manner." Other interests are recognized in Article 4(7-8) of the UNFCCC, including the developing country parties' overriding interest in economic development and poverty alleviation. Also relevant are the interests of scientists in sustaining their programs of research, the interests of program officials in defending or expanding their authorities and budgets, and the interests of employers and employees in protecting or advancing their economic positions. At the local level are diverse interests of community members too specific and numerous for outsiders to understand in the aggregate, although outsiders can gain some dependable insights on a selective basis through direct local contacts.

Thus advancing the common interest of a community requires integrating the multiple interests of community members if possible or balancing them if necessary. "Balancing" means a compromise in which each group can assess what it gained and lost. "Integrating" means a "win-win" outcome in which all the main participants get what they want. Similarly, when the different interests of communities within and across levels, local to international, come into conflict, they must be reconciled as a precondition for joint action. In these circumstances, *politics are necessary to advance the common interest*—if we define "politics" functionally as the giving and withholding of support in making important decisions. We recognize, however, that "politics" has acquired a bad name in popular culture because the giving and withholding of support too often have served special interests. "By definition, a special interest is incompatible with the common interest. It is pursued by some part of the community for its own benefit, at net cost to the community as a whole."[79] In any case, politics are unavoidable when interest groups transform what was once a technical problem, such as global warming, into a political issue. The politics of integrating and balancing multiple interests to advance the common interest of any community, at any level, reframes climate change as an issue in community development. It no longer stands apart as a technical problem amenable to a purely technical solution. In these politics, judgments of appropriateness and validity grounding any specification of the common interest are contestable.

The ultimate objective in Article 2 and other interests in the UNFCCC are an approximation to the common interest of the world's communities; they help define the problems documented earlier in this chapter: The danger of extreme weather events has increased, if aggregated meteorological records and escalating losses are a reliable gauge; insufficient reductions in greenhouse gas emissions portend more losses to come. Underlying these problems is the possibility that those of us concerned about climate change have unwittingly exacerbated them through the mechanism of goal displacement. It is rather easy for us climate change scientists, for example, to justify our proposed policy-relevant research as a contribution toward realizing the ultimate objective or other interests in UNFCCC; to fund and conduct the research; and to produce the requisite reports and publications as a successful *conclusion*. But this turns what was once a means, the research, into an end in itself that effectively displaces the original goal and becomes a substitute for it.[80] Similarly, it is rather easy for an administrator to become preoccupied with a program established or a budget increased—each justified as a response to the threat of climate change and its impacts—or for a negotiator to become preoccupied with arranging a meeting or finding common ground on an international treaty—each similarly justified. Such justifications implicitly or explicitly acknowledge Article 2 and other interests in the UNFCCC as the authoritative formulation of the common interest. But two decades of effort demonstrate that in the aggregate we can produce a plethora of scientific assessments, programs and budgets, and meetings and treaties without doing much to advance the common interest—and perhaps substituting for it.

It is not reasonable to denigrate our many constructive contributions or to take upon our shoulders the full burden of responsibility for disappointing outcomes with respect to the common interest. Public policy decisions on climate change are the responsibility of elected representatives of the public and ultimately the public itself, at each level in democratic systems of governance. But by invoking the common interest in justification of our activities, we assume responsibility for aligning those activities with the common interest in climate change. Otherwise we become de facto special interests. For us climate change scientists, for example, aligning our activities with the common interest means exploiting every opportunity open to us for preparing the minds of decision makers and taking responsibility for any limitations of theory and data that stand in the way.[81] One of these limitations is the relative neglect of politics, including our own interests as scientists, among other factors that explain the disappointing outcomes with respect to the common interest to date. As such, it is not reasonable to defend business as usual as the exclusive approach to climate change science, policy, and decision making. We can focus or refocus on the common interest, coordinate and evaluate our efforts accordingly, and consider what we might do differently and better—if we choose to. That includes opening the established climate change regime to adaptive governance as one approach.

In our judgment, a better approximation to the common interest in response to climate change is to reduce net losses of things valued in the world's many diverse communities, and *not* in the stabilization of concentrations of greenhouse gases in the atmosphere per se.[82] Stabilization of concentrations is one, but only one, means for reducing net losses, and reducing net losses is a somewhat different problem in each community. The many specific values that figure in the multiple interests of community members vary greatly across communities at each scale and are subject to change. However, the values generally include protection of human life and limb, property, other tangible and intangible human artifacts, and the natural environments on which humans depend. Actions proposed to reduce vulnerability to climate change, such as the Toronto Conference Statement, tend to have externalities perceived by various groups to have a positive or negative impact on their other interests—impacts perhaps sufficient to activate their participation in resolving the issue. Valid and appropriate interests may be integrated if possible, or balanced if necessary, to minimize opposition and mobilize support for policy decisions and actions that advance the common interest of the community. We join Michael Shellenberger and Ted Nordhaus in recommending these pragmatic and democratic politics.[83] Specifically, it should not be assumed that all the relevant interests can be reduced to objective standard measures that rise above politics. For example, in the IPCC's Second Assessment Report in 1995, Working Group III provoked analytical and political controversies when it valued a human life in a developing country at one-fifteenth the value of a life in the developed world.[84]

The common interest can be served in the near term through adaptation to extreme events that have damaged or threaten things of value, and in the longer term through mitigation of climate change by reducing greenhouse gas emissions. As strategies for reducing vulnerability to losses, adaptation and mitigation are means, not ends in themselves. Likewise, policy-relevant climate change science, programs, and treaties are means, not ends in themselves. They are properly evaluated according to their contributions to reducing vulnerability to losses from climate change. More emphasis on adaptation in the near term can be justified on pragmatic grounds: Each damaging storm, drought, heat wave, or other disaster identified with climate change tends to motivate, if not force action, providing a window of opportunity to field test promising policy alternatives to reduce the vulnerability of the community impacted. In effect, nature penalizes with severe sanctions past policies, including inaction, that have allowed significant losses from an extreme weather event to occur. And nature in this role serves as a surrogate for the political will to act that has been lacking so far, according to various appraisals, including Gore's. In this sense a disaster is a terrible thing to waste.[85]

To capitalize on such opportunities, it is sufficient to focus selectively on recently damaged or highly vulnerable communities, or communities otherwise motivated to address their own problems. It is neither feasible nor necessary to address or coerce all communities at once, given scarce resources including but not limited to political power and the will to use it. Improvements in policies to reduce vulnerability can be evolved by harvesting experience from policies field tested in selected communities for possible adaptation on a voluntary basis by similar communities elsewhere. Through case studies, important contextual details can be communicated directly from one community to the next without being abstracted and generalized for higher-level authorities. Through informal or formal networks, similar local communities can clarify their common interest in seeking whatever resources may be necessary from higher-level authorities—thereby initiating a process to reconcile the interests of state, national, or international authorities with those of local communities directly impacted by climate change. All of this *has been done* on a small scale, as we show later in this book. It *can be done* on larger scales without a global master plan imposed from the top down.[86] A science-based plan for "Managing Planet Earth" or strategic plan for "Spaceship Earth" is not necessary to reduce vulnerability to climate change, even if there was sufficient knowledge to devise it and sufficient political power and will to impose it.[87]

We focus on disaster-related adaptations in the near term. However, disaster-related adaptations can contribute to mitigation of climate change in the longer term. Perhaps it already has, through news reports and editorial comments linking global warming to natural disasters, including Hurricane Katrina, which devastated New Orleans and other parts of the U.S. Gulf Coast. Similarly, during the prolonged drought in Victoria, Australia, a Melbourne newspaper observed that "[w]ater saving has emerged as the most palpable sign that . . . we finally get it—the climate is changing and we have to adapt."[88] Another example of adverse personal experience as an agent of profound change was reported in *An Inconvenient Truth*. Al Gore's family continued to grow tobacco on its farm in Tennessee long after the Surgeon General concluded from scientific research that smoking caused cancer; they stopped only after Gore's older sister died of lung cancer attributed to smoking. While meteorologists and climate scientists continue to debate the physical and statistical linkages between weather and climate, growing numbers of people have already made the connection and some are prepared to act on it. Major disasters can bring home to ordinary citizens the need for mitigation of climate change, and might do more in this regard than two decades of scientific assessments or promotional politics have done so far. People experience weather, not climate, and they experience it in local places where they live and work. Rayner and Malone anticipated that "[a]ccumulating some experience with adaptation could provide a complementary, even perhaps an alternative, model for pursuing emission reductions."[89]

To summarize our argument, the principal outcomes of the regime to date have been escalating losses from climate-related disasters and only modest reductions in greenhouse gas emissions that portend more losses to come. These disappointing outcomes after two decades of effort are sufficient reason for opening the established climate change regime to alternatives in climate change science, policy, and decision making. Al Gore, among others, has attributed the outcomes to a lack of political will in general and to officials allied with the ExxonMobil's of the world in particular. For those of us concerned about climate change and its impacts, it is not enough to blame political opponents who are indeed responsible in some substantial degree for the disappointing outcomes to date. Introspection suggests that in the aggregate we may have contributed unwittingly to disappointing outcomes in at least three ways: by establishing the climate change regime too narrowly on scientific-management foundations, by framing the climate change problem too exclusively as "globally irreducible," and by substituting such means as scientific assessments, programs and budgets, and meetings and treaties as satisfactory outcomes, thus displacing the common interest in reducing net losses from climate change. We recommend refocusing efforts on the common interest, and evaluating them accordingly, and opening the established regime to effective participation by people vulnerable to extreme weather events and climate change in the places where they live. This would require changes in "our way of thinking," a major factor in humanity's relationship to the earth. But this would also capitalize on opportunities that already exist, including niches of political will around the world, increasing numbers of climate-related disasters that motivate or force action, and experience harvested from exceptions to scientific management.

Our primary purpose is to help open the established regime to adaptive governance and any other approaches that can improve upon the historical baseline to advance common interests in reducing net losses from climate change. Three objectives may be singled out as instrumental to this purpose:

■ The first is to recognize those who have contributed constructive alternatives to business as usual in climate change science, policy, or decision making, in addition to Tennekes, Rayner and Malone, and Cash. Many contributors and contributions have not received the attention they deserve.

- The second is to integrate selected alternatives in the literature, including cases, as proposals for adaptive governance. The practical significance of each alternative is minimized in isolation; together they become mutually reinforcing. Integration is where we hope to make an original contribution beyond field-testing aspects of adaptive governance in Barrow.
- The third objective is to inform and encourage the field testing of alternatives in other communities at any level. This is the most dependable method for falsifying the assumptions of fact and value on which we act, discovering what is important but previously overlooked, and advancing the common interest on a continuous basis.

Progress with respect to these objectives and the primary purpose, we believe, will expand the range of informed alternatives for decision makers, both the public and representatives of the public authorized to make decisions in response to climate change. Adaptive governance is an opportunity for field testing in parallel and in series thousands of alternatives for adapting to those climate change impacts we cannot avoid and for mitigating those we can.

We develop our argument for this purpose and these objectives in the following chapters. They require unusual emphasis on details, including quotations. In particular, the language used by practitioners in many places and by scholars from many disciplines is important evidence bearing on "our way of thinking," including the existence and substance of the two patterns of governance in Box 1.1. Other details are necessary to clarify the political and social consequences in specific contexts. In addition, all details are intended to minimize uncertainty absorption, and thereby help readers assess the extent to which our interpretations and inferences are supported in the chapters that follow.[90] Chapter 2 documents scientific management and exceptions to it in the evolution of the established regime. Chapter 3 tells the story of Barrow as a microcosm of things to come at lower latitudes. This is a body of experience worth harvesting for decision makers elsewhere, and it is an example of intensive inquiry that brings into the picture important details otherwise absorbed and dismissed. Chapter 4 pulls together case material from Barrow and other places and relevant theoretical material to elaborate and support the proposals in Box 1.1. To promote careful consideration of the proposals for adaptive governance, Chapter 5 reframes them as matters of action in the larger context of a transition from the relevant past to possible futures. In the end, the overriding question in climate change science, policy, and decision making remains open: Which interests will we individually choose and collectively decide to serve?

THE REGIME EVOLVES

This chapter documents scientific management, and exceptions to it, in the evolution of the established climate change regime. We focus on major science programs in the first section, and then turn to decisions on climate change policy and decision making in the second. We sketched the historical context of these initiatives in the scientific management tradition in Chapter 1. The details filled in here illustrate considerable investments of time, expertise, effort, and funds consistent with the ideal type of scientific management (Box 1.1) during the last two decades. These details should not obscure the disappointing outcomes documented in the previous chapter: All programs taken together, including those *not* selected here, have made little difference in advancing the common interest *given the magnitude of the task ahead.* In the third section we consider exceptions to scientific management, including case studies of adaptation to extreme weather events and mitigation of climate change on the ground. These exceptions suggest possibilities for opening the established regime to adaptive governance. They also represent bodies of experience available for adaptation to reduce near-term and longer-term losses from climate change.

This account must be highly selective according to our purposes and the two patterns of governance in Box 1.1. We can consider only a handful of initiatives because the climate change literature is vast, and even that literature represents only a small fraction of the experience that could have been harvested. From these initiatives we have selected and often quoted details essential for estimating degrees of approximation to the ideal types, grounding specific interpretations, and assessing general conclusions. The evolution of the regime precludes the closure that would be necessary to tell a complete or simple story: All the many parts of the regime and relationships among them are subject to change on various time scales. It is nevertheless necessary to map the context from which adaptive governance has begun to emerge—the evolution of scientific management in the established climate change regime—to address some basic questions: How did we arrive at this present disappointing state of affairs? Where should we go from here to reduce losses from climate change? How can we get there?

Science

We begin with the evolution of three major scientific programs crystallized by 1990: Phase I of the International Geosphere–Biosphere Programme (IGBP), the Fiscal Year 1990 Research Plan of the U.S. Global Change Research Program (USGCRP), and the First Assessment Report of the Intergovernmental Panel on Climate Change (IPCC). These interconnected programs and their successors typically express assumptions consistent with scientific management that have been challenged elsewhere. The main assumptions are that climate change is an irreducibly global problem that requires international cooperation from the top down; that policy-relevant science accordingly must focus on predictive models of the total earth system that integrate the many factors shaping its behavior; and that science is a necessary foundation for rational policy decisions.

International Geosphere–Biosphere Programme

The IGBP emerged in the mid-1980s as a complement to the World Climate Research Programme (WCRP), which focused on the physical aspects of the climate system. IGBP expanded considerations to include the biogeochemical aspects and global change.[1] As noted in Chapter 1, an Ad Hoc Planning Group met under auspices of the International Council of Scientific Unions (ICSU) in September 1986. Also involved in IGBP planning over the next several years were the government of Sweden, the United Nations Environment Programme (UNEP), the United Nations Educational, Scientific, and Cultural Organization (UNESCO), the Commission of the European Communities, the Organization of American States, the African Biosciences Network, the Commonwealth Science Council, the Third World Academy of Sciences, and various philanthropic organizations. Thus the IGBP depended on a variety of sponsors, and sponsorship has shifted over time. This structure is typical of independent or nongovernmental scientific organizations, including the Advisory Group on Greenhouse Gases (AGGG) in Chapter 1. Unlike the AGGG, however, the IGBP has survived. Prominent in both the AGGG and the IGBP was Swedish scientist Bert Bolin.

In June 1990 the Executive Committee of IGBP released Report No. 12, a full strategic research plan titled *The Initial Core Projects* for Phase 1 covering 1990–2003.[2] The plan accepted the expansion of climate change as an irreducibly global problem requiring nations to make the important decisions from the top down: "The concerns that drive the IGBP are international in character, with causes and effects that transcend national boundaries. . . . All nations have a stake in the consequences of global change; any hope of concerted response strategies requires their involvement in the design and execution of research that must be the basis for recommended policy actions."[3] The expectation that research *must* be the basis for recommended policy actions is a justification that invokes the linear model prescribing but seldom describing the interface between science and policy. Vannevar Bush produced the seminal formulation of the linear model for the United States in 1945. In the first part of the model Bush promoted basic science as essential for progress on major national goals. In the second part he insisted that science itself should be unfettered by practical considerations, which were the exclusive responsibility of government, industry, or other nonscientists.[4] The model is linear because it explicitly excludes feedback from policy and social outcomes to science. Promoters of climate change science often invoke the linear model; we have included it in the ideal type of scientific management (Box 1.1). For the IGBP, the expectation was that this program "will help provide the world's decision makers with input necessary to wisely manage the global environment."[5]

Box 2.1 provides selections verbatim from the Overview of Report No. 12 pertinent to realizing the IGBP's expectation. The objective focused on the total earth system rather than its geographically and culturally distinctive parts; human activities that influence or are influenced by the system were considered exogenous to it.[6] The objective was to describe and understand the total earth system, not to formulate policy recommendations that might reduce human influences on it. The understanding sought was a predictive understanding with expansive time horizons. According to text elaborating the objective, "The primary goal of the IGBP is to develop a predictive understanding of the changing nature of the Earth system. . . . To make this goal achievable, emphasis is placed on a time scale of decades to centuries."[7] The seven key questions for Phase 1 implied a reductive approach to the total earth system. While global changes of interest "transcend the traditional boundaries of scientific disciplines,"[8] some of the key questions are defined more or less along disciplinary lines—for example, programs in ocean biogeochemistry, coastal ecosystems, and terrestrial ecosystems. Separate research on these components would be integrated through synthesis and modeling projects in Regional Research Centers, the second key activity.[9] Report No. 12 served to coordinate research: It gave scientists international guidance to design and international legitimacy to promote projects funded for the most part by their national governments.

In 2006 the IGBP published Report No. 55, the research plan for Phase 2 covering 2004–2013; Phase 1 had been completed in 2003. The research goals were "[t]o analyse (i) the interactive physical, chemical and biological processes that define Earth System dynamics; (ii) the changes that are occurring in these dynamics; and (iii) the role of human activities in these changes."[10] These are similar to the Phase 1 objective (see Box 2.1). The fundamental strategy retained an emphasis on predictive capabilities but added an emphasis on integration as a means to that end. According to the plan's Preface, "we must address over-arching questions that require a systemic—not just a systematic—approach. We must combine research on Earth System components (atmosphere, land and ocean) with research on processes occurring at the interfaces between components, and we must integrate across all of these to develop diagnostic and predictive capabilities for the Earth System."[11] The integration of the three components and the development of diagnostic and predictive capabilities depended primarily on numerical models to simulate the earth system. Human factors beyond exogenous emissions again were omitted, probably because they are difficult to codify in any meaningful way for predictive purposes. Nevertheless there was hope for the "development of simulation tools that capture the richness" of the earth system.[12]

The Global Analysis, Integration and Modeling (GAIM) Task Force developed questions for further research in Phase 2 of the IGBP and reported them elsewhere in 2004.[13] These GAIM 23 Questions were intended to guide earth system science as a whole, not just activities explicitly associated with the IGBP.[14] Question 11 implied the possibility of a mechanical reconstruc-

Box 2.1. International Geosphere–Biosphere Programme

The Initial Core Projects, Phase 1: 1990–2003

Objective

To describe and understand the interactive physical, chemical and biological processes that regulate the total earth system, the unique environment that it provides for life, the changes that are occurring in this system, and the manner in which they are influenced by human activities.

Seven Key Questions

1. How is the chemistry of the global atmosphere regulated and what is the role of biological processes in producing and consuming trace gases?
2. How do ocean biogeochemical processes influence and respond to climate change?
3. How [do] changes in land use affect the resources of the coastal zone, and how [do] changes in sea level and climate alter coastal ecosystems?
4. How does vegetation interact with physical processes of the hydrologic cycle?
5. How will global changes affect terrestrial ecosystems?
6. What significant climatic and environmental changes have occurred in the past, and what were their causes?
7. How can our knowledge of components of the earth system be integrated and synthesized in a numerical framework that provides predictive capability?

Two Key Activities

1. The development of a global Data and Information System that will provide immediate and open access to all researchers, that will provide information needed for earth system models, and that will define and sustain the long-term observations needed to detect significant global changes.
2. The establishment of a set of Regional Research Centers in developing countries where strong synthesis and modeling projects of relevance to overall IGBP objectives and regional priorities will be developed, in close cooperation with existing research networks. Training and exchange programmes will be one of the mechanisms to involve the scientists from the region in Core Project Activities.

Source: The International Geosphere–Biosphere Programme: A Study of Global Change, *The Initial Core Projects*, Report No. 12 (Stockholm, Sweden: IGBP Secretariat, Box 50005, SE-104 05, June 1990), 1–3. Emphasis in original.

---

tion of the earth system: "Is it possible to describe the Earth system as a composition of weakly coupled organs and regions, and to reconstruct the planetary machinery from these parts?" But elsewhere a mechanical reconstruction was called into question when the plan referred to "linking human and environmental domains, understanding decision making and management, [and] handling cross-scale issues" as threads running through several research themes defined in other ways.[15] Normative and operational questions implied the existence of fixed and global answers for environmental management. For example, question 19 asked, "What are the equity principles that should govern global environmental management?" Question 20 asked, "What is the optimal mix of adaptation and mitigation measures to respond to global change?" In view of the world's diversity, if nothing else, perhaps the only working answers acceptable to the world's decision makers (including general publics) are contingent. That is to say, answers depend on the details of natural and human circumstances that differ across particular contexts and are subject to change. In any case, not much had changed from Phase 1. "In terms of overarching questions, the difference between IGBP Phases 1 and 2 is more illusory than real" according to the executive director of the IGBP from 1998 to 2004, Will Steffen.[16]

Phase 2 was informed by the Earth Systems Science Partnership (ESSP) established in 2001 to focus on issues of global sustainability and forge stronger links between IGBP, WCRP, the International Human Dimensions Program (IHDP) and DIVERSITAS (an international biodiversity science program established in 1991). ESSP is relevant here because it is "developing a small set of Integrated Regional Studies (IRS) designed to contribute sound scientific understanding in support of sustainable development at the local level."[17] These studies recognize that "at the regional level, aspects of global environmental change manifest significantly different—yet surprisingly coherent and teleconnected—Earth System dynamics. These are often broadly associated with socio-economic and geopolitical characteristics."[18] Differences in regional context matter, in other words. Work has begun on one project, the Monsoon Asia Integrated Regional Study. However, the frame remains fundamentally global: ESSP was a response to the Amsterdam Declaration on Global Change endorsed by 1,400 scientists from 100 countries seeking a "new system of global environmental science" that will among other things "employ the complementary strengths of nations and regions to build an efficient international system of global environmental science."[19]

U.S. Global Change Research Program

National programs for climate change research evolved concurrently with international scientific programs including IGBP. The largest of these is the USGCRP. According to very rough estimates, USGCRP accounted for half the world's expenditure of $3 billion (USD) per year on global change research in the late 1990s.[20] USGCRP was developed in the late 1980s by the Committee on Earth Sciences (CES). CES published the first U.S. Strategy for Global Change Research in January 1989 and followed up six months later with an "extensive elaboration," the Fiscal Year 1990 Research Plan for the USGCRP. This set the pattern for annual updates that accompany the president's annual budget request in a series titled *Our Changing Planet*. In 1989 CES was composed of a chairman, Dallas Peck, and 15 others representing 10 cabinet or independent agencies plus the Office of Science and Technology Policy, the Office of Management and Budget, and the Council on Environmental Quality in the White House. The proposed budget for the FY 1990 Research Plan was a budget crosswalk, a matrix of budget requests by eight federal agencies for seven research elements. To a large extent the organizational structure of USGCRP is still a budget crosswalk with the elements redefined and additional agencies included over the years.[21] In 1989 an organizational chart placed the USGCRP in international perspective, identifying the ICSU and IGBP among affiliated nongovernmental scientific organizations and WMO, UNESCO, and UNEP as intergovernmental scientific organizations.[22]

CES accepted the global framing in the overarching goal of USGCRP. As stated and emphasized in its FY 1990 Research Plan, the overarching goal was "[t]o gain an adequate predictive understanding of the interactive physical, geological, chemical, biological and social processes that regulate the total Earth system and, hence establish the scientific basis for national and international policy formulation and decisions relating to natural and human-induced change in the global environment and their regional impacts." The policy relevance of the program was justified on the linear model, and more specifically the premise that "effective and rational response strategies to environmental issues can be built *only* on sound scientific information" of the kind sought in the program.[23] The global scope of the program followed from the concept that "[t]he Earth is, after all, one planet and, therefore, contains one interactive 'Earth system'" with complex interplays among the many natural and human systems that drive environmental change.[24]

Like IGBP in Phase 1, CES expanded the scope of climate-related issues to be addressed in USGCRP: "Emphasis in the past understandably has been on needs that are perceived as the most immediate and most local. . . . However, in recent years, the attention of both scientists and policymakers has extended to more global-scale, longer term changes, such as persistent continental-scale droughts, global warming, coastal erosion, and stratospheric ozone depletion. While the impacts of vital concern remain regional, they are recognized to be embedded in the processes of the larger phenomena."[25] For this reason research on *global* change would be necessary to inform decisions at many other levels: "Reliable information and predictions regarding global changes are required at many decision levels within society: individuals (e.g., farmers), industries (e.g., energy and chemical producers), and regulators (e.g., governments)."[26] The reference to predictions as requirements for decisions echoed the overarching goal in which a predictive understanding of the total earth system was intended to establish the basis for national and international decisions. Thus despite acknowledged local and regional concerns, CES affirmed a top-down structure of decision making informed by scientific expertise to predict global changes in the tradition of scientific management (Box 1.1). CES was quite clear that "[t]he goal of the Program is to provide a sound scientific basis for national and international decision making on global change issues."[27]

In the FY 1990 Research Plan CES summarized USGCRP "At-a-Glance" on one page, which is reproduced verbatim in Box 2.2. The first and third points reconfirmed the program's global framing, linear-model premise, and justification as policy relevant. Nevertheless, in the fifth point the scientific objectives of the program—"to monitor, understand, and ultimately predict global change"—were independent of any particular policy goals or action alternatives for realizing those goals. This specification of the linear model raises questions of relevance to decision makers who do not take global change research as basic science, an end in itself. It also facilitates policy applications of global change research by decision makers who cite scientific uncertainties as justification for postponing action on climate change.[28] Several points in Box 2.2 indicated a program formally more comprehensive than Phase 1 of IGBP. USGCRP included social changes among other earth system changes, the effects of human as well as natural phenomena, and Human Interactions as one of seven research activities. Elsewhere Human Interactions were reduced to agents forcing global change, much like impersonal forcing factors such as Solar Influences, Greenhouse Gases, and Aerosols. Apparently external to the program were human responses to global changes in the earth system, a feedback that includes policy decisions affecting future global changes for better or worse.

CES explicitly disavowed policy applications, despite its policy-relevant aspirations and justification: "It is not the role of the program to formulate policies regarding global change, nor does its mandate cover the research required to develop new technologies that might be used to mitigate or adapt to a changing environment."[29] Instead CES expected USGCRP to produce

Box 2.2. The U.S. Global Change Research Program At-a-Glance, 1989

Many global changes can have tremendous impact on the welfare of humans. These events may stem from natural processes that began millions of years ago or from human influence. Responding to these changes without a strong scientific basis could be futile and very costly.

This report presents a comprehensive research plan for the U.S. Global Research Program.

The goal of the Program is to provide a sound scientific basis for national and international decision making on global change issues.

The Program's goals, objectives, research priorities, and strategy are consistent with current national and international global change planning and research efforts.

The scientific objectives of the Program are to monitor, understand, and ultimately predict global change.

The Program is broad in scope, encompassing the full range of Earth system changes, including physical, chemical, geological, social, and biological changes. The Program addresses both natural phenomena, as well as the effects of human activity.

The particular research activities which comprise the U.S. Global Change Research Program are grouped into seven interdisciplinary scientific elements:

1. Climate and Hydrologic Systems
2. Biogeochemical Dynamics
3. Ecological Systems and Dynamics
4. Earth System History
5. Human Interactions
6. Solid Earth Processes
7. Solar Influences

In fiscal year 1989, funding for focused global change research activities total $133.9 million. The President's FY 1990 budget proposes a funding level of $191.5 million, a 43 percent increase for focused programs. This substantial increase will enable the Program to expand and accelerate its research activities in most areas of global change research.

This strategy was developed by a U.S. Federal interagency group, the Committee on Earth Sciences of the Federal Coordinating Council for Science, Engineering, and Technology (FCCSET). The FCCSET is chaired by the Director of the Office of Science and Technology Policy in the Executive Office of the President.

Sources: *Our Changing Planet: The FY 1990 Research Plan: The U.S. Global Change Research Program*, A Report by the Committee on Earth Sciences (July 1989), xiii. This July summary was adapted with few changes from the Executive Summary of *Our Changing Planet: A U.S. Strategy for Global Change Research* (January 1989), A Report by the Committee on Earth Sciences to accompany the U.S. President's Fiscal Year 1990 Budget, 3–4.

"regular, integrated assessments of the current scientific understanding of global change" to clarify future research priorities and provide an updated basis for national and international policy decisions. "Indeed, such periodic assessments will be the *primary output* of the Program with regard to aiding policy decisions."[30] Thus at the interface between basic science and practical applications, and within the constraints of the linear model, were scientific assessments of global change research that excluded the formulation of policy alternatives.

In July 2003 the Bush Administration released the "first comprehensive update" of the original 1989 strategic plan and called it the *Strategic Plan for the U.S. Climate Change Science Program.* The administration established the U.S. Climate Change Science Program (CCSP) in February 2002 to coordinate the USGCRP and the Climate Change Research Initiative. The CCRI was launched in June 2001 and "represents a focusing of resources and attention on those elements of the USGCRP that can best support improved public debate and decision-making in the near term." Implementation of the new strategic plan began with the CCSP report for FY 2004 and 2005 published in July 2004 as part of the series on *Our Changing Planet.* At least nominally, CCSP narrowed the focus to climate change science and moved global change research, the broader concept, into the background. The CCSP Guiding Vision dropped a predictive understanding of the total earth system from the original overarching goal of USGCRP but retained scientific knowledge as the foundation of management information. According to the FY 2004 and 2005 report, "The core precept that motivates the CCSP is that the best possible scientific knowledge should be the foundation for the information required to manage climate variability and change and related aspects of global change."[31] Compared to the USGCRP in 1989, the CCSP mission gave more weight to the application of scientific knowledge in support of decisions.

However, in the spring of 2005 Sen. John McCain, then chair of a Senate committee with climate change jurisdiction, complained about the lack of useful products from CCSP. This prompted CCSP to organize the first Workshop on Climate Science in Support of Decisionmaking in the United States, which convened in the Washington area in mid-November 2005. More than 700 people, mostly scientists, participated in the workshop, which was billed as an opportunity for exploring the intersection of science and action. But the plenary sessions mostly showcased Bush Administration climate change policies and 21 Synthesis and Applied Products (SAPs) planned or initiated by CCSP pursuant to the policy mandate in the Global Change Research Act of 1990 (Public Law 101-606) and the second strategic research plan in 2003.[32] These SAPs are assessments of climate change science focused on questions relevant to CCSP. One skeptic in the audience of a plenary session asked,

"How many assessments do we need?" The implication was that we already knew enough. Another asked, "Is there any leeway to be policy prescriptive?" The implication was that scientific assessments without action alternatives were insufficient. Robert Corell, a plenary speaker, assured the audience that climate change science is not finished, that the real challenge is bridging the gap between knowledge and action. He called this gap "the valley of indecision and delay," a phrase picked up and used by other workshop participants. Another plenary speaker, Roger Pulwarty, mentioned the emergence of adaptive governance in efforts to fill the gap in various other policy domains.[33]

The FY 2006 report on CCSP published concurrently in November 2005 reaffirmed the linear model: "CCSP does not make policy recommendations. Instead, it supports fundamental research that provides necessary objective background understanding for others to analyze policy questions and make policy recommendations."[34] It cited as one of the guiding principles of decision support "[e]arly and continuing involvement of stakeholders."[35] Implementation of this principle would compromise the second part of the linear model by engaging researchers in practical considerations. Elsewhere, however, the FY 2006 report privileged expertise; it put researchers and ultimately departments and agencies in the driver's seat: "The synthesis and assessment products will be generated by researchers in a process that involves review by experts, public comment from stakeholders and the general public, and final approval by the departments/agencies involved in CCSP."[36] Apparently the structure of decision making was still firmly rooted in the tradition of scientific management.

Intergovernmental Panel on Climate Change

National programs such as the USGCRP fund original scientific research providing the basis for scientific assessments. These assessments do not present new scientific research nor do they make policy recommendations. With few exceptions, they do not even formulate alternatives for policymakers to accept or reject in accord with their own interests. Rather, as we have seen, they occupy the turf between climate change science and policy and are exposed to criticism from both sides. The assessments of the IPCC are the most highly publicized and influential worldwide, including in the United States. In this section we rely on IPCC's summaries for policymakers, which are finalized only after line-by-line review by governmental and nongovernmental representatives.

As noted in Chapter 1, IPCC completed its First Assessment Report in 1990 to assist in negotiation of the UN Framework Convention on Climate Change (UNFCCC). Here we focus on the report of Working Group III, also called the Response Strategies Working Group (RSWG), chaired by Frederick Bernthal of the U.S. Department of State. (Bernthal also was a member of the Committee on Earth Sciences that set the pattern for USGCRP in 1989.) The Executive Summary of the RSWG report cited conclusions from Working Group I on Science and Working Group II on Impacts as the appropriate context for its own work. For example:

- "We are certain emissions resulting from human activities are substantially increasing the atmospheric concentrations of greenhouse gases: carbon dioxide, methane, chloroflouor-carbons (CFCs), and nitrous oxide. These increases will enhance the greenhouse effect, resulting on average in an additional warming of the Earth's surface. . . .
- "The longer emissions continue at present-day rates, the greater reductions would have to be for concentrations to stabilize at a given level. The long-lived gases would require immediate reductions in emissions from human activities of 60 percent to stabilize their concentrations at today's levels. . . .
- "In many cases, the impacts will be felt most severely in regions already under stress, mainly the developing countries. The most vulnerable human settlements are those especially exposed to natural hazards, e.g., coastal or river flooding, severe drought, landslides, severe storms and tropical cyclones."

Such conclusions justified Working Group III's work on response strategies. The Working Group added its own conclusion, emphasizing that context matters. "Any responses will have to take into account the great diversity of different countries' situations and responsibilities and the negative impacts on different countries, which consequently would require a wide variety of responses."[37] In an effort to accommodate a wide variety of responses and related considerations, Working Group III outlined "Possible Elements for a Framework Convention on Climate Change." These elements informed decisions on the top-down structure of decision making formalized in the Framework Convention.

The *Policymakers Summary* from Working Group III was approved in June 1990; its Main Findings are reproduced verbatim in Box 2.3. Like IGBP's Phase 1 plans and USGCRP's FY 1990 Research Plan, the first finding accepted the global framing of the climate change issue and a global effort as a require-ment for effective responses. However, the other main findings focused on issues of policy and decision making rather than plans for scientific research to reduce uncertainty. The second finding drew attention to the common responsibilities of industrialized and developing countries, then outlined those responsibilities in the third and fourth findings, respectively. The fifth finding acknowledged the interdependence of environmental protection and economic development, and found it "imperative that the right balance be-tween economic and environmental objectives be struck." The limitation of greenhouse gas emissions and adaptation to climate charge were considered complementary and given equal emphasis in the sixth finding. The seventh finding anticipated the "precautionary principle" and "no regrets" response strategies later accepted in the Framework Convention. These both decouple action from predictive science and challenge the linear model.

The eighth and final finding promoted a "well-informed population" as essential in part to "provide guidance on positive practices." Elsewhere the *Policymakers Summary*, noting that climate change would affect almost every sector, explained that "broad global understanding of the issue will facili-tate the adoption and implementation of such response options as deemed necessary and appropriate." Here, Working Group III did not make explicit *who* would decide which response options were necessary and appropriate nor *how* they would decide. It did promote "public education and informa-tion programmes . . . to encourage wide participation of all sectors of the population of all countries . . . in addressing climate change and developing appropriate responses. . . ."[38] Here perhaps is a glimpse of the need for feed-back from the Peorias of the world. But the document set aside the political difficulties of reconciling conflicting interests reflecting the great diversity of society within and between countries and across levels in the international system. It also set aside the human cognitive constraints on attending to the volume of significant details on the ground as they move up to the national and international levels. The final finding also reaffirmed the conclusion in the Executive Summary on the need for approaches "tailored" to diverse national circumstances.

The Chairman's Introduction sought to separate what was described as the broadly "technical" task of Working Group III from political tasks: IPCC's charge "was to lay out as fully and fairly as possible a set of response policy options and the factual basis for those options. Consistent with that charge, it was not the purpose of the RSWG to select or recommend politi-cal actions, much less to carry out a negotiation on the many difficult policy questions that attach to the climate change issue, although clearly the information might tend to suggest one or another option." The last clause is an important qualification that was and remains an unusual acknowledgement of the biasing effects of information from scientific assessments. That information sheds light on selected parts of a much more complex reality that lies beyond anyone's complete or completely objective understanding, but it leaves most of that reality in the dark. The formulation of response policy options that made this assessment unusual also made it more policy relevant than others, but not policy prescriptive: Chairman Bernthal affirmed that "[s]election of options for implementation is appropriately left to policy makers of governments and/or negotiation of a convention."[39]

IPCC released its Fourth Assessment Report in 2007, during early consideration of a successor to the Kyoto Protocol. Working Group I's report on "The Physical Science Basis" was released in February, followed by Working Group II's report on "Impacts, Adaptation and Vulnerability" in April, and Working Group III's report on "Mitigation of Climate Change" in May. The Synthesis Report was released in November. The main topics followed closely those of the Third Assessment Report in 2001, but with a different division of labor between Working Groups II and III. The Fourth Assessment Report was distinguished from previous assessments by more confidence in the detection and attribution of climate change to human activities: "Most of the observed increase in global average temperatures since the mid-20th century is *very likely* due to the observed increase in anthropogenic greenhouse gas concentrations."[40] This was not quite the certainty based on numerical models expressed in the First Assessment Report in 1990, as quoted above, but in 2007 it was supported by an array of convergent *observations* suggested earlier by *models*. The Fourth Assessment Report was further distinguished by a rising concern about climate extremes and weather disasters. It also reintegrated mitigation and adaptation as compatible response strategies and restored the earlier balance of emphasis in the First Assessment Report. This development merits a brief historical digression.

In 1995 in the Second Assessment Report, Working Group III on the "Economic and Social Dimensions of Climate Change" allocated 13 paragraphs to mitigation and only one to adaptation in its *Summary for Policymakers*, and it suggested they might be incompatible: "[P]ossible tradeoffs between implementation of mitigation and adaptation measures are important to consider in future research."[41] Working Group III supported both alternatives, but chose to emphasize mitigation under resource constraints.[42] Elsewhere at that time support of adaptation measures still implied neglect

Box 2.3. IPCC Working Group III, Response Strategies
Main Findings

1. Climate change is a global issue; effective responses would require a global effort that may have a considerable impact on humankind and individual societies.
2. Industrialized countries and developing countries have a common responsibility in dealing with problems arising from climate change.
3. Industrialized countries have specific responsibilities on two levels:

    a. A major part of emissions affecting the atmosphere at present originates in industrialized countries where the scope for change is greatest. Industrialized countries should adopt domestic measures to limit climate change by adapting their own economies in line with future agreements to limit emissions;

    b. To cooperate with developing countries in international action, without standing in the way of the latter's development, by contributing additional financial resources, by appropriate transfer of technology, by engaging in close cooperation concerning scientific observation, by analysis and research, and finally by means of technical cooperation geared to forestalling and managing environmental problems.
4. Emissions from developing countries are growing and may need to grow in order to meet their development requirements and thus, over time, are likely to represent an increasingly significant percentage of global emissions. Developing countries have the responsibility, within the limits feasible, to take measures to suitably adapt their economies.
5. Sustainable development requires the proper concern for environmental protection as the necessary basis for continuing economic growth. Continuing economic development will increasingly have to take into account the issue of climate change. It is imperative that the right balance between economic and environmental objectives be struck.
6. Limitation and adaptation strategies must be considered as an integrated package and should complement each other to minimize net costs. Strategies that limit greenhouse gas emissions also make it easier to adapt to climate change.
7. The potentially serious consequences of climate change on the global environment give sufficient reason to begin adopting response strategies that can be justified immediately even in the face of significant uncertainty.
8. A well-informed population is essential to promote awareness of the issues and provide guidance on positive practices. The social, economic, and cultural diversity of nations will require tailored approaches.

Source: Intergovernmental Panel on Climate Change, Working Group III chaired by Frederick M. Bernthal, *Policymakers Summary in Climate Change: The IPCC Response Strategies* (Washington, D.C.: Island Press, 1991), xxvi.

of mitigation. By 2007, studies of significant climate change impacts and adaptation practices allowed the Fourth Assessment Report to be more specific and empirical, and support for adaptation no longer implied neglect of mitigation. The expanded assessment of adaptation was prompted by practical considerations, particularly Decision 5 on National Adaptation Programmes of Action (NAPAs) made at the seventh Conference of the Parties in Marrakesh, Morocco, early in November 2001.[43] The Marrakesh Accords intended NAPAs to focus on immediate vulnerabilities in those countries with limited adaptive capacity and to mandate a "participatory assessment of vulnerability to current climate variability and extreme events."[44] Subsequent Conferences of the Parties have recognized the *local* benefits of adaptation activities. Attempts first to consider adaptation and later to fund adaptation activities directed attention to multiple sources of vulnerability, from poverty to weather extremes.[45] Now adaptation to climate change is sometimes considered part of the broader issue of community development.

Returning to the Fourth Assessment Report, Working Group I culminated its *Summary for Policymakers* with projections of future changes in climate. These projections are contingent on emissions scenarios adopted in IPCC's "Special Report on Emission Scenarios" in 2000. The SRES scenarios represent major uncertainties in the levels of the most important greenhouse gases forcing climate changes in alternative socio-economic futures. For example, "The A1 storyline and scenario family describes a future world of very rapid economic growth, global population that peaks in mid-century and declines thereafter, and the rapid introduction of new and more efficient technologies."[46] But neither Working Group I nor the other working groups publicly assessed the SRES scenarios they assumed and used. Working Group I simply stated that "[a]ll [scenarios] should be considered equally sound. The SRES scenarios do not include additional climate initiatives, which means that no scenarios are included that explicitly assume implementation of the United Nations Framework Convention on Climate Change or the emissions targets of the Kyoto Protocol."[47] Thus human activities forcing climate change and human responses to climate change are exogenous to numerical models. This choice allows modelers to focus on what is known about other components of the total earth system, but does not resolve uncertainties in projection of future climate changes influenced by human behavior.

Working Group II's *Summary for Policymakers* "sets out the key policy-relevant findings" of its assessment of impacts and adaptation.[48] The bulk of the summary reviews knowledge of future climate impacts by sectors (e.g., ecosystems, coastal systems, health) and by regions of continental scale excepting small islands. The general conclusions are reminiscent of the First Assessment Report despite the wealth of detailed knowledge accumulated in the intervening years. For example, context matters: "Costs and benefits of climate change for industry, settlement, and society will vary widely by location and scale. The most vulnerable industries, settlements and societies are generally those in coastal and river flood plains, those whose economies are closely linked with climate-sensitive resources, and those in areas prone to extreme weather events, especially where rapid urbanisation is occurring. Poor communities can be especially vulnerable. . . ."[49] Moreover, "It is virtually certain that aggregate estimates of costs mask significant differences in impacts across sectors, regions, countries and populations."[50] The implications of such differences were not formulated as response strategies contingent on the goals of policymakers in the Conference of the Parties, or as active recommendations. Instead, a review of current knowledge about responding to climate change simply noted, for example, that "[t]he array of potential adaptive responses available to human societies is very large, ranging from purely technological (e.g., sea defences), through behavioural (e.g., altered food and recreational choices), to managerial (e.g., altered farm practices) and to policy (e.g., planning regulations)."[51] It acknowledged "formidable environmental, economic, informational, social, attitudinal and behavioural barriers to the implementation of adaptation"[52] without proposing how they might be overcome. It also raised unanswered questions about decision making: "Adaptation measures are seldom undertaken in response to climate change alone but can be integrated within, for example, water resource management, coastal defence and risk-reduction strategies." But who can and should do the integration, where, and how? The conclusion recommended that Working Group II's "judgements about priorities for further observation and research" in the technical summary "should be considered seriously. . . ."[53]

Similarly, the *Summary for Policymakers* by Working Group III provided general conclusions on mitigation, sometimes distinguished by development status or industrial sector (e.g., shipping and construction). It noted that a "wide variety of national policies and instruments are available to governments to create the incentives for mitigation action. Their applicability depends on national circumstances and an understanding of their interactions. . . ."[54] It also noted that "[t]he literature identifies many options for achieving reductions of global GHG emissions at the international level through cooperation."[55] It did not, however, develop this knowledge into strategies or recommendations. Working Group III also raised unanswered questions about decision making: "Decision-making about the appropriate level of global mitigation over time involves an iterative risk management process that includes mitigation and adaptation, taking into account actual and avoided climate change damages, co-benefits, sustainability, equity, and attitudes to risk."[56] But what are the capabilities of the established climate change regime in regard to this iterative, integrative, and adaptive process, judging from its experience? In conclusion, this *Summary for Policymakers* acknowledges gaps in knowledge and suggests that "[a]dditional research addressing those gaps would further reduce uncertainties and thus facilitate decision-making related to mitigation of climate change."[57] Additional research is a consistent interest and recommendation of researchers.

But while most now accept the evidence for climate change itself, some have challenged certain basic assumptions in the four IPCC assessments, USGCRP and CCSP, IGBP Phases 1 and 2, and in other programs like them in the established climate change regime. Here we organize some of the main appraisals around three important assumptions in global change research, including climate change science.

The first of these assumptions is that climate change is an irreducibly global problem that can be managed only through international cooperation from the top down. The corollary is that national governments individually and collectively are the important participants in the decision-making structure. David Cash pointed out one consequence of this assumption: "Traditional centralized assessment efforts . . . forego an opportunity to engage and inform regional and local researchers and decision-makers, increasingly important players in the science and politics of climate change, biodiversity loss, and other large-scale problems."[58] Granger Morgan and his colleagues have clarified why local considerations are important for decision makers: "None of these actors can be expected to make decisions based on globally averaged values . . . Their decisions will be largely influenced by local political context, local preferences and constraints, short-term economic needs, the set of options they have available, and their associated local costs and benefits."[59] Even global changes come together at the local level according to Thomas Wilbanks and Robert Kates: "Global changes in climate, environment, economies, populations, governments, institutions and cultures converge in localities."[60] Convergence of many factors means that each locality is ultimately unique, even though it shares some characteristics with other localities.[61] Convergence also means that the significance of each characteristic is contingent upon other characteristics in the same locality;

significance is not uniform across all localities. Such local considerations are absorbed and suppressed when aggregated statistically, formalized into stable relationships, or abstracted into generalizations for national and international decision makers.

Local decision makers cannot be dismissed as merely passive stakeholders in decisions made elsewhere. Whatever central authorities decide to do can make little difference in reducing near-term and longer-term losses from climate change *unless* people on the ground become active participants, supporting improvements in policy and changing their own behavior. "Climate change is a global problem that requires local action" according to former President Bill Clinton, announcing in 2007 a new global coalition to reduce emissions. "We all know that this is a global problem that requires a successor to Kyoto and national legislation," he said, "but we also know that as you reduce greenhouse gas emissions you must do it place by place, specifically company by company, building by building. The mayors are in a remarkable position to do this."[62] The coalition, organized by the William J. Clinton Foundation, included 5 banks to make loans to upgrade energy efficiency in buildings in 16 of the world's largest cities.

The second assumption is that science relevant to this globally irreducible problem must depend on predictive models of the total earth system that integrate the many factors shaping its behavior. But from the outset climate change modelers themselves understood that even the most comprehensive predictive models of the earth system are only approximations at best. They are forced to omit or simplify what may be important because of computational constraints, limited scientific understanding of process relationships, and incomplete or inaccurate data on initial conditions and parameters. For example, modelers typically omit most human behavior or make it exogenous to predictive models of the earth system because it is relatively difficult to formalize or measure: Humans like other living forms respond selectively to changes in their external environments according to diverse internal predispositions, both genetic and acquired, that are subject to change. Hence, formalizations and measures tend to vary across places and obsolesce over time. Naomi Oreskes and colleagues put the logical consequences of omissions and simplifications for science in rather stark terms: "Verification and validation of numerical models of natural systems is impossible. This is because natural systems are never closed and because model results are always non-unique." In other words, a closed model cannot capture all the dynamics of the open system it represents, and alternative models can produce the same results. The implication is that "[m]odels can be confirmed by the demonstration of agreement between observation and prediction, but confirmation is inherently partial. Complete confirmation is logically precluded by the fallacy of affirming the consequent and by incomplete access to natural phenomena. Models can only be evaluated in relative terms, and their predictive value is always open to question. The primary value of models is heuristic."[63]

Similarly, but from a policy perspective, Steve Bankes concluded that "the principal result of the increasing use of computer models seems to be not a marked improvement in the quality of decision making, but rather a growing sensitivity to the shortcomings of models." He affirmed the uses of models to explore the implications of assumptions and hypotheses for policy purposes. (Exploratory implications are heuristics for observations on the target system, not substitutes for making those observations.) Bankes went on to criticize "[b]uilding a model by consolidating known facts into a single package and then using it as a surrogate for the actual system. . . . When insufficient knowledge or unresolvable uncertainties preclude building a surrogate for the target system, modelers must make guesses at details and mechanisms." Such guesses are unavoidable in modeling the behavior of humans or other living forms; this helps explain the tendency to exclude their behavior from climate system models. Moreover, "There is a strong tendency to model in detail phenomena for which good models can be constructed, and to ignore phenomena that are difficult to model, producing systematic bias in the results."[64] This helps explain the focus on continental or larger scales in global climate system models, the primary tools presently employed in simulating the earth system. In the absence of a stronger climate change signal, focusing on large scales is necessary to reliably detect the effects of factors forcing global climate change (e.g., changes in greenhouse gases, aerosols from natural and human sources, and solar flux) and to separate them from local factors that may be independent (e.g., land-use changes). Local factors become more important at smaller scales.[65]

The third assumption is that scientific assessments are necessary foundations for rational policy decisions in accord with the linear model, and unfettered by other practical considerations. That scientific assessments are either necessary or sufficient is questionable in view of energy conservation programs that reduce greenhouse gas emissions and predate major scientific assessments of climate change.[66] Likewise, that assessments are unfettered by practical considerations is questionable in view of technical difficulties in policy modeling that leave scientific assessments incomplete and laden with judgments of facts and values. Bankes explained how these technical difficulties can open Pandora's Box in a policy decision process: They "can interact with psychological or bureaucratic tendencies to produce a host of problems, including using models to rationalize institutional prejudices, poor models driving out careful thinking, and tending to emphasize the aspects of a problem that can best be simulated. The result can often be that models provide an illusion of analytical certainty for problems that are not well understood, or in the worst cases provide scientific costume for points of view that are self-serving."[67] Climate models are not exempt from these problems. For example, climate models have historically omitted or empirically approximated poorly understood processes, utilized parameters that have no analog in observations, and relied on parameter choices that are nonunique. Thus, it is always possible to find something to criticize in a climate model. This is one reason among many why scientific assessments are not above politics; they are politically contested.

Victor Baker wrote what could become a fitting epitaph for the linear model, after documenting failure at the science-policy interface in the New Orleans hurricane and flood disaster of August 2005. "Underlying this failure is the misrepresentation of science in a mythical status founded on the failed notions of logical positivism but supported for the advantages provided to various political ends. Among the flawed notions are (1) that science can aspire to certainty in the prediction of specific hazards, (2) that these predictions provide the authoritative basis for action because of the a priori rationality of their generation, and (3) that the detachment of science from the development of policy is essential to ensuring a good outcome."[68] Similarly, consider Herman Karl and his collaborators at MIT on a popular version of the linear model often invoked by interest groups competing on the same issue. "The concept of 'decision based on sound science' is predicated upon the presumptions that science is a neutral body of knowledge immune from value judgments, science can predict with certainty and clarity what will happen in the physical world, and policy making is a [completely] rational process. None of these is true." In the real world of environmental policy, they continue, "decisions are unavoidably based on a range of values along with the interests of a great many stakeholder groups. Science cannot be separated from these values and interests. For many of our very complex environmental problems... decisions based on sound science must integrate social science, natural science, and stakeholder concerns."[69]

Finally, the corollary of the linear model is that scientific assessments are policy relevant in the sense that they are in fact used to make climate change policy decisions. We have already contended that prominent scientific assessments do not engage the interests of an important group of climate change decision makers, people on the ground in local communities worldwide. In the next section we turn to the influence of scientific assessments on climate change policy and decision making by central authorities. As noted in Chapter 1, scientific assessments did indeed put global warming on national and international policy agendas in the 1980s, and each IPCC assessment attracts attention. But during the last two decades climate science in the aggregate has contributed little to policy decisions. Scientific uncertainty in particular has been used primarily to defer action on mandatory policies enforced by severe sanctions to reduce greenhouse gas emissions in the United States and in the Conference of the Parties to the UNFCCC.[70] Progress in climate change mitigation has been very modest, and for the most part selective and voluntary by nation and other entities.

Policy and Decision Making

Although scientific assessments have put climate change on various policy agendas, we have found few specific substantive links between scientific assessments and subsequent decisions by central authorities in the history of the climate change regime. Central authorities base their decisions on other considerations primarily. Here we review international decisions under the UNFCCC, regional decisions represented by the European Union's Emissions Trading System, and national decisions represented by policies in the United States.

International Decisions

Ordinary policy decisions, such as targets and timetables for the mitigation of greenhouse gas emissions, are distinguished from constitutive decisions that prescribe structures and processes for making ordinary policy decisions. Constitutive decisions, in other words, are decisions about decision making.[71] The most important constitutive decisions at the international level are formalized in the UN Framework Convention on Climate Change (UNFCCC). IPCC's First Assessment Report (FAR) succeeded in structuring negotiations that culminated in the Framework Convention. Annex I in the *Policymakers Summary* from Working Group III specified "Possible Elements for Inclusion in a Framework Convention on Climate Change."

The "Executive Summary" of Annex I highlighted three general issues for negotiation:

- "the political imperative of striking the correct balances: on the one hand, between the arguments for a far-reaching, action-oriented Convention and the need for urgent adoption of such a Convention so as to begin tackling the problem of climate change; and, on the other, between the cost of inaction and the lack of scientific certainty";
- "the extent to which specific obligations, particularly on the control of emissions of carbon dioxide and other greenhouse gases, should be included in the Convention itself or be the subject of separate protocol(s)";
- "the timing of negotiations of such protocol(s) in relation to the negotiations on the Convention."[72]

In addition, one of three specific issues concerned institutional powers: "Views differ substantially on the role and powers to be created by the Convention, particularly in exercising supervision and control over the obligations undertaken."[73] Elsewhere the FAR raised the possibility of economic sanctions for the enforcement of agreements. "This would require an international convention to establish a system of agreed trade or financial sanctions to be imposed on countries not adhering to agreed regimes. Many contributors expressed considerable reservations about applying this approach to greenhouse gas emissions because of the complexity of the situation."[74]

Regarding the first general issue above—striking politically correct balances—negotiators of the UNFCCC, primarily diplomats, selected urgent adoption over either inaction or a far-reaching, action-oriented convention. But they also authorized action on the precautionary principle in Article 3(3) of the UNFCCC: "The Parties should take precautionary measures to anticipate, prevent, or minimize the causes of climate change and mitigate its adverse effects. Where there are threats of serious or irreversible damage, lack of full scientific certainty should not be used as a reason for postponing such measures." In addition, the Preamble recognized "various actions to address climate that can be justified economically in their own right"; these so-called "no regrets" policies make sense economically or in other ways regardless of anthropogenic climate change. Regarding the second and third general issues, negotiators included Article 4(2), the nonbinding provision to reduce greenhouse gas emissions to 1990 levels by 2000 in industrialized (Annex I) countries, and the Conference of the Parties subsequently negotiated the separate Kyoto Protocol at the end of 1997. (Both policies were introduced in Chapter 1.) But the negotiated targets and timetables for emissions reductions had little to do with the reductions needed according to IPCC assessments. Early in 1998, Bert Bolin, the first chair of the IPCC, observed that "[t]he Kyoto conference did not achieve much with regard to limiting the buildup of greenhouse gases in the atmosphere. If no further steps are taken during the next 10 years, CO_2 will increase in the atmosphere during the first decade of the next century essentially as it has done during the past few decades. Only if the new cooperation among countries succeeds will the Kyoto conference represent a step toward the ultimate objective of the convention. . . ."[75] Bolin's expectations about the first decade of the century turned out to be correct (see Fig. 1.1).

The question of new cooperation raises the specific issue posed by Working Group III in the FAR. The Conference of the Parties has not yet agreed on a system of trade or financial or other sanctions sufficient to enforce compliance with even the modest targets and timetables in the Kyoto Protocol. Consider the official description of new sanctions to enforce compliance by parties to the Kyoto Protocol, part of the Marrakesh Accords: "If a Party fails to meet its emissions target [for the first commitment period, 2008–2012], it must make up the difference in the second commitment period, plus a penalty of 30%. It must also develop a compliance action plan, and its eligibility to 'sell' [emissions credits] under emissions trading will be suspended."[76] However, a country not in compliance is unlikely to have emissions credits to sell to another country. A compliance action plan is not necessarily a sanction nor an improvement over previous plans that fell short; it can function as an interim substitute for action among other things. The major sanction carries over the unmet commitment from the first period to the second and makes the second period commitment more stringent through the penalty. It is doubtful that these formal sanctions augmented by informal sanctions such as national pride or embarrassment are sufficiently severe to enforce compliance. But, ultimately, this is an empirical issue to be informed by the relevant facts as they become available during the first commitment period. These facts include emissions averaged over those five years, as reported by each party and verified by others, and enforcement actions actually taken by the Compliance Committee established in the Marrakesh Accords.[77]

Other international policy decisions may be mentioned briefly. The Kyoto Protocol established three innovative "mechanisms" to help industrialized countries in Annex I meet target and timetable and other commitments.[78] Joint Implementation (Article 6) authorizes an Annex I country to apply emissions credits earned through projects in another Annex I country toward meeting its commitment under the Kyoto Protocol. Credits typically can be earned at lower cost in countries in transition to a market economy, the central and eastern European countries once part of the Soviet bloc. Eligible projects might reduce emissions (e.g., through energy efficiency improvements) or remove greenhouse gases from the atmosphere by sinks (e.g., through reforestation). The Clean Development Mechanism (Article 12) is designed to assist in the sustainable development of non–Annex I countries. It authorizes Annex I countries to implement reduction or removal projects in developing (non–Annex I) countries and apply net credits toward meeting their Kyoto Protocol commitments. With certain restrictions, Emissions Trading (Article 17) authorizes Annex I countries to buy and sell credits earned through emissions reductions or removals by sinks at home or abroad. For calculating credits, the rules to be invoked by nations and applied by the Compliance Committee are complex enough to open many loopholes. Accounting tricks make it possible to generate credits without significantly affecting trends in the atmospheric concentration of CO_2 or other greenhouse gases.[79] The extent to which these loopholes are exploited by accounting tricks is an indication of the effective interests of participants.

In the absence of significant reductions in greenhouse gas emissions or fewer losses from climate-related extreme events, Bodansky's summary appraisal of the regime in 1995 still seems valid: "The emerging climate change regime—with the UN Framework Convention on Climate Change (FCCC) at its core—reflects the substantial uncertainties, high stakes and complicated politics of the greenhouse warming issue. The regime represents a hedging strategy. On the one hand, it treats climate change as a potentially serious problem and, in response, creates a long-term, evolutionary process to encourage further research, promote national planning, increase public awareness, and help create a sense of community among states. But it requires very little by way of substantive—and potentially costly—mitigation or adaptation measures." Elsewhere Bodansky explained that "[t]he basic problem is that few countries have been willing to make difficult political decisions to limit emissions." The convention "reflects a carefully balanced compromise among the often conflicting positions of states."[80] At about the same time Boehmer-Christiansen concurred that "[t]he policy outcome so far . . . is a weak, research intensive framework treaty which reflects a political balance of power rather than any firm direction derived from science."[81]

And Ungar suggested that acceptance of the UNFCCC was predicated on a "no regrets" strategy in which actions would be justified on grounds other than climate change.[82] Bodansky concluded that "[t]he FCCC . . . is capable of evolution and growth, should the political will to take stronger international action emerge."[83] But the necessary political will has not emerged in more than a decade and a half, and the regime has not yet passed the field test established by its overriding objective, Article 2 of the UNFCCC.

Among later appraisals, Granger Morgan challenged the assumption that international agreements are essential. Writing in 2000 he considered it "unlikely that all the world's major states will simultaneously agree to a serious program to curtail emissions of CO_2 and other greenhouse gases. Fortunately, a universal top-down framework is not the only route to a global regime for managing CO_2. Norway, the Netherlands, and others have begun to take unilateral action." (By 2006, Norway had reduced its greenhouse gas emissions by 28.7% and the Netherlands by 2% from their 1990 baseline levels.[84]) In the spirit of adaptive governance, he concluded that "the world can learn from these efforts and begin to move, in a progressively more coordinated way, toward a more sustainable future."[85] The expectation is that progress is possible outside mandatory, legally binding international agreements by harvesting experience from voluntary actions taken by nations in a decentralized international system. However, others insist that voluntary approaches are not enough. In connection with the voluntary purchase of carbon offsets, Stephen Schneider contended that "[v]olunteerism doesn't work. It's about as effective as voluntary speed limits. No cops, no judges: road carnage. No rules, no fines: greenhouse gases. We're going to triple or quadruple the CO_2 in the atmosphere with no policy. I don't believe [voluntary] offsets are just a distraction. But we'll have failed if that's all we do."[86] More generally, in advance of the meeting of the Conference of the Parties in Kyoto in December 1997, Jessica Tuchman Matthews contended that "[t]he only thing we know for absolute certain is that voluntary programs won't work."[87]

In retrospect, we know with confidence that international agreements enforceable and enforced by trade or financial or other severe sanctions are not yet politically feasible. Under these circumstances, it would be prudent to exploit every opportunity to improve on voluntary alternatives. Voluntary and mandatory approaches, like adaptation and mitigation, or scientific management and adaptive governance, are *not* mutually exclusive in practice. However, there are well-known tendencies toward the dichotomization of alternatives in promotional politics.[88]

The cornerstone of the European Union's (EU) efforts to meet its collective commitment under the Kyoto Protocol is an Emissions Trading System (ETS) adopted by the European Parliament and European Council in October 2003 (Directive 2003/87/EC).[89] ETS is a cap-and-trade system, one that puts an absolute limit (or cap) on total emissions and relies on the trading of allowable emissions to reduce overall compliance costs efficiently. It creates a market, in other words. Trading depends on scarcity, a shortage of allowances; theoretically, the price of an allowance would be zero and no trading would occur without a shortage under the cap. The primary unit traded is the EU Allowance (or EUA), which allows one metric ton of CO_2 emissions in a year. The addition of 10 central and eastern European countries to the EU in May 2004 enhanced opportunities for efficient compliance. The collective emissions of those countries declined 33% from 1990 to 2006, making EUAs available for sale to those needing more allowances.[90] In addition to efficient emissions reductions, another purpose of the ETS in the first trading period, 2005–2007, was learning by doing. This was expected to put the EU on a path to meet its collective commitment under the Kyoto Protocol in the second trading period, 2008–2012.

Under the Kyoto Protocol the original 15 member states of the EU agreed to reduce their collective emissions of greenhouse gases 8% below 1990 levels by 2012. The ETS is structured hierarchically from the EU down through member states to large industrial installations within the states. The EU's member states, now numbering 27, must submit at least 18 months before a trading period begins a National Allocation Plan (NAP) for approval by the European Commission. At the national level, "No allowances can be distributed before a plan has been accepted and approved by the European Commission."[91] The NAP must specify the state's allocation scheme and the amount of allowances to be distributed to each industrial installation. The total amount of allowances must be consistent with the Kyoto Protocol target of the member state. Scarcity was limited in the first trading period: A minimum of 95% of the total allowances had to be allocated free, leaving 5% or less of the allowances to be auctioned. Most states distribute allowances on an annual basis. Thus, in the first trading period of three years, one-third of the allowances permitted by the European Commission were distributed for each year. Each member state issues allowances to each installation's account in an electronic National Registry maintained by the member state. At the next level down are the industrial installations covered by the ETS. In the first trading period, coverage was limited to industrial point sources of CO_2 emissions—10,282 in 2005, 11,186 in 2007—involved in the production of electricity and refined oil products; coke, iron, and steel; lime and cement; glass and ceramics; and paper and pulp. This included about 40% to 45% of all greenhouse gas emission from EU's member states. It excluded CO_2 emissions in the commercial, residential, and transportation sectors of member states' economies and emissions of greenhouse gases other than CO_2.[92]

"Each installation under the EU ETS has to have a greenhouse gas permit before it can operate."[93] The essential part of the permit prescribes rules for reporting CO_2 emissions for each year within three months of that year's end. The report must conform to the reporting rules and be verified independently. Within four months of year's end, the operator must transfer an amount of allowances equal to the verified emissions. This final step is called "surrendering" the allowances; they are used only once. "If an operator of an EU ETS installation is short of allowances he will have to buy allowances from an installation that has more allowances than needed to cover the verified emissions. . . . A company that does not surrender enough allowances will be fined [and] is still obliged to surrender the missing allowances."[94] The fines are €40 per missing EUA in the first trading period. Thus verified emissions and allowances for each installation each year are reconciled in the National Registry, and each registry is connected to the EU's Community Independent Transaction Log, which monitors all transactions of allowances. Transactions are further complicated by an amendment to the ETS directive in November 2004 allowing under certain restrictions the use of emission reduction credits from the Clean Development Mechanism and Joint Implementation in addition to EUAs to cover verified emissions.

What did field testing reveal about these formal provisions of the ETS? Most of the NAPs for the first trading period were approved by the EC before the end of 2004, but the last from Greece was not approved until June 2006, nearly midway into the first trading period. The EC found revisions necessary in NAPs that attempted either to allocate allowances in excess of national emission targets or to adjust allocations after EC approval. Neither is allowed under the ETS rules. In May 2006 the EC released emissions data and compliance status for the 21 member states with active national registries during the first year, 2005.[95] The data show a surplus, not a shortage, of EUAs. Allocated allowances of 1.8295 billion metric tons for 2005 exceeded independently verified emissions of 1.7853 billion metric tons for installations operating in these countries in 2005. The EU's Environment Commissioner called the difference, an excess of 44.2 million EUAs, an "over-allocation" of allowances. Amid reports of overallocation, the price of allowances dropped by about two-thirds in April and May 2006 from a high of nearly €30. Monthly trading volumes were modest, not exceeding 1.6% of first-period allocations.[96] By the compliance deadline for 2005, the end of April 2006, 8,980 installations accounting for more than 99% of allocations had reported their 2005 emissions. At same time, "a total of 849 installations were identified as not having surrendered a sufficient number of emission allowances," but many of them fulfilled their surrender obligations in the next two weeks.[97] The official results were similar for 2006 and 2007, at the end of the first trading period.[98]

Appraisals near the middle of the first trading period were mixed at best. In April 2006, Climate Action Network Europe reported the results of evaluations of NAPs by its members, several hundred nongovernmental environmental organizations scattered across Europe. First, they found that "[e]missions limits set by Member States for the first phase were a major disappointment . . . they need to be strengthened considerably" for the second trading period. In particular, "the majority of them are not in line with their corresponding national climate change targets or energy strategies. Out of 25 countries, only two Member States (Germany and the United Kingdom) have asked the participating industry sectors to reduce their emissions over historic levels (based on the information provided for respective base years— mainly 2000–2002). All other Member States allow for increases." Second, they found that national allocation mechanisms based on past emissions and other rules often were problematic. "While good examples exist, which make the carbon price an important factor in future investment decisions, many Member States have, unfortunately, developed rules that are either not clearly rewarding of emission reductions or even give perverse incentives to continue investing in high-emission technology or running more polluting plants." Finally, "NGOs identified major problems with the transparency of NAP development processes and the methodologies that were decided." In particular, the processes often privileged participation by industrial associations over environmental organizations, and "various data sources employed in the NAPs were not accessible and therefore unverifiable."[99]

In May 2006 the EC emphasized learning by doing in anticipation of the second trading period: "The new 2005 emissions data gives independently assessed installation-level figures for the first time and so provides Member States with an excellent factual basis for deciding upon the caps in their forthcoming national allocation plans for the second trading period, when the Kyoto targets have to be met."[100] In February 2007, PricewaterhouseCoopers noted differences in reporting and compliance standards and procedures in a survey of major emissions trading schemes worldwide. These differences "add complexity and cost, and are likely to increase risk—risk of non-compliance, of unintended misreporting, of fraud and, ultimately, of market failure. This issue is perhaps most marked in the EU scheme." On the EU specifically: "Although most Member States require third-party verification, EU-wide standards for verification and the accreditation of verifiers have not been implemented."[101] At a meeting in early June 2007, countries that administer ETS reportedly acknowledged that "the system was shadowed by some major flaws, including a government-credit allocation plan that allows companies to profit by lobbying for additional pollution permits." The steel sector, for example, sold excess permits from large government allocations. "'Though it has been a success, we have undergone a steep learning curve, and we have seen some windfall profits being made by power companies,' said Barbara Helfferich, a spokeswoman for the European environment commissioner, Stavros Dimas. 'We are considering auctioning up to 100 percent of credits,' she said, 'and the EU is determining whether there should be a mandatory level of auctioning. The commission would complete its review by the end of the year [2007],' she said."[102]

On January 23, 2008, the commission submitted proposals for the second trading period. What seems clear so far is that the EU has reduced the annual average cap on total emissions for the second trading period by about 10%, keeping the price of an EAU above €20 in the first quarter of 2008. However, based on nearly complete reports from industries covered by the ETS, a carbon-market research and consulting firm reported that total CO_2 emissions increased by 1.1% in 2007, but still fell short of total allowances permitted for that year.[103] Official figures confirmed the overallocation of allowances but put the increase in emissions in 2007 at 0.8%, or at 0.68% when adjusted for changes in the number of installations covered.[104] An appraisal in May 2008 discounted emissions reductions in favor of the structure and process of the ETS: "Its primary goal was to develop the infrastructure and to provide the experience that would enable the successful use of a cap-and-trade system to limit European GHG emissions during the second trading period, 2008-2012. . . . The trial period . . . was never intended to achieve significant reduction in CO_2 emissions in only three years. In light of the speed with which the program was developed, the many sovereign nations involved, the need to develop necessary data, information dissemination, compliance and market institutions, and the lack of extensive experience with emissions trading in Europe, we think that the system has performed surprisingly well."[105] Meanwhile, the press reported in April 2008 that European countries, in response to record-high oil and natural gas prices and other factors, "are slated to build about 50 coal-fired plants over the next five years, plants that will be in use for the next five decades." The "clean coal" technology to be used will reduce local air pollution but have no effect on carbon emissions. Technology for carbon capture and storage is still under development.[106] This reinforces earlier concerns about ETS's "suitability for directing long-term investment toward a low-carbon future—the ultimate goal of any climate change program."[107]

Like the Framework Convention, the EU's Emissions Trading System has not yet passed the field test established by its primary goal, reductions in greenhouse gas emissions. Among other aspirations of scientific management, EU–ETS was established by EU's central authorities on the advice of economists as an efficient means of meeting Europe's collective and national commitments to reduce greenhouse gas emissions under the Kyoto Protocol. Large numbers of industrial installations were to implement substantive and procedural decisions made at the top of the decision-making structure. But so far as we can tell there are still "significant differences between Members with respect to participant definitions, industry level emissions caps and allocations, and enforcement."[108] Such diversity frustrates central control on behalf of efficient reductions in emissions. And implementation of formal provisions has not been above politics. Learning by doing, a purpose of the first trading period, recognized a possible gap between formal provisions and what actually happens on the ground. But learning by doing must be followed by action on the lessons learned to terminate failed policies and put EU member states on a path to meet their individual and collective commitments under the Kyoto Protocol in the second trading period. This requires more political power and the will to use it than has been evident so far.

National Decisions: The United States

At the national level, the United States acted to fulfill its voluntary commitment under Article 4(2) of the UNFCCC to reduce greenhouse gas emissions to 1990 levels by 2000. In October 1993, President Bill Clinton and Vice President Al Gore announced the Climate Change Action Plan (CCAP), a collection of nearly 50 new or expanded programs. CCAP did not require a predictive model of the total earth system, new technologies, or severe sanctions for compliance. Instead it relied on modest incentives and existing technologies to encourage voluntary participation in "no regrets" programs justified by savings in energy costs, if nothing else. For example, the Green Lights program helped business partners such as Johnson & Johnson, a manufacturer of medical supplies, reduce emissions and energy costs through more efficient lighting in buildings. The Cool Communities program helped cities, military bases, and other federal facilities reduce emissions by strategically planting trees and lightening the surfaces of buildings and pavement, thereby saving air-conditioning costs. Participation was rational and cost effective from the standpoint of partners at least. However, Congress cut the president's budget request for CCAP by 46% to $184 million in FY 1995, leaving CCAP with a budget an order of magnitude less than USGCRP's budget for research in the same fiscal year. This is one reason why CCAP fell short of its goal, although it did produce modest reductions in greenhouse gas emissions compared to estimates of what emissions would have been without CCAP.[109] Direct references to CCAP disappeared in the Bush Administration, but some component programs appear to have survived. In April 2006, the Government Accountability Office (GAO) identified 20 voluntary climate change programs in the U.S. government, in addition to the Environmental Protection Agency's Climate Leaders program and the Department of Energy's Climate VISION program.[110]

The Clinton Administration never submitted the Kyoto Protocol to the Senate because of insufficient support for ratification. In advance of the meeting of the Conference of the Parties in Kyoto, the Senate in mid-1997 approved by a vote of 95-0 a resolution (S. Res. 98) urging the administration not to sign any climate pact that exempted developing countries from emissions targets or that would result in serious harm to the U.S. economy. Shortly before conclusion of negotiations in Kyoto in December 1997, Senate Republican leaders declared the Kyoto Protocol "dead on arrival."[111] The Bush Administration rejected the Kyoto Protocol in March 2001.

In mid-June 2001, in a speech from the Rose Garden, President Bush "acknowledged that the world has warmed and that greenhouse gases have increased, largely due to human activity, but emphasized that the magnitude and rate of future warming are unknown."[112] In February 2002, President Bush announced a policy to reduce the energy intensity of the U.S. economy by 18% in 10 years. (Energy intensity is the ratio of greenhouse gas emissions to economic output.) A 14% reduction was expected in any case without this policy.[113] In July 2005, the United States announced the Asia-Pacific Partnership on Clean Development and Climate with Australia, China, India, Japan, and South Korea to "complement, but not replace the Kyoto Protocol." One purpose was to create "a voluntary, non-legally binding framework for international cooperation to facilitate the development, diffusion, deployment, and transfer of existing, emerging and longer-term cost-effective, cleaner, and more efficient technologies among the Partners. . . ."[114] It remains to be seen whether the Asia-Pacific Partnership will improve energy intensity, a focus of its efforts, and whether it will be complementary to the Kyoto Protocol. Of the six members, only Japan had signed and ratified the protocol. But all six joined with the fifteen other member nations in APEC (Asia-Pacific Economic Cooperation) to sign the Sydney Declaration in September 2007. The declaration explicitly and officially deferred to negotiation of a successor to the Kyoto Protocol. Meanwhile, consider columnist Thomas Friedman's summary appraisal of American climate change policy in 2007: "Climate change is not a hoax. The hoax is that we are really doing something about it."[115]

As a partial exception, consider the aftermath of Hurricane Katrina that devastated New Orleans and other areas of the U.S. Gulf Coast in late August 2005. Within one year, the Bush Administration proposed and the Congress approved four emergency appropriations bills totaling $125 billion for hurricane relief and reconstruction in the devastated areas. As the first anniversary of the disaster approached, the *National Journal* reported that "[h]uge sums have flowed into the region to pay for debris removal, levee repair, emergency food and shelter, and a giant federal relief operation. But reconstruction money is arriving much more slowly, and with many more strings attached."[116] An open question is the extent to which vulnerabilities to future hurricanes will be reduced or increased in New Orleans and other devastated areas through private and public policy decisions. Another open question is the extent to which the experience of Katrina will be used to adapt relevant policies to reduce net losses from future extreme weather events in the United States generally. Nevertheless, it is already clear that Hurricane Katrina forced action from an administration and a Congress otherwise adamantly opposed to action on climate change, opening a window of opportunity to improve climate-related policies and policy outcomes. As noted in Chapter 1, a disaster is a terrible thing to waste.

Except for the Climate Change Action Plan, U.S. climate change policy has relied primarily on more research to support *future* decision and action, thereby deferring action on *existing* science and technology. As the Congressional Research Service put it, "Concerted investment in science and technological research has been the cornerstone of the federal climate change strategy since the 1960s, in the interest of reducing scientific uncertainties and lowering the costs of technology solutions."[117] At a White House research conference in April 1990, President George H. W. Bush echoed and reaffirmed the course set for the USGCRP by the Committee on Earth Sciences in 1989. Citing "the many uncertainties that abound" and the "diametrically opposed points of view" of two unnamed scientists, the president concluded: "What we need are facts, the stuff that science is made of, a better understanding of the basic processes at work in our whole world, better Earth system models that enable us to calculate the complex interaction between man and our environment." He requested from Congress a 60% increase in USGCRP's FY 1991 budget "to reduce the uncertainties surrounding global change, to advance the scientific understanding we need if we are to make decisions to maximize benefits and minimize the unintended consequences."[118] The administration of George W. Bush maintained the priority of research after the turn of the century. In FY 2006, appropriations for the major research programs, CCSP and CCTP, were $1.709 and $2.773 billion, respectively. Each exceeded the budgets for the other two components of the federal strategy in the same fiscal year: $1.084 billion in Energy Tax Provisions that may help reduce greenhouse gas emissions and $241 million for International Climate Change Assistance.[119]

Concern about the policy relevance of scientific research can be traced back at least to the U.S. Global Change Research Act (Public Law 101-606) enacted in November 1990. The act provided legislative authority for USGCRP, which had been established by executive order in 1987. But it also included a policy mandate for USGCRP "to produce information readily usable by policy makers attempting to formulate effective strategies for preventing, mitigating, and adapting to the effects of global climate change" (Section 104). In addition, the purpose of the Act referred to USGCRP as a program to "assist the Nation and the world to understand, assess, predict, and respond to human-induced and natural processes of global change" (Section 101b). The mandate "to respond" went beyond the three scientific objectives of the FY 1990 Research Plan. The act also directed the committee that succeeded CES to "consult with actual and potential users of the results of the Program to ensure that such results are useful" (Section 102e). Consultation taken seriously would fetter with practical considerations the basic science on which integrated assessments were to be constructed. In these respects, the act challenged expectations based on the linear model and expressed in the FY 1990 Research Plan.

The policy provisions of the U. S. Global Change Research Act of 1990 were initiatives of the House Committee on Science, Space, and Technology (later renamed the House Science Committee). Rep. George E. Brown,

Jr., was a long-time member of that committee and became its chair after the elections of November 1990. In 1992 he criticized both the neglect of policy goals by climate change scientists and the exploitation of that neglect by politicians: "Scientists tell us their data are objective, unburdened by ethical or moral implications. That leaves us politicians free to apply the data in any way we see fit. And we do. Scientific uncertainty has become an operational synonym for inaction on global environmental issues, and the debate over global change has thus become an impediment to action on a wide range of issues critical to our survival." We consider these outcomes to be consequences of linear-model aspirations, unintended for the most part. Rep. Brown also rejected better predictions as the point of climate change research: "The immediate challenge for science and technology must not be viewed as the need to reduce scientific uncertainty about climate warming. This is a hollow ambition. It is too easy to support and too unlikely to bear fruit."[120] Brown insisted that "scientists and policy makers must work together to make certain that research goals stay focused on policy goals, that 'scientific excellence' defines a path for achieving these goals, rather than an excuse for political inaction...."[121] Neither research focused on policy goals nor political action was the direction set in plans for USGCRP in the early 1990s and carried over into the CCSP a decade later.

Others have complained about the lack of policy relevance of USGCRP, but with little obvious influence judging by research plans and products and by the persistence of complaints over the years. In 1993, the first year of the Clinton administration, critics included the president's science advisor in testimony before the Senate Committee on Energy and Natural Resources on March 30; witnesses in hearings of the House Committee on Science, Space, and Technology on May 19 and November 16; and the Office of Technology Assessment (OTA), a research arm of Congress. In a major report published in October, OTA concluded that "there is no mechanism in USGCRP for making the link between policy and science."[122] The successor to CES responded in the next annual research plan for USGCRP: "Most significantly, new research in FY 1995 will result in better connection of research to policy making."[123] Yet complaints persist. As previously noted, Sen. John McCain complained about the lack of useful products from CCSP in the spring of 2005, leading to the first Workshop on Climate Science in Support of Decisionmaking in the United States in November 2005. In December 2006, 24 members of the House of Representatives, including two Republicans, wrote to the acting director of CCSP urging him to comply with the letter and spirit of the Global Change Research Act of 1990. Among other things the letter complained that the first of the 21 Syntheses and Applied Products released in April 2006 "contained no explicit policy-relevant information."[124]

In September 2007, the National Academy of Sciences announced the results of an evaluation of the first four years of climate change research under CCSP. CCSP commissioned the evaluation by the National Research Council, which concluded among other things:

- *"Discovery science and understanding of the climate system are proceeding well, but use of that knowledge to support decision making and to manage risks and opportunities of climate change is proceeding slowly."* In particular, 19 of the 21 Synthesis and Applied Products were still in progress. "Also, only a few small programs . . . have been initiated to identify and engage decision makers."
- *"Progress in understanding and predicting climate change has improved more at global, continental, and ocean basin scales than at regional and local scales.* Information at regional and local scales is more relevant for state and local resource managers and policy makers, as well as for the general population, but progress on these smaller spatial scales has been inadequate."
- *"Our understanding of the impact of climate changes on human well-being and vulnerabilities is much less developed than our understanding of the natural climate system.* Progress in human dimensions research has lagged progress in natural climate science, and the two fields have not yet been integrated in a way that would allow the potential societal impacts of climate change and management responses to be addressed."[125]

Evidently, CCSP has sustained the priorities of the FY 1990 Research Plan for USGCRP, its older component. In 1989 CES emphasized adapting those initial plans and priorities to future research, not to decision support or feedback from policymakers: *"The degree to which future research requires that these analyses be updated, and hence the degree to which the annual versions of this document change, will be a measure of the success of the U.S. Global Change Research Program."*[126] In November 2008, the National Academies "launched a new, congressionally requested study called America's Climate Choices. Experts representing various levels of government, the private sector, and research institutions will serve on four panels and an overarching committee. Five consensus reports will be released in 2009 and 2010."[127]

Let us step back from such details to reconsider the evolution of the mainstream climate change regime as a whole. Under the UNFCCC, the formal structure of decision making in the regime is still centralized in the Conference of the Parties. However, this is a matter of authority, not control. (Authority is "formal power [or] the expected and legitimate possession of power."[128] Control is effective power based on various resources—e.g., knowledge, skill, respect, trust, loyalty, and wealth, as well as authority—including the will to use them.) The resources necessary for the Conference of the Parties to control concentrations of greenhouse gas emissions in the atmosphere were withheld in negotiation of the UNFCCC. Similarly, the parties to the Kyoto Protocol lacked the political will to prescribe sanctions severe enough to enforce even modest targets and timetables for emissions reductions on themselves. Thus, compliance has been effectively voluntary. The dispersion of control is indicated by the diversity of policy outcomes: Among Annex I parties to the UNFCCC, Norway, as noted, decreased its greenhouse gas emissions by 28.7% from 1990 to 2006; but over the same period, the United States increased its emissions by 14.0%, Spain and Canada by more than 50%, and Turkey and Sweden by more than 100%.[129]

These outcomes also reflect the reconciliation of priority interests in each party's climate-related policies. Stabilization of concentrations of greenhouse gases has seldom been a priority interest in competition with economic prosperity or national security, for example. In any case, national climate-related policies have not been based primarily on IPCC or other scientific assessments. Nor are such assessments necessary for rational policy decisions and action: The precautionary principle and "no regrets" policies decouple action from scientific assessments in principle; the U.S. Climate Change Action Plan decoupled them in practice. Indeed, scientific uncertainty in contested assessments has been used to justify more scientific research as a substitute for action in the United States at least. National climate-related policies have been based on politics. Politics can serve common interests or special interests. But like the weather, politics are inevitable in important climate-related decisions—if there is much at stake.

In short, if the aspirations of scientific management helped shape the climate change regime, scientific management has not been an achievement of the regime. Climate science is still contested; national climate-related policies have not risen above politics on an "objective" scientific foundation; and effective control remains dispersed among the separate parties, despite creation of a single central authority. But the future evolution of the mainstream regime is not determined. It is contingent on new insights, choices, and decisions, among other changes in circumstances.

Exceptions

Initiatives consistent with adaptive governance have emerged on the periphery of or outside the established climate change regime. These are exceptions to scientific management. As such, they tend to be isolated in separate niches, often lack resources sufficient to realize their potential, and are less well documented compared to the major programs already considered. Nevertheless, they merit selective consideration to illustrate more concretely aspects of adaptive governance (Box 1.1) and to introduce experience worth building upon in opening the established regime. Recall that adaptive governance is not an end in itself, but one approach to reducing near-term and longer-term net losses from climate change.

Toward Case Studies

Case studies in support of policy decisions are an integral part of intensive inquiry for adaptive governance. In the United States, the Office of Global Programs (OGP) in the National Oceanic and Atmospheric Administration (NOAA) took a step in that direction by initiating the Regional Integrated Sciences and Assessments (RISA) program. Beginning in 1995, RISA launched regional pilot programs to support "integrated research across a range of disciplines to expand decision-makers' options at the regional level. It does this in a manner cognizant of the context in which decision-makers function and the constraints they face in managing their climate-sensitive resources."[130] By FY 2002, RISA supported five regional pilot programs, each taking a somewhat different approach, at a total cost of $3.3 million per year. The charter of a House Science Committee hearing in April 2002 claimed that RISA was unique: "Other than [RISA], there is currently no structure or process within USGCRP to identify potential users, understand their needs, and connect them to the research agenda." Thus, RISA challenged the priority of unfettered basic research in the linear model. The charter also observed that "RISA has been called a step in the right direction by some . . . while others view it as a model that could guide larger efforts within USGCRP."[131] At those hearings, the former staff director of the House Science Committee, Radford Byerly, Jr., acknowledged RISA's scientific excellence and continuing scientific contributions. He emphasized, however, that the program did not ask the right questions, for several reasons:

- "The first reason is that the program operates under [a flawed] assumption, explicit in program documents and in the authorizing legislation, that if we do good research and accurately predict future climate, then making policy for climate problems will be easier."
- "Second, and this is most important, a prediction will not tell us what to do in terms of mitigation and adaptation. . . . 'What to do' involves politics and policy. Politics is not a dirty word. In a democracy it is how we resolve conflicts of values. . . ."
- "Third, the assumption that prediction will make decisions and solutions easy extends into areas, that is, politics and policy, in which climate scientists have little expertise. This assumption may be unconscious and therefore unexamined."[132]

Byerly concluded that we already knew enough for some decisions without further improvements in scientific understanding and predictive capability. For other decisions we needed to know what users want. His major recommendation was to authorize a program to involve users directly in planning and evaluating climate change research. RISA was designed to move beyond the assumptions singled out by Byerly and toward improvements in decisions on the ground. The extent to which it may have succeeded is unclear, in part because appropriate metrics for a full appraisal from this standpoint are lacking. There is anecdotal evidence that RISA has achieved some visibility in regional decision processes.[133]

RISA was featured in a box in *Our Changing Planet* for FY 2006 under the banner "Decision Support Research: Bridging Science and Service." When the FY 2006 report was published in November 2005, RISA had eight regional pilot programs under way, but the budget had been reduced to about $1.0 million in FY 2004 and in FY 2005. The president's budget requested an increase to $1.8 million for FY 2006,[134] which is about half the amount appropriated for FY 2002. These figures must be interpreted with caution, however. They do not include other funds available to RISA pilot programs through other parts of NOAA's budget or other federal agencies, or through state funds leveraged by federal funds. On the one hand, the descriptions in the box emphasized that "[f]indings from RISA activities and the development of experimental decision-support tools are proving highly valuable in a range of practical settings." On the other hand, the examples cited emphasized scientific input—model-based climate information, water forecasts, a decision-support system, and a university curriculum—over the particular policy decisions on which the practical utility of the input depends. Moreover, regional studies were presented as means to larger theoretical ends: "Regionally based programs, such as RISA, utilize studies of the application of climate information and rely on human-dimensions research to strengthen the theoretical foundations of decision support."[135] Pending more detailed and systematic appraisal, RISA appears to be a small but potentially important work in progress.

It is worth noting that one of the eight RISA teams, the Climate Impacts Group based at the University of Washington coauthored *Preparing for Climate Change: A Guidebook for Local, Regional, and State Governments* with King County, Washington, in 2007. The *Guidebook* reiterates the rationale for participation by decision makers from the bottom up in adapting to climate change: "Preparing for climate change is not a 'one size fits all' process. Just as the impacts of climate changes will vary from place to place, the combination of institutions and legal and political tools available to public decision makers are unique from region to region. Preparedness actions will need to be tailored to the circumstances of different communities."[136] The *Guidebook* was produced in association with the International Council for Local Environmental Initiatives (ICLEI)—Local Governments for Sustainability and its Climate Resilient Communities Program, which was launched in the fall of 2005 with funding from NOAA. The *Guidebook* makes the connections between its chapters and ICLEI's five milestones for preparation for climate change: "1: Initiate your climate resiliency effort. . . . 2: Conduct a climate resiliency study. . . . 3: Set preparedness goals and develop your preparedness plan. . . . 4: Implement your preparedness plan. . . . 5: Measure your progress and update your plan."[137] Clearly, much uncertainty in local experience must be absorbed in such generalized guidelines. In applying them, most of the work lies in making the connections with the details of unique circumstances in each community. The value of such guidelines depends on their contribution to decisions that actually increase resilience or reduce net losses to climate change, community by community.

The U.S. National Assessment of the Potential Consequences of Climate Variability and Change was another step toward opening USGCRP to case studies in support of policy decisions. Beginning in 1998, experts came into direct contact with stakeholders in a process that included 20 regional workshops and produced assessments of climate change impacts for 16 regions (combined into 9 mega-regions) and 5 sectors in the United States. One of the mega-regions was Alaska; one of the sectors was Coastal Areas and Marine Resources. In an overview report forwarded to the president in November

2000, the National Assessment Synthesis Team (NAST) of experts officially recognized the necessity of adaptation: "[T]he planet and the nation are certain to experience more than a century of climate change, due to the long lifetimes of greenhouse gases already in the atmosphere and the momentum of the climate system. Adapting to a changed climate is consequently a necessary component of our response strategy."[138] NAST concluded that "impacts of climate change will be significant for Americans" and that context matters: "[t]he nature and intensity of impacts will depend on the location, activity, time period, and geographic scale considered. For the nation as a whole, direct economic impacts are likely to be modest. However, the range of both beneficial and harmful impacts grows wider as the focus shifts to smaller regions, individual communities, and specific activities or resources." The headline summed it up: "Large Impacts in Some Places."[139] The assessment focused on climate impacts, not identifying or analyzing strategies for adaptation. Recognizing this limitation, NAST recommended that "the next assessment should undertake a more complete analysis of adaptation. In the current Assessment, the adaptation analysis was done in a very preliminary way, and it did not consider feasibility, effectiveness, costs, and side effects. Future assessments should provide ongoing insights and information that can be of direct use to the American public in preparing for and adapting to climate change."[140]

But the next assessment was blocked. In October 2000, the Competitive Enterprise Institute (CEI), Sen. James Inhofe, and others sued the Clinton Administration to stop production and use of the National Assessment, alleging that it violated certain laws including the Federal Advisory Committee Act. That suit was settled the following September when the Bush Administration agreed to characterize the National Assessment as undertaken by a third party, and not an official position of the administration. In August 2003, CEI sued again after the Environmental Protection Agency (EPA) included National Assessment findings in the EPA's 2002 *Climate Action Report to the U.N.* That suit was settled early in November when the administration inserted in the USGCRP's Web site a clarification and disclaimer that "[t]he National Assessment Overview and Foundation Reports were produced by the National Assessment Synthesis Team, an advisory committee chartered under the Federal Advisory Committee Act, and were not subjected to OSTP's Information Quality Act Guidelines."[141] In June 2005, whistleblower Rick Piltz claimed that "[a] decision was made very early on by this president [Bush] to deep-six the national assessment. Any reference to the National Assessment is continually removed from any document or report."[142]

In the IPCC's Fourth Assessment Report, Working Group II moved toward intensive inquiry in Chapter 17, "Assessment of Adaptation Practices, Options, Constraints and Capacity." Freed from the constraints of a *Summary for Policymakers*, the authors went beyond the point that context matters in *general*. They elaborated the point and pursued its logical implication: case studies documenting how multiple factors interact to shape adaptation outcomes in a *specific* context. Context was clearly an important theme in Chapter 17; it used variations on the word 20 times in 17 pages of text, tables, and boxes. Elaborations of the contextual point included the importance of cross-level linkages, the limitations of national indicators, and related points as documented in the following selections (emphasis added):[143]

- Cross-level linkages: "Most of this literature also argues that there is limited usefulness in looking at only one level or scale, and that exploring the regional and local *context* for adaptive capacity can provide insights into both constraints and opportunities."
- National indicators: "It has been argued that national indicators fail to capture many of the processes and *contextual* factors that influence adaptive capacity, and thus provide little insight on adaptive capacity at the level where most adaptations will take place."
- Vulnerability and adaptive capacity: "A significant body of new research focuses on specific *contextual* factors that shape vulnerability and adaptive capacity, influencing how they may evolve over time."
- Lessons: "The lessons from studies of local-level adaptive capacity are *context*-specific, but the weight of studies establishes broad lessons on adaptive capacity of individuals and communities."
- Technology transfer: "Existing or new technology is unlikely to be equally transferable to all *contexts* and to all groups or individuals, regardless of the extent of country-to-country transfers."
- Human cognition: "It is increasingly clear that interpretations of danger and risk associated with climate change are *context* specific and that adaptation responses to climate change can be limited by human cognition."
- Human perspectives: "Individuals' interpretation of information is mediated by personal and societal values and priorities, personal experience and other *contextual* factors."

Where context matters in these and other ways, research must be centered on the particular case. The significance of each factor depends on the other factors in the same context; and the other factors in the context must be considered comprehensively. Selecting a few factors according to a theory or model, or reducing their significance to a stable relationship or standard measure, absorbs but does not reduce or eliminate uncertainty about the context. The results tend to be misleading or irrelevant. Case studies, however, concretely clarify how context matters.

The most prominent case in Chapter 17 centered on Tsho Rolpa, a glacial lake high in the mountains of Nepal. (This case was not mentioned by name in the *Summary for Policymakers,* which deleted case-specific local information for national and international policymakers.) As explained in Box 17.1, warmer temperatures over several decades melted ice in nearby glaciers causing the accumulation of a large volume of water in the lake by 1997. If the moraine dam containing the lake were to burst, about a third of the water would be released catastrophically, posing a major risk to people, a hydropower plant under construction, and presumably other things of value downstream. "These concerns spurred the Government of Nepal, with the support of international donors, to initiate a project in 1998 to lower the level of the lake through drainage. An expert group recommended that . . . the lake should be lowered three metres by cutting a channel in the moraine. A gate was constructed to allow for controlled release of water. Meanwhile, an early warning system was established in 19 villages downstream in case a Tsho Rolpa [flood] should occur despite these efforts. Local villagers were actively involved in the design of the system, and drills are carried out periodically. In 2002, the four-year construction project was completed at a cost of US$3.2 million."[144] This helped reduce risk but left additional drainage to be done.

Table 17.1 listed nearly a dozen more specific cases of adaptation identified by region, countries, and references in the literature, by climate-related stress (e.g., sea level rise, drought), and by adaptation practices.[145] Such a listing can signal to policymakers elsewhere facing similar problems the existence of experience that might be adapted to their own contexts. The level of detail they need depends on the degree of their interest and concern. For example, a policymaker interested in and concerned about the risk of a catastrophic flood from a glacial lake outburst in another area might need additional information beyond Box 17.1 to address practical questions: What motivated people at various levels to act? Aside from the rising waters of Tsho Rolpa, concern has been traced back to the most catastrophic glacial lake outburst in the area, which apparently serves as the local historical baseline for assessing risks and progress. On August 4, 1985, a flood surge of 10 to 15 meters caused four or five deaths, wiped out 14 bridges, 30 houses, and an almost-completed hydroelectric plant. The disaster certainly impressed

Nawa Jigtar, a monk from the village of Ghat: "If it had come at night, none of us would have survived."[146] But a reasonably dependable understanding would require a more comprehensive understanding of the particular policy process, as suggested by these generic questions: What were the alternatives available to reduce vulnerability? Who supported and opposed them? Who signed off on this alternative? Who paid for it? And so on.

The Tsho Rolpa case study in Chapter 17 is a valuable precedent for future IPCC assessments that include concrete working solutions for the consideration of policymakers on the ground. In addition, the National Assessment and the RISA programs are valuable precedents for intensive research centered on climate-related problems in diverse subnational communities, including collaboration with policymakers having local knowledge of those communities.

Action on Adaptation

Consider in more detail three case studies of action on adaptation to extreme weather events and climate change, located in west Greenland, the Pacific islands, and southeastern Australia. Each is an exception to scientific management and an approximation to the ideal type of adaptive governance (Box 1.1).

In response to climate-related changes in the North Atlantic, Inuit communities on the west coast of Greenland adapted their economies twice in the twentieth century as a matter of practical necessity, not scientific predictions. The first transition was from seal hunting to cod fishing; the second was from cod to shrimp fishing. This case brings into focus the many complex interactions involved in adaptation to climate changes at the regional and local levels over the long term, clarifying what it means to say that context matters. This account is based on the published research of Larry Hamilton and his collaborators.[147]

In west Greenland gradual temperature increases beginning in the latter half of the nineteenth century became significant by 1920 and peaked around 1930. Meanwhile the warm Irminger Current flowing around the southern tip of Greenland reached farther north along the west coast. This current pushed the ice pack farther north, and along with it prime seal hunting areas. Overhunting also contributed to declining seal harvests that could no longer support the growing Inuit population. However, the warm current also allowed cod to migrate into west Greenland waters after 1920 and spawn off southwest Greenland. "By 1930, fishing had replaced seal hunting as Greenlanders' most important economic pursuit."[148] Fishing pressure on mature cod breeding stocks intensified after World War II, but Greenlanders' cod catch was only a fraction of the catch from larger and better-equipped international fishing fleets. During the 1960s, "it became clear that conditions were changing. West Greenland waters were cooling; cod began appearing later each year, and not as far north."[149] The total cod catch peaked in the early 1960s and declined steeply for the next few decades. The decline prompted investments to deploy more efficient technologies that further intensified pressure on dwindling cod stocks. "Falling sea temperatures after 1960, and the particularly cold winters of 1982–1984, could have produced codfish declines even without overfishing. But the interaction between climate and fishing proved deadly."[150] By 1992 the cod catch was insignificant everywhere in west Greenland waters, even in the south.

According to Hamilton and colleagues, "The same environmental pressures—cooling waters and overfishing—that doomed the cod also made shrimp more abundant."[151] The shrimp could survive in the colder waters, and there were few cod or other predator fish left to control the shrimp population. In the 1960s, the shrimp fishing industry began to expand from its origins in Disko Bay in the north. Its catch more than quadrupled from the 1970s to 1995, replacing the cod catch; in 1995 the shrimp industry accounted for nearly three-quarters of Greenland's exports. "The cod-to-shrimp transition appears a roughly even exchange at the national level. Its local consequences, however, were not even at all."[152] Northern municipalities in west Greenland around Disko Bay, where cod were long gone, maintained shrimp production in the period from 1988 to 1998. Sisimiut was one of two mid-coastal municipalities that "lost cod but made offsetting gains in shrimp. Most of the southern municipalities, on the other hand, lost cod and did not gain shrimp. Paamiut, a cod-fishing specialist, experienced the greatest fall of all."[153] These different economic outcomes had significant social consequences. "Municipalities with rising landings all gained population [from 1988 to 1995]. Most municipalities with declining landings experienced low or negative population growth. . . ." Community well-being sometimes is measured by crime rates, which tended to be relatively low in places with high landings of fish and relatively high in places with low landings.[154]

To explain these uneven consequences, Hamilton and colleagues told the tale of Paamiut in the south and Sisimiut 500 km farther north up the west coast. Sisimiut had an enterprising tradition that left it less specialized and better able to adapt. "In 1928, Sisimiut resident Elias Kleist and five partners pooled resources and became the first Greenlanders to purchase their own fishing vessel, the 36-foot *Nakuak* ('strong').[155] This enterprise successfully capitalized on a local canning factory built in 1924 by Danes who had been fishing halibut in the area since about 1910. But as halibut began disappearing under fishing pressure from European vessels, "[t]he manager of the canning factory, Martin Hansen, recognized that shrimp provided one alternative. Four former halibut boats began fishing for shrimp in 1935, and the canning factory was converted accordingly."[156] Kleist and his crew capitalized on the opportunity in 1938; they began fishing for shrimp near Sisimiut while their neighbors ignored this nontraditional resource. Other developments in Sisimiut, including establishment of a shipyard in 1931 and a technical school in 1948, "supported the new industry and paint a picture of an enterprising, politically connected town."[157] When cod landings in Sisimiut fell sharply in 1968, Sisimiut was prepared to shift toward shrimp and other species, and then to maintain its early dominance over would-be competitors as concentrations of shrimp extended down the coast with cooling waters. "Sisimiut was landing, and its workers were thus profiting from, shrimp caught all along the southwest coast."[158] After a new snow crab fishery was organized at Nuuk in 1989, "Sisimiut residents invited Canadian fishermen to teach them crab fishery techniques and began local manufacture of traps. Sisimiut became the center of this new fishery."[159] Thus it sustained the enterprising and diversifying tradition.

Also an early site of commercial fishing, Paamiut developed more slowly and became more specialized under Danish rule and then Home Rule when Greenland gained limited sovereignty in 1979. After World War II, "the Danish government began a lengthy, capital-intensive effort to rationalize the fishing industry. Doing so required not only reorganizing and expanding Greenland's fish catching and processing capacity, but also constructing a modern social infrastructure (including housing, transportation, power, education, training and health services) around it."[160] Some planners were concerned about overdependence on a single species, cod, and encouraged diversification after 1950. Nevertheless, "government planners in the late 1950s identified Paamiut as geographically ideal for industrial development because of its location in the open-water district and access to cod. The government made substantial investments in a new fish processing plant, where production began in 1967, and in large trawlers that could supply this plant with fish, beginning in the early 1970s. For a time, Paamiut's plant was the largest in the North Atlantic."[161] But the economic value of cod landings in Paamiut dropped precipitously as the cod population collapsed. Neither the shrimp nearby nor other resources were sufficient to support large-scale industrial processing in competition with the established shrimp fishery in Sisimiut. "Snow crab were found around Paamiut as well, but production there lagged behind that of Sisimiut [and others] until the government transferred a processing vessel . . . to Paamiut in 2000. This vessel provided welcome economic activity, although it also conformed to the historical pattern of top-down development in Paamiut, contrasting with more local initiatives in Sisimiut."[162]

According to Hamilton and colleagues, "That Sisimiut would prosper during the cod-to-shrimp transition was by no means assured, but rather emerged from the interactions among various kinds of capital—in a path-dependent fashion."[163] This capital consisted of the natural capital of a geographically concentrated source of shrimp, the physical capital of early investment in shrimp processing plants and trawlers, the human capital of a diverse economy and workforce, and finally the social capital of an "enterprising spirit and social cohesion."[164] The latter included Sisimiut's ability to influence from the bottom up central government decisions supporting the fishing industry. Paamiut was relatively passive and dependent on decisions from the top down; the main office of the Greenland Technical Office in Copenhagen took the initiative to construct the processing facility in Paamiut. Hamilton and colleagues generalized these conclusions: "First, socially important environmental changes result not simply from climate change, but from interactions between climate, ecosystem, and resource usage. . . . Second, environmental changes affect people differently through interactions with social factors." In particular, "Social networks and cohesion (social capital), as well as skills (human capital), investments (physical capital), and alternative resources (natural capital), shape how the benefits and costs are distributed."[165] No one of these factors alone tells the story; rather, outcomes depend on how they interact in particular contexts down to the local level. That is why context matters.

The Pacific Islands case involved the 1997–1998 El Niño, one "so intense that scientists have since labeled it 'The El Niño of the Twentieth Century.'"[166] The El Niño–Southern Oscillation (ENSO) occurs on a time scale of about four to seven years. It arises from interactions between the upper ocean and the atmosphere in the equatorial Pacific. It is manifest most clearly in surface temperature changes in the eastern Pacific Ocean that influence a range of atmospheric processes around the Pacific and beyond, including extreme weather events. In the El Niño or "warm" phase, characterized by warmer-than-normal sea surface temperatures in the eastern equatorial

Pacific, rain-producing clouds in the western Pacific tend to shift eastward leaving drought to prevail in the Pacific Islands. The La Niña or "cool" phase is characterized by colder-than-normal sea surface temperatures in the eastern equatorial Pacific. During "neutral" conditions, the upper ocean in the Pacific is on average neither warm nor cool. In the context of the 1997–1998 El Niño, the Pacific ENSO Applications Center (PEAC) successfully field tested a model for research in support of action on adaptation. It was ahead of its time; equivalent models are still only envisioned in most research programs, although there are exceptions.[167] This account is based primarily on lessons learned by leaders of PEAC and documented by them in 2000.[168]

PEAC began in the early 1990s as a pilot project to test the feasibility of integrating climate variability research, forecasts, and application services "end-to-end" on an operational basis. Participants credited Eileen Shea in NOAA's Office of Global Programs (OGP) for securing initial grant support. PEAC was established in August 1994 as a joint venture of the University of Guam, the University of Hawaii, the Pacific Regional Office and the Climate Prediction Center of the U.S. National Weather Service, NOAA OGP, and the Pacific Basin Development Council. It focused on providing seasonal to interannual forecasts of the El Niño–Southern Oscillation and related information products for the U.S.-affiliated Pacific islands: American Samoa, the Commonwealth of the Northern Mariana Islands, the Federated States of Micronesia, Guam, Hawaii, the Republic of the Marshall Islands, and the Republic of Palau.

OGP initially explored how large-scale coupled ocean–atmosphere models could produce ENSO-related forecasts that could be turned into useful information products for Pacific islanders and other potential forecast users around the globe. "However, the spatial resolution of large-scale models . . . did not meet the needs of the people the Center was intended to serve."[169] PEAC supplemented the global-scale model forecast information with empirically based statistical models that "would allow the Center to forecast rainfall on specific islands using historical rainfall data." This adapted research to the different interests of different island communities. It produced "simple guides that describe rainfall and tropical cyclone activities expected under 'normal,' El Niño, and La Niña conditions."[170] Meanwhile, on the applications side, PEAC "conducted workshops, focus group meetings, and local briefings about ENSO in all of the client jurisdictions during 1995 and 1996. . . . From these briefings PEAC identified the concerns of participants on potential impacts of El Niño and La Niña events, and elicited information about the specific kinds of ENSO forecast information needed."[171] Research and applications came together in the *Pacific ENSO Update*, a quarterly newsletter distributed in hard copy and by Web site beginning in August 1996. These resources were in place in February and March 1997 when PEAC found indications of a developing El Niño warm event in several coupled atmosphere–ocean models it consulted regularly. "By May, it was clear that an ENSO event was developing very quickly."[172]

PEAC began alerting clients through the newsletter and informing "government officials that drier than normal conditions could be expected beginning in late 1997 and running through May or June 1998."[173] Working with other members of the PEAC scientific team, Charles (Chip) Guard and the Water and Energy Research Institute (WERI) at the University of Guam produced seasonal forecasts of the percent of normal rainfall for Guam, Micronesia, and Palau, at the request of officials in those places. The forecast for the Marshall Islands was qualified because of mixed signals in the historical record. "By October 1997, PEAC issued the first quantitative rainfall forecasts in the *Pacific ENSO Update*."[174] A month earlier, Guard and a colleague had begun traveling across the affiliated Pacific Islands to brief government officials on PEAC's forecasts and to suggest immediate preparations for impacts to come. In January 1998, Alan Hilton, Cheryl Anderson, and Mike Hamnett gave similar briefings in American Samoa. "The personal briefings were later identified as a key component of issuing the forecasts, gaining an understanding of the situation, and motivating people to action."[175] PEAC experienced some difficulties, including skepticism about PEAC's ability to forecast rainfall up to a year in advance and about the expected drought when briefings occurred during rainstorms in several jurisdictions. In December 1997 Supertyphoon Paka brought rain and destruction to many of the Micronesian Islands, but by January significant rainfall deficits began to spread and eventually impact all the islands. Drought in American Samoa was delayed until April because of its location.

Despite such difficulties, PEAC succeeded in catalyzing action in response to its forecasts. The Marshall Islands, state and national governments in the Federated States of Micronesia, Palau, the Northern Mariana Islands, and Guam developed task forces and mitigation plans similar to those already in place on Yap in Micronesia. The task forces "mounted public information campaigns to inform the public about what to expect from the El Niño, to explain measures that could be taken to conserve water and prevent outbreaks of diseases, and to warn of the increased wildfire risk and actions to reduce risks."[176] All the jurisdictions served by PEAC in Micronesia eventually established El Niño or drought task forces. Consider a few specific actions taken in advance.

> In Palau, the Department of Public Works surveyed the water distribution system in Koror and completed repairs on about 80 percent of the system before the onset of the drought. Throughout the [Federated States of Micronesia], people repaired water catchment systems. In Yap, local vendors were able to supply new household catchment tanks to meet the demand that developed in response to the public information campaign. In the Marshalls, local hardware and building supply companies ordered new catchment tanks, but, unfortunately, they did not arrive until after the drought was underway.[177]

Rainfall forecasts also helped prepare for wildfires associated with drought. For example, "The Government of Guam decided that fresh water from their main reservoir should be conserved, and used brackish water to fight fires. In Yap, firefighters worried about the long-term result of increasing the salt content in the garden areas, so they used water pumped from an old quarry."[178] Most of the islands imposed restrictions on water consumption during drought, but their policy responses differed according to their circumstances.

PEAC was the catalyst pulling together participants and resources scattered throughout a distributed decision-making structure. PEAC put the scattered task forces in contact with each other and provided useful information before and during the drought: "The task forces maintained weekly contact by using PEACESAT satellite teleconferencing, where Chip could regularly inform them of updates and provided technical information through WERI, such as specific catchment design requirements and estimates of water needed per person to withstand the drought."[179] PEAC sought external resources to help the affiliated Pacific Islands cope with the drought. As early as November 1997 Chip Guard recommended that agencies of the U.S. government provide new wells and pre-position reverse osmosis units in certain areas. "PEAC staff in Hawaii actively consulted and discussed these recommendations with federal officials, trying to impress on them the fact that the cost of providing disaster assistance could be reduced significantly should plans be implemented before water needs became critical. Unfortunately, response to these opportunities required a US presidential disaster declaration before the agencies could take action against the impending disaster."[180] That declaration came near the peak of the drought in March 1998. It allowed the Federal Emergency Management Agency to provide assistance to the Marshalls and Micronesia, including six reverse-osmosis units for Yap.[181]

It would be useful to supplement this account with case studies from sources on the ground in the U.S.-affiliated Pacific islands, where most of the action on policy took place. Nevertheless, it seems clear that PEAC's outside assistance was important, if not necessary to help islanders reduce their losses from the 1997–1998 El Niño. In doing so PEAC demonstrated the value of adapting research to the different needs of decision makers on the ground, engaging them directly in their own policy processes, and mobilizing external resources in support of their local policy decisions where needed. Note that PEAC accomplished all this in an advisory capacity; it was policy relevant, not policy prescriptive. Local governments made and implemented the decisions, not PEAC, and agencies of national governments decided to support them. PEAC made the transition from a research pilot project to operational funding around 2001, and it continues to be supported by the National Weather Service as part of the NOAA Climate Program (formerly OGP) and by other related NOAA programs. PEAC still publishes the *Pacific ENSO Update*. Through Cheryl Anderson and others, PEAC's experience in connection with the 1997–1998 El Niño helped inform the emergence of the Pacific Regional Integrated Science and Assessment. The Pacific RISA is being implemented by the East–West Center in Hawaii in collaboration with Anderson and other University of Hawaii colleagues along with partners in the National Weather Service and other NOAA programs and offices.

The southeastern Australian case is centered on the city of Melbourne in the state of Victoria. In response to three of the most severe droughts on record, Melbourne dramatically reduced per capita water consumption and institutionalized permanent water-saving rules accepted by the public. In doing so it demonstrated how climate-related extreme events can help people on the ground adapt water policies for sustainability. This account is based primarily on public documents and interviews with former Deputy Premier and Minister for Water for Victoria John Thwaites.

Drought is a natural part of life in Australia, the driest inhabited continent on earth. But dry conditions generally have been relieved within a year or two as part of the ENSO cycle. Historically, ENSO has imposed a four- to seven-year cycle of drought and heavy rain on the northern and eastern parts of the continent, thereby influencing human settlement patterns and, more recently, norms of water usage. The influence of climate change on the frequency and intensity of the El Niño (dry) and La Niña (wet) events remains uncertain.[182] Nevertheless, three of the most severe droughts in the instrumented record have occurred in the last two decades, in 1994, 2002, and 2006, with average annual rainfall deficits in Melbourne of around 180, 250, and 210 millimetres, respectively, compared to the 1855–2008 average.[183] The drought of 1994 was part of an unusually persistent, multiyear El Niño event. During 10 months from March to December 1994, rainfall was in the lowest 10% of observations over the last century across most of Australia. The drought effectively ended with very heavy rain and flooding over northern Victoria in January 1995. Another period of below-average rainfall occurred from February to April 1995 before higher rainfall returned throughout Victoria. However, the ample rainfall typical in previous decades across agriculture-intensive southeast Victoria has never fully returned.

Just prior to the drought of 1994, water professionals in the state authority, Melbourne Water, were exploring innovative approaches to water management. This included water-sensitive urban design, a radical idea that grew out of academic conversations on waterway health at that time, according to water engineer Tony Wong.[184] These innovations were developing as the state government was disaggregating Melbourne Water into "a wholesale water, drainage and waterway authority and three new retail water supply and sewage businesses."[185] This process dispersed the cohort of water professionals, but they maintained working relationships through an informal network that spanned government, private sector, and academic boundaries. They have continued to provide leadership and innovations for a more comprehensive approach and have helped resolve political differences.[186] Following the 1994 drought, the network promoted a more comprehensive approach to urban water management; as it began to gain acceptance, opportunities emerged to field test ideas. For example, the Melbourne Water executive began to encourage its staff to actively use existing authority to subject approval of new developments to drainage and water-quality conditions.

Support for comprehensive urban water management became widespread in the public arena after the 2002 drought. The El Niño event that year was weak but had a very strong impact in Victoria. The drought ranked in severity and areal extent with the most extreme droughts of 1902 and 1982–1983 (both El Niño periods) and exacerbated the effects of eight years of ongoing rainfall deficits. The drought also coincided with exceptionally warm conditions: maximum temperatures established by a wide margin new records for autumn, winter, and spring in the period since 1950. University researchers and the Australian Bureau of Agricultural and Resource

Economics estimated the drought nationwide cost more than 40,000 jobs (and perhaps as many as 70,000), decreased agricultural output by 30%, decreased economic growth from 3.8% to 3.1%, and created a 50% increase in the trade deficit.[187] Government committed drought relief totaling $728 million (AUD) over three years to farmers and businesses in affected areas. Because of severe economic consequences, extensive bushfires in eastern Victoria, and widespread water shortages, people began to recognize that this drought was unusual.[188] Another rather dry year in 2003 reinforced the impression that present conditions were not only an exception to the historical cycle but manifestations of permanent climate change.[189]

After the 2002 election, water became "one of a few key areas of emphasis for the government," according to John Thwaites. "The fact that water fell under the portfolio of the Deputy Premier (me) indicated the importance of water to us and to the community." The process for constructing a strategy for the state was "a classic process of scientific report, followed by a green paper, followed by extensive consultation, followed by the white paper...."[190] The scientific report led by researcher Nancy Millis was *Water Smart City*, an intensive integrated assessment of local rainfall patterns, geography, demographics, infrastructure, and options for responding to expected changes.[191] This report led to a green paper in 2003, *Securing Our Water Future*, which generated a consultation process that included "600 submissions, public meetings, farming community meetings, business groups, environment groups, right up to the highest level," according to Thwaites.[192] Subsequently, the Victorian state government released the white paper in 2004, *Our Water Our Future*. The priority in urban areas was water supply; water quality and maintenance of environmental flows were priority concerns in rural areas. As noted by Thwaites, "the consequences of running out are dramatic, to say the least." The white paper proposed a *Central Region Sustainable Water Strategy* to integrate water resource planning, including rivers, reservoirs, aquifers, recycled water, storm water, and seawater, for Melbourne and the surrounding areas. The white paper was the basis for a series of amendments to the 1989 Water Act—collectively, a portfolio of more than 100 actions to improve water conservation, more efficient use of water, and river health throughout Victoria.

A goal of the amendments was to reduce urban water demand in Melbourne by 15% by 2010. Provisions enacted to meet the goal included the Stormwater and Urban Water Conservation Fund to "support local scale innovative water sensitive urban development initiatives, stormwater conservation and water recycling initiatives across Victoria."[193] Another provision reformed water pricing with a block tariff system: each additional block of water use cost more per unit; the penalty for high blocks of water use was quite severe.[194] Other provisions provided rebates for water-saving devices such as more efficient shower heads and established a major campaign to change water-consumption behavior. The public campaign included research on public attitudes concerning water, education on water conservation, and advertising. Through these and other provisions, the Victorian approach exceeded its goal: It achieved a 22% reduction in urban demand in Melbourne by 2006. [195] Explaining this success, Thwaites noted that "community input was integral to the successful implementation of the Strategy's initiatives." Moreover, "you can never do it with just ads. Financial incentives and penalties need to be part of the strategy. You can be fined, but no one has ever been fined. Community pressure does the job."[196] Community pressure is a manifestation of the consent of the governed, as well as a substitute for the application of penalties for noncompliance that were approved in the amendments.

Another effort to secure Melbourne's water supply picked up on the *Central Region Sustainable Water Strategy* presented in the 2004 white paper. The effort began with a discussion paper, released for comment in October 2005, which attracted about 80 comments during the two-month submission period. An independent panel chaired by freshwater ecologist Peter Cullen was appointed in February 2006 to consider the public comments and make recommendations. The draft strategy was released in April 2006. In the Foreword, Thwaites stated that "[p]riority is given in the Strategy to conserving water and demonstrating its efficient use before considering alternative supplies and additional augmentations."[197] Alternative sources of water, including recycling, storm water treatment, and desalination, were future possibilities for ongoing investigation. However, the context was evolving. The extreme failure of winter and spring rains in the drought of 2006 produced inflows to the Melbourne catchments 30% lower in 2006 than the previous driest year and less than half the 1997–2006 annual average.[198] As Thwaites noted, "We got it wrong. No one realized how bad it was going to get."

Thus the final strategy, released in October 2006, had to account for reductions in streamflows during the previous decade that were somewhat more severe than had been projected for 2055 as a result of medium climate change. The strategy highlighted an "adaptive management approach" with "a diversity of options" to minimize the risk of running out of water. It relied on 112 separate actions, including the introduction of on-the-spot fines for breaching water restrictions (Action 3.4) and the establishment of an environmental entitlement for the Yarra River in the city of Melbourne of 17,000 megaliters (Action 4.39a). Action 3.27 required the government to complete a feasibility study of desalination options for Melbourne, and if feasible to develop a business case. In June 2007, Premier of Victoria Steve Bracks announced plans for a desalination plant to increase water supply. At a capital cost of $2 billion (AUD), it would supply about one-third of Melbourne's water, 150 gigaliters per year, with annual operating costs of $70–100 million (AUD).[199]

Bracks invoked a consensus on climate change to justify the cost: "[P]eople in Victoria realize that water prices have to go up to account for new infra-structure because of drought, because of climate change. We have to find new water and there is a cost to that." Thwaites noted that "the desalination plant is about diversity of supply. At this stage conservation won't do it." Nancy Millis, who led the first study in a *Water Smart City*, commented that "[t]he prolonged drought across Australia reminds us that dams are of short term value if populations increase and rainfall seriously declines. Alternate sources of supply exist. . . . The technologies to convert these sources of water to water 'fit for purpose' are now well established and have been safely used for years in a number of countries."[200] Don Bursill, another water re-searcher, said that a "desalination plant is a sensible approach as part of an overall strategy."[201] However, political opposition to this highly visible target emerged rather quickly. And Liberal Party planning spokesman Matthew Guy said the government was treating its planning and environmental pro-cesses "with contempt."[202] Jorg Imberger, of the Centre for Water Research, called it "a technological fix."[203] The desalination plant project was conceived in the tradition of scientific management as the best instrument to realize a single target efficiently through a discrete decision. It remains contested.

The policy process begun in 2002 was transparent and open to partici-pation by anyone interested, and at the same time it was coordinated by like-minded water professionals scattered across sectors but centered in Melbourne Water. Thwaites observed that "Melbourne Water has been a critical part of our success. . . . Retailers were initially more skeptical, be-cause they are judged by corporate performance (like energy suppliers), but the message got through. Increasingly they saw that sustainability is the correct way forward." Thwaites also noted that "extensive consultation sup-ported public acceptance of the provisions of the Act. It was a transparent process." It was also focused intensively on local water problems through scientific reports, green papers, and white papers, which through public consultation brought expert and lay perspectives to bear on water problems highlighted by droughts. The policy process was procedurally rational in the face of multiple uncertainties, including weather, climate, and the feasibility and effectiveness of policy alternatives. The series of strategies implemented and proposed contributed to adaptation through learning by doing. Leaving aside the proposed desalination plant, each strategy consisted of a portfolio of relatively modest policies. The number of policies meant that progress did not depend on planning projections about the effectiveness of any one of them. The relatively modest scale and consequences of each policy made it easier to resolve differences among interest groups, and to experiment: The policies that failed field testing were relatively easy to terminate, freeing resources to build on those that succeeded.

Certain aspects of procedural rationality are understood as "good public policy" by one practitioner, Mike Waller, who has held senior positions in the U.K. Treasury and in Australian government, including head of microeconomics for Prime Minister Bob Hawke. As chair of Sustainability Victoria, Waller commented on the issue of democracy raised in a roundtable discussion on the cultural and political challenges of climate change at the University of Melbourne in August 2008: "In responding to climate change, we need to keep in mind three principles of good public policy." While editing notes on his comment, Waller distinguished between resilience and experimentation in the third principle, resulting in these four:

- "Don't build another fundamentalism. By that I mean that we should not focus solely on climate change or assume we understand all the connections in a complex interacting system (e.g., demographics, social/cultural attitudes, peak oil, technological change)."
- "Provide an integrated narrative—it isn't all about climate change. People have other, more immediate concerns—equity, security, maintenance of livelihood, and the like. We have to empower people to make their lives. Give them knowledge and power and there is willingness and ability to change."
- "Adopt measures that build social, economic and environmental resilience—this means addressing change at all levels and accepting the need for less focus on pure efficiency in order to build in a cushion to deal with unexpected changes and pressures."
- "Experiment and be ready to accept failure. Democracy, for all its faults, can change direction. If we conduct many experiments at once, we can keep what works and discard what does not work. We must embrace failure—allow our politicians to fail—and learn from it."[204]

There is nothing here advocating a discrete, comprehensive policy to realize a given target efficiently, as the one best way that rises above politics on a scientific foundation. Instead, Waller's practice-based principles converge on adaptive governance as we understand it (Box 1.1).

Looking back over these cases of action on adaptation in Australia, the Pacific islands, and Greenland, it is reasonable to conclude that there is no one-size-fits-all solution for reducing net losses from climate change. Instead, context matters: The effectiveness of any policy initiative depends on particular circumstances. Among these circumstances are innovators and leaders like Elias Kleist in Sisimiut, Chip Guard in Guam, and John Thwaites and the water professionals in Melbourne. These are real people, not the cardboard caricatures sometimes substituted for scientific or managerial purposes. Real people do not have full or fully objective information on the particular context. Nevertheless, despite uncertainties, they choose to act (or not act) on their limited understandings. Science improved those understandings in some cases, but that depended on integration with local knowledge. Their actions inevitably are matters of trial and error in some degree; sometimes they pay off more or less as expected, and sometimes they do not. The diversity of circumstances, trials, and outcomes on the ground frustrates centralized decision making from the top down in the tradition of scientific management. But that diversity is an asset in adaptive governance. Among other things, diversity provides opportunities for policy innovation and learning by doing in many communities at the same time—and, as we have shown here, for harvesting experience on what works, and for diffusing that experience to inform choices and decisions elsewhere. This basic pattern turns up in cases of action on mitigation in the next section, and the Barrow case in Chapter 3.

Action on Mitigation

Exceptions to scientific management in the established climate change regime also include local, state, and private sector initiatives to mitigate greenhouse gas emissions. There are diverse and multiple paths to emissions reductions, even among the local communities of primary interest here. The general tendency, however, is consistent with adaptive governance: Progress is based on the experience of leading communities, which is diffused through networks to other local communities for possible adaptation on a voluntary basis. Sometimes communities influence from the bottom up higher-level decisions to support what worked on the ground.

The village of Ashton Hayes in Cheshire, U.K., cut its CO_2 emissions by 21% in the few years after May 2006, the baseline for annual household surveys conducted by the University of Chester.[205] The village's project, Going Carbon Neutral, demonstrates a working, partially field-tested model for other communities to consider. Ashton Hayes resident Garry Charnock organized the project after attending a debate on climate change sponsored by Greenpeace and featuring the government's Chief Scientific Advisor, Sir David King.[206] Charnock and Roy Alexander, a village resident and technical director of the project based at the University of Chester, coauthored *A Practical Toolkit for Communities Aiming for Carbon Neutrality*. Our account is based primarily on the *Toolkit*, the project's Web site, and its 16-minute film.

The *Toolkit* outlines the steps taken and lessons learned in Ashton Hayes as guidance for other communities. It begins with promoting carbon neutrality from the bottom up: "[F]irst try out your ideas and thoughts on some trusted friends within the community. If you find they support your idea, take the next step and chat to respected individuals in your community such as landowners, leaders of clubs and societies, the church and most of all the primary school teachers. If you get a warm and generally supportive response go outside your village and engage your local environment representatives such as your Town & County Councils and explain to them that people within your community are keen to help and do something about climate change."[207] The Carbon Neutral project team sought critical democratic authority and administrative support from the Ashton Hayes Parish Council. The project obtained that support in mid-November 2005, after promising the council that "our team would take responsibility for running it and we would generate the finance it needed without using monies from the Parish Council Precept." The Parish Council insisted on three things in return for its support. First, one team member would take the vacant seat on the Parish Council; Charnock took that seat and responsibility for running the project. Second, the team would test "the idea in an open forum to gauge the level of village support"; the project launch two months later served that purpose. And third, the team would "share any experiences of our journey with other communities"; that was done through the *Toolkit*, talks with other villages in the United Kingdom, and other channels.[208] Support from the Parish Council catalyzed interest from the Cheshire City Council and the news media, and generated further interest and understanding among residents of Ashton Hayes.

To prepare for the project launch that would gauge the level of village support, the team gained crucial support from the Energy Saving Trust and primary schools. As the team reported, it also "identified all the business leaders and owners in the village and asked if they would consider providing money or support-in-kind for the launch. After a quick series of phone calls (many via personal contacts) we raised £3,500 and several businesses also offered to send staff and exhibition stands to the launch."[209] The project called them "sponsors" and gave them publicity through its Web site and other channels. The project identified 13 business sponsors with local connections, including Technical Editing Services (TES), a communications company founded by Charnock and his wife; the RSK Group, one of the United Kingdom's leading environmental consultancy firms; and Shell Global Solutions.[210] It also identified 15 governmental, educational, and other nonprofit organizations that provided technical support. Evidently, these organizations and business sponsors shared a common interest in the project. Publicity for the launch included signs, direct mail to homes, and press kits. The project was covered by local newspaper, radio, and TV in advance of launch night, and by the BBC NW 6 o'clock news and Granada TV on launch night.

Launch night was January 26, 2006, a cold winter night; nevertheless, 400 people turned out and heard various short presentations, less than 10 minutes each, explaining what the Going Carbon Neutral project planned to do in coming months—for example, a baseline survey of CO_2 emissions in Ashton Hayes by Alexander's students at the University of Chester. Charnock and Alexander observed that "[t]his was a remarkable turnout from a community of just 1,000 people. Previous village meetings had never drawn more than 40 or 50 people."[211] In the *Toolkit* they reported what they learned from participants in the launch: "[A] number of people were very concerned about climate change but were anxious about taking individual action (such as installing wind turbines or solar panels) as they thought they might be considered rather cranky." Richard May summarized their feelings in his comment: "Having a community-wide project 'gave everyone permission' to take action. We also learned that people respected the people running the event because most of them had lived in Aston Hayes for years. People were also pleased that they were not being 'sold' ideas, just being asked to do whatever they could to stem climate change." In addition, the BBC's World Service Radio covered the launch, which "reassured village residents that this was a worthwhile project and that the outside world would support rather than ridicule our efforts at addressing climate change."[212] Deference and respect for local people, their personal contacts, and their identifications with the local community are resources that have been underestimated in the climate change regime.

Activities toward achievement of the carbon neutral goal are reported in less detail than promotion of the project. They included continuing communications through the Ashton Hayes primary school, news media, and the project Web site. These sources provided answers to the question, "What can we do?" In the project's film, resident Richard Holland answered for himself and apparently others: "Basically, try to cut down on the amount of energy that we use. Don't waste energy is the main thing. Don't leave things switched on. I don't use the car unless I absolutely have to. Think about energy, what they are using, all the time. Think whether or not they actually need to use that energy." In addition, "In the first five months, some residents have implemented energy-saving changes, by, for example, installing loft insulation, using the tumble dryer less, and switching to energy-saving lightbulbs."[213] Residents also have installed solar thermal panels and planted trees, to offset carbon emissions or for a renewable source of fuel. The *Toolkit's* extensive section on measuring carbon footprints advises that "[i]t is rare to get anything right the first time. Inevitably there will be mistakes and it is important to recognise and learn from these as you reflect on each stage of the process."[214] Mike Waller and like-minded advocates of good public policy would agree.

With the rise in project activity, the "team of active volunteers expanded rapidly to over 30 people. . . . To keep the decision making fast, efficient and free of 'red tape' we allocated members to one of 5 'autonomous' teams that would manage key aspects of the project."[215] Four of the autonomous teams were for Technology/Design, Carbon Neutral Clinics, Carbon sinks, and Conferences & Exhibitions. They report to the fifth team at the center, the Media team.[216] The original project team adopted and retained "simple project guidelines that we named the 'Big Rules' . . . [that] help us all steer the same path and avoid conflicts in the village."[217] In management of the project, it appears that each small accomplishment put the project in a position to gain more support, including financial support. For example, "Having business sponsors also attracts the interest of other organizations, such as the Cheshire Community Council, which donated £1000 for a project computer when they saw the wide range of community participation in our project." The project was awarded a two-year £26,500 grant from the Climate Challenge Fund of the Department for Environment, Food, and Rural Affairs to help the project spread the word about its activities. "The University of Chester has also secured knowledge transfer funds and hopes to win a Green Gown award for the project."[218]

The project's self-appraisal concluded that "[i]t is difficult to pinpoint any single element of the project that has been crucial to success. Synergy seems to be the key." Among the critical factors, "The 'Big Rules' we adopted . . . have helped to prevent conflict and avoiding engendering feelings of guilt or fear associated with climate change. . . . The idea is to encourage people to participate in whatever way they can without pointing the finger or criticizing anyone's lifestyle. There is also no doubt that the immense media coverage we have enjoyed has also energised the population. . . ." Other critical factors explaining progress to date were the support gained through the primary school and the technical support and long-term commitment of two nearby universities, Chester and East Anglia. In addition, project collaborators considered whether their experience in Ashton Hayes was relevant elsewhere. As Garry Charnock reported, "We began to wonder if Ashton Hayes was perhaps a 'special' community and if it would be hard to replicate this enthusiasm elsewhere." They concluded that "[w]e are probably better-off than the average UK community but not markedly so. . . . But we have discovered that we are not special. Over the past year we have given talks to almost thirty communities around the UK and have come to realise that every one of them is similar to us. . . . And they are all keen to do something about climate change. However, each community faces a different challenge. This is why it is important to share our ideas, try out new and sometimes innovative concepts and shape our experience, together."[219]

Those involved in the Ashton Hayes project seem to understand the rationale for decentralized decision making (Box 1.1), judging from these voices in the project's film:

■ On expanding participation, Derik Bowker, resident and member of the Cheshire Community Council: "It's brought a variety of people in that wouldn't normally get involved in community projects, and it's been very beneficial for other areas within the Chester district, because they've come on board and joined in as well." Narrator Marie Friend added that "[w]e have been delighted with the response from over 20 communities around the UK, and as far afield as Castlemaine in Australia."

■ On learning across communities, Heather Barrett, Carbon Neutral co-ordinator, Castlemaine [Victoria], Australia: "I think it's really useful for Ashton Hayes and Castlemaine, and any other town in the world that's trying to do this same sort of thing, to reach the goal of being carbon neutral because we can learn from each other, we can learn so much. Our experience will be different because we've got different government, we've got different solar energy, we've got all sorts of different things in our communities, but we can learn so much basically about behavior change, what motivates people—that will be the same all over the world."

- On bottom-up strategies, Joan Fairhurst, member of the Chester City Council: "Ashton Hayes is really breaking new ground, and it's because it's a bottom-up effort, it's a community effort, rather than being told to do something. Government is very good, isn't it, at actually saying, you know, we should all be doing this, and then not doing it themselves. It's much more positive, I think, if it's the community who is saying this is what we feel we can actually do, and do it in a way that the local community benefits very significantly."

- On integrating common interests, Tracy Todhunter, Carbon Neutral group: "It gives me hope that small people in small communities can make a huge difference to climate change. That we can actually act together, work together to reduce our energy consumption and our greenhouse gas emissions, by all working further on small things that are cumulative. It's also been good . . . that people are getting to know each other, they are working together, they are talking to each other. . . ."[220]

Such perspectives, reflecting experience on the ground and hopes for the future, are also resources that have been underappreciated in the climate change regime.

The Danish Island of Samsø, located in an arm of the North Sea, demonstrates another model for climate change mitigation. According to reporter Elizabeth Kolbert, a few of its residents began to reconsider the island's energy system in the late 1990s. "By 2001, fossil-fuel use on Samsø had been cut in half. By 2003, instead of importing electricity, the island was exporting it, and by 2005 it was producing from renewable sources more energy than it was using."[221] Samsø has achieved carbon neutrality and more, by averting more CO_2 emissions from fossil fuels than it releases into the atmosphere. Our sources trace the immediate origins of Samsø's effort back to a contest organized by the Danish Ministry of Environment and Energy to promote innovations in renewable energy through planning. A nonresident engineer in consultation with the mayor of Samsø drew up a plan to wean the island from fossil fuels and submitted it to the Ministry. Samsø won the contest in October 1997 when the Ministry deemed its plan "most likely to succeed," but received essentially nothing except designation as Denmark's "renewable-energy island." On Samsø "the general reaction among residents was puzzlement. . . . One of the few people on the island to think the project was worth pursuing was Søren Hermansen."[222]

Intrigued by the concept of a "renewable-energy island," Hermansen became the project's first employee after federal money was found to fund a single position. Not much happened initially. According to Hermansen, "There was this conservative hesitating, waiting for the neighbor to do the move."[223] As a resident of the island, Hermansen understood this tendency and learned how to make good use if it. Kolbert reported, "Whenever there was a meeting to discuss a local issue—any local issue—Hermansen attended and made his pitch. He asked Samsingers to think about what it would be like to work together on something they could all be proud of. Occasionally, he brought free beer along to the discussions [in accord with local custom]. Meanwhile, he began trying to enlist the support of the island's opinion leaders."[224] He appealed to multiple interests: "When I go out to explain to people why moving to renewable energy is a good thing, it helps to make the economic argument about saving oil costs, selling wind power and getting Government money. But there is a good feeling among Danes about self-sufficiency and the environment I can play on, too. Danes have a romantic attachment to the idea of leaving land unharmed."[225] The strategy of building interest and support incrementally from the bottom up paid off on Samsø as it did in Ashton Hayes: "As more people got involved, that prompted others to do so."[226] Thus, broad support was an outcome of getting started on multiple initiatives appealing to a variety of particular interests; broad support was *not* a precondition for getting started on any one initiative.

At the outset in 1998, the project arranged many public meetings on wind turbines to harvest energy from winds that blow continuously off the North Sea. Soon it became "obvious that there was no shortage of potential investors. In particular, farmers with potential wind turbine sites on their land were keen to invest in their own wind turbine." Moreover, the initiative's scheme of "spreading ownership also greatly improved citizen acceptance for the erection of these wind turbines."[227] Samsingers constructed and operate 11 large land-based wind turbines, each about 230 feet tall. Together they produce 26 million kilowatt hours per year, "which is just about enough to meet all the island's demands for electricity." In addition, the island has 10 offshore turbines that are even larger, about 300 feet tall. Together they produce about 80 million kilowatt hours per year, enough to power about 20,000 homes at normal rates of consumption in Denmark. Kolbert reported that the offshore turbines "were erected to compensate for Samsø's continuing use of fossil fuels in its cars, trucks, and ferries . . . in the aggregate, Samsø produces about ten per cent more power than it consumes." The turbines are owned individually and collectively by many of the island's 4,300 residents plus some nonresidents. "At least four hundred and fifty island residents own shares in the onshore turbines, and a roughly equal number own shares in those offshore." Community ownership continues to be important. According to Hermansen, "We care about the production, because we own the wind turbines. Every time they turn around, it means money in the bank. And, being part of it, we also feel responsible."[228] Much of the wind power is sold to off-island consumers by utility companies. They were required by the Danish government to offer 10-year fixed-rate contracts to buy electricity. That reduced risk to investors in Samsø's wind turbines, who expected the investments to be repaid in about eight years.

In 1998 the Samsø municipal council made participation in district heating systems voluntary for all existing homes, but mandated participation by new buildings in areas with existing or planned district heating systems.[229] Samsø has district heating plants in 4 of its 22 villages. Those in Ballen, Tranebjerg, and Onsbjerg burn straw, while another in Nordby burns wood chips from fallen trees. The plants heat water, which is piped underground to deliver space heat and hot water to homes nearby—260 of them in Ballen and neighboring Brundby, for example. The burning of biomass fuels releases CO_2 into the atmosphere, but the CO_2 would have been released anyway, by burning straw in the field or by letting fallen trees rot. These are CO_2-neutral biomass fuels because new plant growth absorbs the same amount of CO_2 that is released when they are burned. In contrast, burning fossil fuels releases CO_2 that otherwise would have remained sequestered in the ground. Behind the Nordby district heating plant is "a field covered in rows of solar panels, which provide additional hot water when the sun is shining."[230] Moreover, "Samsø is experimenting on a modest scale with biofuels: a handful of farmers have converted their cars and tractors to run on canola oil. [Kolbert and Hermansen] stopped to visit one such farmer, who grows his own seeds, presses his own oil, and feeds the leftover mash to his cows." Perhaps because these innovations appeared to have been successful, Kolbert "asked Hermansen whether there were any projects that hadn't worked out. He listed several, including a plan to use natural gas produced from cow manure and an experiment with electric cars that failed when one of the demonstration vehicles spent most of the year in the shop. The biggest disappointment, though, had to do with energy consumption. . . .

Essentially, he said, energy used on the island has remained constant for the past decade."[231] In contrast, reducing energy use was the major achievement in Ashton Hayes.

Samsø's own evaluation 10 years after the initial 1997 plan covers changes in a range of locally important value outcomes: self-sufficiency through ex-ploitation of local resources; the production and consumption of heat and electricity; transportation; investments, savings, and employment in the local economy; communication and renewable energy tourism; local involvement and participation; and environmental benefits. Investments in the project paid off for the local economy. As a net energy exporter, Samsø no longer has to import energy that cost the local economy about $10 million (USD) in 1998. The investments were overwhelmingly local: Over the 10-year period, $65 million of the total investment of $75 million (USD) came primarily from local investors; the remaining $10 million came from subsidies by the Danish government and European Union.[232] "Both the total investment and the total subsidies have been much lower than originally predicted" in the initial plan in 1997. The difference was attributed in part "to the many projects not yet effectuated and the projects abandoned, especially because their economic feasibility was too poor."[233] Thus, the initial plan was not a straitjacket but a guide that was adjusted in the course of implementation. The evaluation emphasized local participation and leadership as explanatory factors: "The mobilization of the local populace and the spheres of cooperation established between local participants and interest groups has been exceptional, and a central factor to bear in mind when explaining why the project has achieved such good results. The Samsø Energy Company's staff have been acknowl-edged and credible 'whips' in touch with reality and what was feasible."[234]

In his interview with Kolbert, Hermansen was not sure why the renewable-energy-island effort had made progress "because different people had had different motives for participating. 'From the very egoistic to the more over-all perspective, I think we had all kinds of reasons.'" When Kolbert asked about the lessons from Samsø for other communities, Hermansen replied, "We always hear that we should think globally and act locally . . . I under-stand what that means—I think we as a nation should be part of the global consciousness. But each individual cannot be part of that. So 'Think locally, act locally' is the key message for us."[235] What Kolbert added is also relevant to understanding the local motives and broader relevance: "The residents of Samsø that I spoke to were clearly proud of their accomplishments. All the same, they insisted on their ordinariness. They were, they noted, not wealthy, nor were they well educated or idealistic. They weren't even terribly adventure-some. 'We are conservative farming community' is how one Sam-singer put it. 'We are only normal people. . . . 'We are not some special people.'" Kolbert concluded that "Samsø transformed its energy system in a single decade. Its experience suggests how the [global] carbon problem, as huge as it is, could be dealt with, if we were willing to try."[236] Meanwhile, in Samsø, "The process has not stopped, but continues to develop and still generates many new initiatives and products."[237] It also continues to attract visitors of many kinds from around the world who, like Kolbert, are interested in what has been accomplished there.

Toronto became one of the first cities to set targets and timetables to reduce its greenhouse gas emissions. For Toronto's initiative, Henry Lambright and his collaborators credited Tony O'Donohue, an influential conservative politician who was deeply impressed and concerned by what he learned at the World Conference on the Changing Atmosphere in Toronto in 1988. The Toronto City Council unanimously adopted three recommendations of an advisory committee in January 1990. These declared as a target reducing Toronto's CO_2 emissions by 20% compared to 1988 emissions by 2005; authorized a Toronto Atmospheric Fund (TAF) eventually endowed with $23 million (CND) to help finance demonstration projects; and created a city Energy Efficiency Office to support energy-efficiency improvements in city-owned buildings and vehicle fleets and street lighting, and to review new building proposals for energy and water use. In addition to O'Donohue's political leadership, climate change action in Toronto was supported by "the long tradition of environmental consciousness in the city" and by "widespread recognition that economic and other benefits will accrue from initiatives in the name of climate change."[238] Thus the precautionary principle and "no regrets" were part of the motivation.

Toronto Mayor David Miller was invited to an international climate conference in May 2007 in New York, where he formally announced a new plan for Toronto that included a greenhouse emissions reduction target of 30% by 2020. "Emissions from municipal operations have dropped, but the [original] overall goal fell by the wayside," in part because Toronto, like other cities, lacks control over some of its environmental problems.[239] Progress in Toronto is more apparent in innovative programs; one uses deep water in Lake Ontario to cool buildings downtown, for example. Toronto leaders take some pride in their innovations and international leadership role. "Key elements of the ambitious conservation plans proposed for both New York and London are direct copies of programs developed in Toronto, according to Phillip Jessup, TAF Director. 'Toronto can shine because we have developed and piloted programs that have been picked up by other cities,' he said. 'But Toronto can learn as well. There is a lot we can learn from Berlin and some other cities.'"[240]

Toronto and many other local initiatives on mitigation are affiliated with ICLEI.[241] ICLEI was headquartered in Toronto City Hall when it was established in 1990 under auspices of UNEP and the International Union of Local Authorities to represent local governments in the United Nations. One of ICLEI's early projects was the Urban CO_2 Project. In 1991 it involved Toronto and 12 other major North American and European cities working together to reduce their CO_2 emissions.[242] The project was expanded into the Cities for Climate Protection (CCP) campaign in January 1993. By 2006 CCP operated in 27 countries and included 546 local governments with about 8% of the world's urban population (or 243 million people) and about 20% of all *urban* greenhouse gas emission.[243] In its 2006 progress report, ICLEI International emphasized its original aspiration to organize and influence from the bottom up: "[L]ocal governments engaged with ICLEI are speaking in concert, much louder than they could on their own. . . . ICLEI will ensure that we speak even louder, to influence other levels of government and to continue to prove: *Local action moves the world.*"[244] ICLEI's "basic premise is that locally designed initiatives can provide an effective and cost-efficient way to achieve local, national, and global sustainability objectives."[245] ICLEI USA's office elaborated these aspirations in its review of 2006: "Our work targets and impacts the places where most people live, where most energy is consumed, and where most greenhouse gas emissions are produced. These are also the places where solutions to global warming will be designed, implemented and led, city by city, day by day." An affiliate of ICLEI USA, the U.S. Mayors Council on Climate Protection, aspires to have mayors "work together to share best practices, develop regional solutions, and inform the federal government about effective policies local governments need to keep improving."[246]

In accord with its aspirations, ICLEI has attempted to stimulate the sharing of successful innovations among cities. Most of ICLEI's 2006 progress report described particular programs in particular communities and their benefits. Examples included more efficient lights, water boilers, and timers in municipal buildings in Ekurhulni, South Africa; a mandate to cover 60% of water heating needs through solar heating in new and retrofitted buildings in Barcelona, Spain; a light rail system in Calgary powered by 12 windmills in southern Alberta, Canada; and a facility that generates electricity from landfill gas in São Paulo, Brazil. Comparable examples of decisions influenced at higher levels of government are missing from ICLEI's progress report, despite ICLEI's original aspiration. Beyond cases, it is difficult to evaluate the role of Toronto or the CCP campaign in the international network. Among other things, measures of greenhouse gas emissions and related benefits and costs are seldom standardized, if they are measured at all, and even standardized measures are difficult to interpret or explain because outcomes depend on multiple supporting factors and constraints that differ considerably from place to place. Information sent out from networks centered on Toronto or CCP is only one of many factors that might help explain outcomes elsewhere.

However, in 2006 ICLEI International claimed that CCP participants over the last 10 years had realized reductions in greenhouse gas emissions equivalent to approximately 60 million tons of CO_2 per year, or 600 million tons for the decade. "Together CCP Campaign participants annually save an estimated USD $2.1 billion in associated fuel costs. These savings do not take into account other community benefits—like improved air quality and public health, new product markets and job opportunities."[247] The recent spin-off from CCP, the U.S. Mayor's Climate Protection Agreement, was modeled on the Kyoto Protocol and formalized in June 2005. John Bailey of the Institute for Local Self-Reliance conducted a preliminary outcomes appraisal based on 10 cities that had signed it. He concluded that "it is unlikely that more than one or two of our ten cities and quite possibly none will reduce their GHG emissions 7 percent below 1990 levels by 2012. Overall emissions increases ranged from 6.5 percent to 27 percent from 1990 baseline measurements. An exception was Portland, Oregon, which reports a tiny 0.7 percent increase above the 1990 baseline."[248] Nicholas Kristof was more enthusiastic in 2006, especially about Portland. "In 1993, the city adopted a plan to curb greenhouse gases, and it is bearing remarkable fruit: local greenhouse gases are back down to 1990 levels, while nationally they are up 16 percent. And instead of damaging its economy, Portland has boomed. . . . So it's time to abandon the old self-defeating notion that curbing greenhouse gases is too costly to be effective. Portland and other localities are showing there's plenty we can do inexpensively. . . ."[249]

Harriet Bulkeley and Michele Betsill conducted a process appraisal of CCP based on six case studies conducted from 1998 to 2002. They concluded that CCP had the most impact in Denver, CO, in the United States, and Newcastle, New South Wales, in Australia, where "climate change considerations have been integrated into the institutional structure of local government."[250] CPP had mixed impacts in Newcastle and Leicester, U.K., and limited impacts in Cambridgeshire, U.K., and Milwaukee, WI in the U.S. These impacts depended on multiple interacting factors. Those cited as key were leadership from committed officials and politicians, internal and external funding, legal authorities, how the climate change mitigation issue was defined and understood with respect to other community interests, and the political will to address conflicts among interests.[251] Sometimes mitigation was integrated with other community interests; sometimes it was co-opted by them. CCP had the most impact in those places "that were best able to capitalize on both the material and non-material resources offered by the network. For example, Denver and Newcastle (NSW) have benefited, financially and politically, from sharing their expertise across the network. Officials in each of these cities value the CCP programme not merely as a source of information, but also as a vehicle for demonstrating environmental leadership."[252] This appraisal tends to confirm the diversity of participation and outcomes at the local level in the CCP network. Once again, context matters.

Clearly, local action on mitigation is no panacea. (Neither is national or international action for that matter.) But the sources cited above indicate that it can help. And such leaders as Toronto, Denver, Newcastle, and Portland represent bodies of experience available to be tapped for other communities. Bulkeley and Betsill downplay the role of information in their cases, apparently to correct theoretical expectations in the literature on global environmental governance. But they tend to overlook the practical value of information on field-tested mitigation alternatives for other communities as they become motivated and able to act. Information alone is not sufficient, nor is any other single resource. The lesson appears to be that outcomes depend on the interaction of multiple factors in each community, but that networks can diffuse knowledge and information that helps expand the range of informed choices and motivate decisions and action on the ground. It is not necessary for every community to reinvent the wheel.

Like local communities, some states have taken the initiative on the mitigation of greenhouse gas emissions. In the United States, in the absence of significant action at the federal level, 42 states had conducted greenhouse gas inventories, 30 states had climate action plans under way or completed, and 12 states had set targets for greenhouse gas emissions by 2007. "However, a smaller but growing number of states have implemented or are creating mandatory emission reduction programs."[253] One of the more significant is a partnership of nine Northeast and mid-Atlantic states begun in 2003. Called the Regional Greenhouse Gas Initiative (RGGI), it would establish a cap-and-trade system to limit CO_2 emissions by power plants but is expected to face legal challengers and has not solved the "leakage" problem: Emissions reductions in RGGI states could be offset by emissions increases in states that export electric power to RGGI states.[254] Recent reports indicate that the system has been expanded to 10 states covering 233 power plants. When the cap goes into effect at the beginning of 2009, the total allowed carbon dioxide emissions to be auctioned are expected to exceed the plants' total projected emissions.[255]

In 2004 California issued regulations to limit the fleet-average greenhouse gas exhaust emissions of motor vehicles beginning in 2009. More stringent caps each year after that would reduce fleet-average emissions to 30% below 2009 emissions by 2016. At least 12 states are likely to follow California's lead. But the state's legal authority to regulate motor vehicle emissions depends on a waiver denied by the federal EPA in December 2007, when new federal legislation required increases in corporate average fuel efficiency standards for cars and trucks. California's attorney general called the denial "absurd," citing "special topographical, climate and transportation circumstances, which require tougher air quality standards than those set nationally." The EPA administrator defended denial of the waiver: "The Bush administration is moving forward with a clear national solution, not a confusing patchwork of state rules."[256] In 2006 California passed a Global Warming Solutions Act (AB 32) that "directs the California Air Resources Board (CARB) to develop and implement a statewide program that would reduce the state's greenhouse gas emissions to 1990 levels by 2020."[257] The act gives CARB broad authority; it also addresses the leakage problem by covering all electric power consumed in California. Also in 2006 California enacted legislation (SB 1368) that eventually will limit, if not effectively prohibit, the use of coal-generated electricity in California, provided the legislation survives legal challenges. California's track record enhances its credibility: Per capita electricity use in the state was roughly constant, at 7,000 to 8,000 kilowatt hours per capita, for 27 years through 2001. For the same period, per capita electricity use in the United States as a whole grew 50% to about 12,000 kilowatt hours per capita, a rate of about 2% per year.[258]

According to the Congressional Research Service, "The states' motivations may be as diverse as the actions themselves. Some states are motivated by projections of climate changes, while others expect their policies to provide economic opportunities or other co-benefits, such as improvements in air quality, traffic congestion, and energy security. Another driver behind state action is the possibility of catalyzing federal legislation."[259] Multiple interests have been mobilized on behalf of climate change mitigation. An interest in catalyzing federal legislation evidently prompted the governors of

California, Utah, and Montana to appear in televised ads in November 2007 demanding that Congress act on climate change.[260] Moreover, the possibility of a patchwork of different climate change regulations across the 50 states "is causing some industry leaders to call for a federal climate change program. If Congress seeks to establish a federal program, the experiences and lessons learned in the states may be instructive."[261] Thus, both instructive experience and political pressure can move from the bottom up in a federal system that allows states to serve as "laboratories of democracy."[262] Notice also that demands for the reduction of greenhouse emissions frustrated within the core of the climate change regime do not necessarily dissipate. Sometimes they are expressed in state and local initiatives.

The evolving context also includes voluntary initiatives by corporations to mitigate their greenhouse gas emissions and advance other interests at the same time. Johnson & Johnson, previously mentioned as a participant in the U.S. Climate Change Action Plan, reduced greenhouse gas emissions in its plants worldwide, motivated in part by a corporate environmental tradition, access to disinterested technical advice on efficient lighting technologies from the Green Lights program, and savings in lighting costs realized from initial implementation of those technologies.[263] British Petroleum, under the leadership of Sir John Browne, withdrew from the Global Change Coalition opposed to action on climate change in 1996, supported the Kyoto Protocol, and then set its own target and timetable for emissions reductions in 1998. In 2000 it rebranded itself, shortening its name to BP, marketing itself as moving "beyond petroleum," and adopting an environment-friendly logo. By 2002 "BP had not just met its target—to reduce its emissions of greenhouse gases by 10 percent below 1990 levels—it had exceeded it, done so eight years ahead of schedule and with no net economic costs."[264] BP realized a net gain of $600 million (USD) from energy efficiency. More recently, Frito-Lay, a manufacturer of potato chips, has embarked on an ambitious plan to take its huge plant in Casa Grande, AZ, "off the power grid, or nearly so, and run it almost entirely on renewable fuels and recycled water." The concept is called "net zero." "Frito-Lay officials maintain that trying net zero provides a hedge, particularly if the most pessimistic predictions about climate change and the availability of water and petroleum hold true." The track record supports some degree of confidence: "Since 1999, Frito-Lay has reduced its water use by 27 percent and electricity by 21 percent, cutting $55 million a year in utility costs."[265]

Corporations and industrial sectors are sometimes called "communities of interest" because they are specialized to the pursuit of particular business interests. Functionally, environmental nongovernmental organizations, professional associations, and even government agencies with limited mandates and jurisdictions also are specialized to the pursuit of particular interests. For public policy purposes, these organizations are better conceived as interest groups that cooperate and compete in place-based communities from local to global levels. The common interest is defined only for place-based communities and depends on the integration of pluralistic interests where possible and trade-offs where necessary. Thus, the question is whether or to what extent the interests of these groups are aligned with the common interests of the communities in which they operate; no single interest, including the environmental interest, is equivalent to the common interest. In assessing the role of interest groups of any kind addressing climate change issues, it is important to distinguish between the proclamation of targets and timetables, other goals and projections, and new or modified institutions on the one hand and the outcomes realized by such means on the other hand. The substitution of symbols (including symbolic action) for actual results is a recurring tendency in the evolution of an established regime and exceptions to it. But actual outcomes are a better guide to understanding the past and shaping the future on behalf of common interests.

The actions on mitigation reviewed here bring into the picture complex realities discounted or overlooked in the mainstream quest for mandatory international targets and timetables to reduce greenhouse gas emissions. These complexities include the diversity of contexts, the multiplicity of relevant interests in each context, and the variety of policies implemented to reduce emissions. Some successful policies have diffused through ICLEI and other networks to inform decisions elsewhere. Adaptation to climate change is even more complex because there is no equivalent to emissions reduction, the quantifiable single target shared in mitigation efforts.

These complex realities, combined with human cognitive constraints, imply decentralization of climate change science, policy, and decision making: Because no one can understand the global problem completely or completely objectively—or even come close—it is prudent to factor the global problem into simpler parts, community by community. In addition, control over policy responses to climate change is dispersed among and within nations. Decentralization makes it feasible to bring people in local communities into the process of evolving better policy responses adapted to their own circumstances. In contrast to remote policymakers, they can contribute local knowledge but cannot avoid taking responsibility for the direct consequences of their own decisions—short of emigrating to a different community.

Moreover, the exceptions indicate that the political will and other resources necessary to address adaptation and mitigation problems already exist in niches here and there. These niches can be expected to grow as the adverse impacts of climate change become more obvious in more communities—assuming, as we do, that the impacts projected by mainstream climate science turn out to be dependable.

This chapter has reviewed scientific management, and exceptions to it, in the evolution of the established climate change regime. The details indicate that scientific management and adaptive governance are different but *not* mutually exclusive approaches to simplifying the complex realities of climate change for purposes of understanding and action. Aspects of each can co-occur in specific examples that closely approximate one or the other of the ideal types (Box 1.1). Each of the exceptions on the ground occurs within a larger framework of lawful climate-related policies at least national in scope. But the mainstream quest for mandatory international targets and timetables for emissions reductions cannot avoid the dispersion of control, barriers to technically rational policy arising from politics and scientific uncertainties, and contested scientific assessments of climate change and its impacts. The aspirations of scientific management go a long way toward explaining how we arrived at the present disappointing state of affairs. But scientific management is not an achievement of the regime. The exceptions consistent with adaptive governance that have already emerged on the periphery of or outside the regime suggest where and how we might go from here to more effectively reduce losses from extreme weather events and climate change.

BARROW AS MICROCOSM

3

This chapter tells the story of Barrow, AK, as a microcosm of things to come at lower latitudes, as signs of climate change become ever more obvious there. The primary purpose is to harvest a body of experience on climate change adaptations for decision makers elsewhere. In the process we also illustrate intensive inquiry as proposed in Box 1.1, inquiry centered on a single case considered comprehensively and integratively. This brings into the picture important details on the ground that must be omitted in the extensive analysis characteristic of scientific management. In the first section we review adaptations in the long history of Barrow and the North Slope from the earliest settlers to the eve of rapid modernization in the region. This culminates with the story of the great storm of October 3, 1963, the most damaging on record and in living memory in Barrow, and the most influential baseline for policy planning there. In the second section we review Barrow's vulnerability to big storms based on trends in human and natural factors, and show how those factors came together in extreme events that caused damage in Barrow. The third and final section focuses on various policy responses to these extreme events in Barrow over the last several decades, highlighting both scientific management and adaptive governance. The latter includes initial steps toward organizing a network of Alaskan Native villages to exchange policy-relevant information and influence federal policy from the bottom up.[1]

As previously noted, climate change is not an issue in Barrow or the North Slope; the signs have been obvious for all to see. Ronald Brower, Sr., of Barrow recently surveyed the signs that are important for subsistence life in the community.[2] Because of temperature increases, Brower said, "My ice cellar this year had two inches of water. . . . We need to look at alternative means of storing our wild-country foods." Caribou are among the sources of wild-country foods adversely impacted. "Normally we get snow at the end of August, but this year not until the beginning of October," he said." Late-season rains and the ensuing ice coat freeze the vegetation that caribou need to sustain themselves and their young. "The caribou are used to digging in snow. Now they have to dig in ice," Brower said. Once-abundant lemmings have decreased dramatically in number, affecting the food chain for other wildlife. The positive impacts include growing abundance of species once rare in the Arctic. "Now we are catching king salmon in our nets up in Barrow. We are catching king crab 40 miles south of Barrow." But Brower considered this a mixed blessing. He was concerned that more seafood meant more commerical fisheries and state and federal policies to deal with their impacts on Arctic villages. Brower was also concerned that more big-game trophy hunters meant fewer animals for subsistence hunters who depended on them for food. In short, Brower said, "We are experiencing things in one lifetime that should take five or six generations. . . . We are making do with less (subsistence food) and trying to make the most of it." Other signs of climate change important to modern life in Barrow are considered below.

As an example of what we call intensive research, this chapter takes a bottom-up perspective on larger issues of climate change science, policy, and decision making. The details from Barrow cannot be replicated exactly, even in other Native villages on the North Slope coast of Alaska. But decision makers elsewhere can capitalize on various similarities that have been observed in other local communities and can be expected to emerge elsewhere. As signs of climate change become more obvious at lower latitudes, Barrow is one body of experience on which to build. With Norman Chance, author of an ethnography of development in Arctic Alaska, we believe that "the initial step in solving large problems [involving the whole of humanity and its future] is to break them into more manageable units in the context of particular historical circumstances. Once these elements are grasped more clearly, then the various parts can be rejoined and the whole issue seen with greater understanding."[3] In this frame of reference the details in Barrow, complex relationships among them, and similarities with other local communities may expand the range of informed choices and decisions for policymakers concerned about climate change elsewhere.

Historical Contexts

Recent and impending adaptations to climate change by people in Barrow are best understood in the context of their ancestors' responses to interacting challenges in the region from the earliest times. In addition to climate change, the interacting challenges include changes in the raw materials available in the natural environment and changes in population, indigenous group relations, colonization, and modernization in the human environment. This section constructs the context from the earliest times through the colonial era culminating with the great storm in 1963.

Explorations and Adaptations

The first Americans crossed the Bering Strait by land or sea at least 12,000 years ago (Fig. 1.2). To them, according to Norman Chance, "the momentous crossing between Siberia and Alaska, occurring gradually over many generations, was simply an exploration of new hunting territory and nothing more."[4] But little is known about them from the few Paleo-Arctic archaeological sites that have been found. Studies in several disciplines suggest that "the ancestors of present-day Alaska Natives can be traced to two migrations occurring 10,000 to 5,000 years ago." The first was basically an inland population that "moved widely through the interior and northwest coast of British Columbia early in the postglacial period. A second, maritime-oriented group arrived later, perhaps 7000 to 6000 years ago."[5] This second group diverged into Aleut and Eskimo sometime prior to 4,000 years ago when the archaeological record becomes clearer. Specialized adaptations of the Eskimo to local ecologies included the Ipuitak, the Old Bering Sea, and the Birnirk cultures. The people of these cultures "resolved many problems of subsistence," allowing them "to devote increasing amounts of time to developing the technical and artistic skills for which they became known."[6] They constructed small semisubterranean houses that retained heat efficiently. The cold-trap entrance led down into a long underground passage that turned up to a trap door in the floor of the dwelling; the construction was supported by whalebones and covered with sod, and it was adapted to temperatures as low as −50°C in winter. "As their technical skills increased, they modified their clothing, allowing them to remain out-of-doors for longer periods."[7] Caribou hides with hollow hairs were especially good insulators for retaining body heat. They also developed a complex hunting technology known for its flexibility, adaptability, and openness to innovations.

As early as 2,000 years ago, camps existed on the North Slope coast. This coastal region is predominantly low-lying, wetland tundra, dotted by numerous thaw lakes. Offshore islands and shoals moderate the local influence of polar pack ice on the coast. Most of these islands are sand and gravel barrier islands bounding shallow lagoons, while others are relics of earlier coastal retreat processes and lie farther offshore. The islands, and the mainland coast where unprotected by islands, have always been subject to considerable erosion by wave action. Among the camps on the coast was Ukpeaġvik, the Iñupiaq name for "the place where we hunt snowy owls," located on the bluffs at the southern edge of present-day Barrow (Fig. 3.1). (Barrow was settled much later, in the colonial era, on lower ground for easier access to ships.) Down the coast from Ukpeaġvik were other important prehistoric sites including Ulġuniq (or Wainwright) and, at the western end of the North Slope coast, Tikigaq (or Point Hope). Up the coast about 16 km from Ukpeaġvik on the Point Barrow sand spit was Nuvuk, now abandoned and vulnerable to erosion. At the base of that spit was Pigniq, where summer hunting camps exist today.[8] Pigniq is considered the type site, or model, of Birnirk culture.

"Beginning with the Birnirk peoples (ca. A.D. 400–800) up to the present, whaling has been the focal point of local people's subsistence pursuit on Alaska's Arctic coast from Point Hope to Point Barrow."[10] Other aspects of culture in this coastal region were adapted to sustain the whaling orientation; this is the main point of Glenn Sheehan's monograph, *In the Belly of the Whale*, which we rely on here. The Birnirk peoples were sparse, mobile, and dispersed in the interior and along the coast. They began the whaling orientation in suitable locations by developing technology to increase the harvest, including more efficient large *umiat* (boats) and smaller *qayaqs* (kayaks) made with skins carefully sewn together.[11] However, they lacked the concentrations of people necessary for consistently successful whale hunts. After an initial strike, an injured whale was more likely to escape from a few whaling crews than many, and landing and butchering a whale still require the cooperation of many people. Some Birnirk people incorporated whaling into their subsistence way of life without modifying that way of life appreciably. Others concentrated on other game, and some left no evidence of

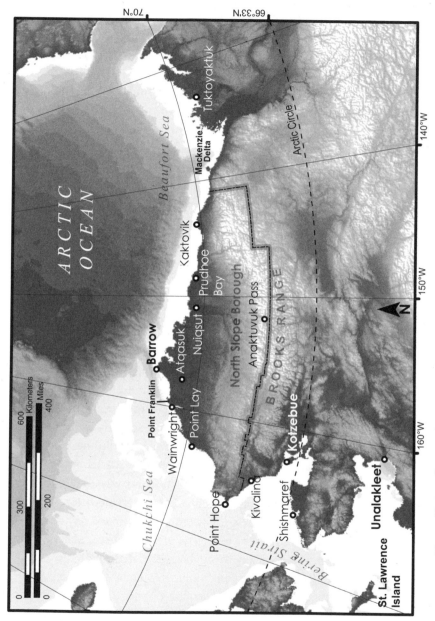

Fig. 3.1. Native villages in the region. Sources: Leanne Lestak and Eric Parrish, NOAA ETOPOS data.[9]

whaling at all. Around A.D. 800, small groups of more mobile Thule began to emerge and succeeded the Birnirk roughly a century later. "Differences between Birnirk and Thule are that Thule invariably made use of whales, Birnirk did not always, and Thule peoples appear to have been more numerous."[12] The Thule shared "the same language, the same world view, [and] the same cultural preconceptions" but differed through adaptations to their local environments.[13]

More dependent on whales for subsistence, the Thule were also more consistently successful at whale hunting. Their growing population supported more crews to improve the local whale harvest, but also increased competition for limited local resources including whales and other game species. By A.D. 1000 this competition helped motivate the Thule to expand their range thousands of miles eastward along the coast into Canada and Greenland, much like their ancestors expanded their range from Siberia across the Bering Strait. Expansion also was encouraged by the pan-Arctic climate warming that occurred about the same time as part of the medieval warm period. During that warming, "The Atlantic and Bering Sea bowhead [whale] populations may have mingled in expanded summer feeding grounds, providing good hunting in most coastal areas."[14] However, "Around A.D. 1200, the climate deteriorated, becoming cooler and more modern. Whale hunting became impractical over most of the Thule's geographic range."[15] Cooling left more summer ice floes that interfered with open ocean hunting in several ways: by restricting whales' summer feeding grounds, by providing barriers for protection from whaling crews, and by making hunting more dangerous for crews. Cooling also restricted lead hunting geographically in the spring when whales concentrate in open leads—cracks in sea ice—as they migrate to summer feeding waters. "Although leads generally parallel the entire coast, they only approach land and landfast ice closely and predictably near points that jut into the ocean and affect current flow. Whaling was now confined to these places." Moreover, "As cooling progressed, the formerly long hunting season was confined to the time it took the migration to pass any one point. Scheduling was critical as hunters had to be present at particular times instead of hunting opportunistically through an extended whaling season."[16]

Forced to adapt to changing conditions, descendants of the Thule in Alaska evolved a more complex social structure to sustain the whaling orientation. One adaptation beginning around A.D. 1200–1400 was settlement in large permanent villages at Point Barrow, Point Hope, and other points on the Alaskan coast where migrating whales concentrated in accessible open leads in the spring. With permanent settlements, people began to centralize ice cellars for storage of harvests once scattered among camps. "Whaling gave a permanent food surplus, but over time other vital game-derived material became relatively less available locally as populations grew. At some point, people could only increase supplies of raw materials by gaining access to resources outside local catchments."[17] Thus, another adaptation was trade, primarily with caribou hunters inland but also coastwise trade for raw materials from walrus, seals, and other marine species distributed unevenly along the coast. Trade supplied coastal villages with "hides for lines and covers, and ivory and bone for hunting gear"; it also supplied inland caribou hunters with "oil for eating, lighting, heating and preservation, as well as meat and blubber for eating."[18] Without trade, a single failure of the annual caribou hunt could jeopardize the survival of these small inland groups. Similarly, coastal villages concurrently became more dependent on inland caribou hunters, and more specialized: "Relying upon trade rather than their own caribou hunting put them under obligation to trading partners and made them dependent upon them. At the same time, trade supported the primary whaling subsistence focus by freeing time that otherwise would be diverted to inland hunting."[19] Trading relations included other Native peoples over a vast area. Chance notes that trade fairs occurred east of Ukpeaġvik during summer months at the mouth of the Colville River (Nuiqsut) and on Barter Island where the village of Kaktovik is located.[20]

Another adaptation was a more complex decision-making structure than the earlier egalitarian structures. The whaling crew captains (*umialit* in the plural, *umialik* in the singular) owned and distributed the huge whaling surplus, the basis of the growing trade network.[21] Distribution of the whaling surplus to crew members also sustained the voluntary support necessary to keep crews intact. But personal leadership qualities also helped increase the surplus available.[22] "Given the importance of a successful hunt, choosing the most knowledgeable individual to lead the effort was far more effective than limiting the selection to a member of one's family. Once harvested, the game was divided among individual participants" in the hunt, and then redistributed within their families.[23] Some whaling captains eventually capitalized on their wealth and power to expand control beyond the hunt and their crews. They increased their power through ceremonial centers (*qargich* in the plural, *qargi* in the singular). Archaeologist Glenn Sheehan led the very first excavation of a *qargi*, at Mound 34 in Ukpeaġvik; his monograph emphasized its whale symbolism and whalebone structure. "Eskimo whalers entering the *qargi* were entering the belly of the whale figuratively, and to the best of their construction techniques, literally."[24] Sheehan also summarized its significance in the communal life of permanent settlements: "Work, re-ligious and social activities came to be focused upon a place and a role, the *qargi* and the *umialik*. More food was distributed for a longer time of each year by *umialik*. New elements of social control were manifest: communal religion took a permanent physical locus and role; redistribution gained extra-familial aspects and the traditionally flexible definition of family was stretched to help larger groups remain cohesive; retribution passed beyond the polity and became warfare; more labor was directed for longer periods of time; and more free time and socialization fell under the direct overview, if not oversight, of the religious/political leaders."[25]

After about A.D. 1400, as the population of permanent coastal villages continued to increase, "Whaling continued to provide a food surplus. . . . However, the supply of various non-whale game species within easy travel distance of each village was limited. Demand for raw materials, most of them game-derived, outstripped local village game catchments."[26] Further expansion of trade apparently was insufficient. To obtain more raw materials, people moved out to settle permanent daughter villages in outlying coastal areas but returned to the main village for spring whaling. "Relations were strong from one whaling village to another, between each whaling village and its daughter settlements, and between whaling villages and the caribou hunters of the interior."[27] But territorial expansion eventually brought them into conflict, and the whaling captains who were increasingly in control engaged "in political relations with other groups that extended to large scale warfare."[28] Indigenous warfare significantly decreased the population after about 1700; one major village was abandoned about that time, for example, and Tikigaq (Point Hope) lost many people early in the nineteenth century. Just prior to European contact on the North Slope in 1826, "Eskimo whalers lived in sedentary villages of from 300 to 600 people, with up to 1300 more in the supporting region around each village."[29] Following contact, newly introduced diseases decreased the population and perhaps the need for in-digenous warfare, which ceased shortly thereafter.[30]

In any case, conflicts between large permanent settlements, like conflicts within them, were tempered by older kinship linkages including comarriage (spouse exchange) and adoption (baby exchange) and by older coopera-tive economic linkages including trading, hunting, and sharing the subsis-tence harvest. The economic linkages were based on mutual dependence, as Chance explained in the case of sharing: "When a local family had little food and a neighbor had more, a request for assistance would carry more weight if the one without had been generous in the past. Thus, in times of need, sharing across family lines was common, each local family knowing it could count on another's offer of surplus food when the occasion arose."[31] According to Robert Spencer, the primary principle of North Alaska Eskimo culture was the obligation of the individual to cooperate both with kin and within the hunting group or crew, "a body working together under the banner of an *umealiq* and pledged to the concerted activity of the hunt, whether on the sea, for whale, walrus, or *ugruk* [seal], or inland, in the intensive caribou drive and impound."[32] Similarly, in Chance's assessment, "For the Iñupiat prior to their encounter with the West, people rather than things were the crucial resource."[33]

Contact and Colonialism

The first significant contact with Europeans on the North Slope occurred in August 1826 when Thomas Elson and William Smythe from Capt. F. W. Beechey's mapping expedition arrived in Ukpeaġvik. They named the place after Sir John Barrow of the British Admiralty. In his journal Beechey noted the results of the extensive trade network among Eskimos: "The inhabitants of Point Barrow had copper kettles, and were in several respects better supplied with European articles than the people who resided to the southward. . . . The copper kettle in all probability came from the Russians"[34]—no doubt indirectly, through a series of trades with Eskimos to the south. Direct contacts were sporadic for a few decades. Then in July 1848 the Yankee bark *Superior* sailed through the Bering Strait and into the Chukchi Sea off the west coast of the North Slope. There it found the Iñupiat in *umiat* apparently eager to trade, and large numbers of bowhead whales that were more profitable than smaller right whales and sperm whales. This was a major discovery for the Yankee whaling industry at its height, supplying oil for a variety of uses and baleen for buttons and corset stays. Demand for these products was subject to changing fashions and substitute products, including crude oil discovered in Pennsylvania in 1859. Commercial whaling in other oceans had already impacted whale stocks in the Arctic in the early 1850s, when Capt. Rochfort Maquire of the H.M.S. *Plover* overwintered in Barrow. To maintain profits amid declining stocks, Yankees hunted whales more aggressively and turned to walrus, which were decimated by the early 1880s. Whaling ships became a familiar sight on the North Slope, but "the first whaling ships visited Barrow in 1854 and next ones did not visit for another 15 years." "By about 1870, whalers and traders began to deliberately and frequently contact villagers."[35] The first to settle in Barrow was a Yankee whaler, Charles Brower, who arrived in 1886 and established the first store in 1893. Brower's Café still stands near the beach in what is now called Browerville.[36]

The Iñupiat Eskimos traded whalebone, caribou meat, and fur clothing for ammunition, lead, flour, black tobacco, molasses, and whiskey, which "disrupted and demoralized village life." They also found paid employment in various capacities. But contacts with outsiders decimated the Iñupiat population. According to Chance, "Some observers blamed the effects of alcohol brought to the villages by traders, but a more realistic explanation focused on the loss of sea mammals by a Native population with limited other resources on which to draw. Overhunting by Natives supplied with rifles by Americans and the British was similarly responsible for a severe decline in the Western Arctic caribou herd. . . ."[37] Disease from contacts took its toll along with starvation. "In 1900, more than 200 inland Iñupiat trading at Point Barrow died of influenza following the visit of a whaling ship. Two years later more than 100 Barrow Iñupiat perished in a measles epidemic." In addition, "more than one-third of the town's 600 Native population died within a week" after arrival of the Spanish flu pandemic of 1918.[38] Thus, Barrow and other coastal villages failed to maintain population, despite in-migration from inland Iñupiat who depended on whaling and shared the whaling orientation.[39] But commercial whaling in the Arctic died out by World War I, as prices paid for whale products declined, reflecting changes in fashions and competition from substitutes in world markets.

Meanwhile, in 1867 the U.S. government bought Alaska from Russia for $7.2 million "without consulting the original occupants of the region or obtaining title through purchase or treaty. . . ."[40] They became subject to U.S. laws and regulations, including those pertaining to Native Americans at lower latitudes. The plight of the Iñupiat and interests in "civilizing" and assimilating them prompted coordinated actions by missionaries and the U.S. Bureau of Education, the agency given jurisdiction. To replace depleted caribou and "instill a new entrepreneurial spirit" the Bureau established a program that initially loaned and later gave reindeer to Native Alaskan apprentices.[41] The apprentices were provided with food, clothing, housing, and instruction as their herds increased; at the end of the apprenticeship they were expected to sell surplus reindeer meat to support their families. The first reindeer were imported by U.S. government agents to supply whalers marooned in the Barrow area in 1897–1898, but 125 of the reindeer were turned over to villagers at a time when they were interested in alternatives to the declining whaling industry. Reindeer flourished with little competition for forage from depleted caribou herds, but the caribou eventually came back on their own. Nearly all the Barrow reindeer were lost by 1939. The government tried again in 1943, borrowing more reindeer from another village. However, "By 1952, the last of the remaining Barrow reindeer had disappeared. Similar losses were reported at other reindeer herding villages on the Slope."[42]

Ill-advised changes in management figured in the program's demise. The consolidation of individual herds into a central herd with corporate management motivated dissatisfied herders to "lose" many animals before consolidation. The cancellation of wages for apprentices reduced their dependence on government but also contributed to decline and loss of reindeer herds. Having learned to expect compensation for their labor from whaling ships and other outsiders, apprentices did less herding without wages. This allowed reindeer herds to become wilder, easier prey for wolves, and scattered beyond their home ranges. Owners also left herds unattended to continue hunting and to trap foxes around remaining herds when prices for pelts were relatively high before and after the Great Depression. Animals from the mixing and interbreeding of reindeer and caribou were fair game for hunters, despite questions of ownership. Basically, the program itself seems to have been ill conceived and inconsistent with realities on the ground. According to Chance, "on the North Slope, which had no market for reindeer meat, the hoped-for entrepreneurial self-reliance was simply not possible. Instead, reindeer herding was viewed by the Barrow and other North Slope Iñupiat largely as an extension of their earlier subsistence hunting practices."[43] Toward the end of the program, reindeer herding had difficulty competing with an increasing number of jobs that paid cash in Barrow.

Meanwhile, missionaries arrived on the North Slope in 1890. Their intention was "to replace Iñupiat religion with a more humane Christian one and to restore a moral basis for a society that [missionaries] saw as defiled and demoralized by whalers and traders."[44] In Barrow, the Presbyterian mission operated the government school for Native children for the first four years, supervised the reindeer apprentice program for the federal government until 1904, and provided medical care and staffed a clinic that was transferred to the Indian Public Health Service in 1936. "There is no question that the offering of such educational, medical, and economic services helped considerably in converting the Iñupiat to Christianity." Conversion also was assisted by resymbolizing the hostile spirits in Native fatalism as the devils in Christianity and by cultivating hope of relief from them through miracles, song, and prayer. One observer found in 1908 that "almost every Iñupiat he met along the northern coast had become converted to this 'nativized' Christianity."

But the influence of missionaries also had adverse consequences for converts' traditional culture. Wooden structures were promoted as cleaner and more hygienic than sod houses, but they were draftier and more difficult to heat; they depended on scarce driftwood for fuel. Keeping the Sabbath, a substitute for traditional taboos, reduced the time available for fishing and hunting, especially during the six-week spring whale hunt in Barrow. This reduced the collective harvest, weakened the pattern of community-wide sharing, and encouraged people to buy food in the cash economy.[45] Evening church meetings at the missionary were considered morally preferable to Native dancing and unsupervised young people at *qargich*, which were suppressed by missionaries. Church-sanctioned and government-legalized marriages increased criticism of traditional marriage and divorce, which were more flexible. Patriarchical church teachings altered the role of women: "Once an active co-partner in subsistence, sharing, and socialization, Iñupiat women found that under the banner of being good wives and mothers, they carried considerably more household responsibilities and had far fewer freedoms than previously."[46] Similarly, outside influences altered the socialization of Iñupiat youth: "Before the arrival of school teachers and missionaries, Iñupiat socialization was largely focused on learning how to survive. A child's education entailed continual observation mixed with regular instruction tempered by practical experience."[47] With the Organic Act of 1884 the federal government sought to assimilate Alaska's Iñupiat youth through education, but the effort was less pronounced on the North Slope where there were fewer alternatives to subsistence hunting.[48] Even in the 1950s, "new teachers were encouraged to change cultural practices . . . and promote in their place new forms of behavior and thought more conducive to life in a civilized society."[49] This included instruction in English only; punishment for using Iñupiaq words in school was resented.

Meanwhile, government jobs in construction and maintenance attracted Iñupiat from villages across the North Slope to Barrow. In 1944, the federal government constructed an airstrip several miles northeast of Barrow to support the U.S. Navy's plans to explore for petroleum in the area. Near the airstrip the Navy built a large construction camp as a base of operations in 1946; it evolved into the Naval Arctic Research Laboratory (NARL).[50] Government employment opportunities also included the radar station constructed nearby early in the Cold War, part of the Distant Early Warning (DEW) Line. Iñupiaq elder Kenneth Toovak, for example, recalled government employment as a young man operating a bulldozer on the beach to remove gravel for various building projects. A young nurse from Seattle stationed in Barrow, Grace Redding, recalled that most of the employment was at the camp, the Weather Bureau, the Bureau of Standards, other agencies, and the schools and hospital. "In '58 it was really a very primitive village," she said.[51] The population of Barrow had increased from 363 in 1939 to 951 in 1950, and to 1,314 in 1960.[52] In February 1963 local support for modernization had proceeded far enough that the Barrow City Council informed U.S. Sen. Ernest Gruening by letter that Barrow residents desired "assistance in a step-by-step improvement plan in the present village location, the realignment of houses, the establishment of well-defined streets and alleys, and the laying out of sewage, gas distribution lines, and power lines." They requested his help in expediting a town-site survey by the Bureau of Land Management. That survey was completed later in the year, just after the great storm of October 1963, according to a story in the *Tundra Times* in December 1963. The headline proclaimed, "Improvements Coming to Barrow; but Eskimos Will Keep Independence."[53]

"Still," in Chance's assessment of the period immediately following World War II, "while some Iñupiat were becoming affluent [in these government jobs], most continued to rely on hunting and fishing to supply their basic dietary needs."[54] In Sheehan's assessment, subsistence hunting and fishing were among the earlier cultural forms that were *not* displaced by the emergence of more complex forms under the powerful *umialit* and their *quagich*. The more complex forms were constructed on the earlier ones as adaptations to increases in Iñupiat population and related changes; they largely disappeared after the population was decimated through outside contact.[55] By the time Robert Spencer conducted his field work in Barrow in 1952 and 1953, and before new archaeological data became available, any remnants of the complex forms were difficult to perceive. Thus, he emphasized survival of the earlier forms and concluded that "much more of the aboriginal patterns of life remained than was expected. . . . Money and modern industrialized technology, it is true, had made inroads and changes: both modern education and Christian missionizing had effected modifications of some patterns." However, "The Eskimo of North Alaska had solved the problem of living together by establishing a strong cooperative kinship bond. This remains, despite the collapse of some other social forms, and . . . continues to provide the keystone of the social structure, a nuclear point around which all other institutional aspects constellate."[56] In 1958 Harold Kaveolook put it more succinctly: "We always try to help each other. That is the Eskimo way."[57]

The story so far demonstrates that the Iñupiat survived and adapted to an unforgiving environment for centuries before the arrival of Europeans,

Americans, and modern science and technology on the North Slope. They relied instead on continuous detailed observations of themselves and their changing environments, and the accumulation of knowledge and insight from those observations. This exemplifies "the science of the concrete." Anthropologist Levi-Strauss considered it "no less scientific and its results no less genuine [than the exact natural sciences]. They were achieved ten thousand years earlier and still remain at the basis of our civilization."[58] In addition, the Iñupiat were pragmatic. They incorporated Western technologies and ideas selectively, according to what served their own needs. At the same time they sustained their older cultural forms, including the whaling orientation, family and kinship ties, and sharing the harvest. But they lost the more complex forms organized around the powerful *umialit* and their *quagich*.

In the most charitable interpretation, colonialism in the Arctic was inhumane even when well intentioned. The imposition of outside practices from the top down was often ill informed, inept, and destructive. Remnants of this early history continue to influence contemporary events.

The Great Storm

The great storm of October 3, 1963, was another challenge to the survival and well-being of people in Barrow. As noted, it is still the most damaging on record and in living memory in Barrow, even though it did not produce record winds at Barrow. The story is worth telling in some detail. Among other things, it shows how an understanding of losses depends on a comprehensive array of interacting factors that cannot be reduced to weather, climate, geography, geology, or human factors alone. It shows the priority of protecting family, friends, and other people in emergency responses to a disaster, the importance of improvisations in protecting them, and the benefits of reserves, redundancies, and outside assistance in expediting recovery. However, such generalizations do not capture the storm as experienced by survivors, some of whom still participate in local decisions to reduce storm damage.

The great storm originated as a low-pressure system along the Arctic front over Siberia around 145.6°E longitude late on October 1. Over the next 24 hours it traversed Siberia to the coast of the East Siberian Sea and continued northward over the Arctic Ocean on a track typical for such systems. However, shortly after 9:00 P.M. (all times Alaska Standard Time) on October 2, the storm turned eastward, steered by a strong flow in the upper atmosphere known as the jet stream in both popular and scientific circles. It began to deepen rapidly, reaching a central low pressure of an estimated 976 millibars (mb) at 11 A.M. on October 3 while located in the Beaufort Sea around 490 miles north of Barrow. At this time, the winds at Barrow had already shifted from southerly to westerly and were reaching 40 miles per hour, with gusts recorded up to 60 miles per hour. The depression continued to intensify as it tracked eastward and the winds at Barrow turned west-northwesterly.[59]

The strongest winds at Barrow were reported between 1:00 and 3:00 P.M. on October 3 with gusts possibly as high as 75 or 80 miles per hour. The highest official observation from the National Weather Bureau (now the National Weather Service) was 55 miles per hour for wind sustained for one minute. These extreme winds were reinforced by a strong Aleutian high to the south with a central pressure of 1030 mb. Between 3:00 and 4:00 P.M. wind strength at Barrow started to diminish, but the low-pressure system continued eastward and may have reached its maximum intensity in the Canadian archipelago early on October 4. (Reports differ on the later stages of the system.) As the cold sector of the depression passed over Barrow, it produced a blizzard during the next few days. Guy Okakok, then the *Tundra Times* correspondent in Barrow, wrote that "I am 60 years old now . . . and I have never seen the winds as strong as we had that day on October 3. High winds and high water everywhere."[60]

The long fetch, or open water, to the north allowed these winds to build up a storm surge and extensive wave action. The fetch was about 350 miles from the shore at Barrow to the area where the Arctic Ocean was at least 50% covered by ice. Greater concentrations of sea ice are known to dampen the generation of wind-driven surge and waves; the extent of any moderating effect from the lesser concentrations of ice floating below the storm is unknown.[61] In any case, the sea level at Barrow started to rise around 5:00 A.M. on October 3. Seawater covered the runway at NARL and contaminated the freshwater lake by 10:00 A.M. The peak occurred between 2:00 and 4:00 P.M. with water as deep as 3 feet in some locations and currents that were very strong and dangerous. Based on the reported water depth, still water lines, and beach elevation, this translates to a storm surge of 12 feet. The unusual height of the surge was caused primarily by the long duration of the strong northwesterly winds. The long fetch allowed the formation of waves on top of the storm surge that could have reached as high as 15 feet.[62] In a nearly contemporary account, Pedro Schafer described the storm as "unique in its violence and consequences. No such disastrous event has been mentioned either in the records of the observatory at Barrow, dating back to 1901, or in the tales of the local eskimo population."[63]

The storm surge and wave action caused extensive erosion of the shoreline and flooding of near-shore areas in and around Barrow. The storm "moved over 200,000 cubic yards of sediments, which is equivalent to 20 years' normal transport," according to geologists James Hume and Marshall Schalk.[64] They compared the estimated sediment transport during the storm surge on October 3 with the accumulated transport during the years 1948 to 1962, a period that included another major storm in 1954. The result of sediment transport was shoreline retreat, leaving steeper and higher elevations in cross sections of land perpendicular to the beach. They estimated that the bluffs on the southwest edge of Barrow retreated as much as 10 feet during this storm, exposing large masses of permafrost to thawing. The permafrost subsequently thawed, causing further shoreline collapse. As a result the shoreline retreated as much as 33 feet in some parts of the bluff, according to estimates based on aerial photographs.[65] Thawing permafrost destabilizes Arctic shorelines and makes them more vulnerable to both normal and storm-induced sediment transport.

Because of coastal geography, flooding from the storm surge and waves was more extensive several miles northeast of Barrow around NARL (or the camp) and the nearby Distant Early Warning (DEW) Line radar station (Fig. 3.2). In the spring of 1964 Hume and Schalk reconstructed the extent of flooding by surveying the line of debris frozen in place immediately after the storm, and drew on eyewitness accounts of flooding at NARL. They concluded that "[n]ear the Camp, the water flowed over the beach and down the back slope into the Camp area. A temporary lake was created which had an elevation of 9.2 feet above sea level. The lake extended about ¾ of a mile inland directly behind the Camp." The depth of this temporary lake ranged from about 1 to 3 feet. It might have been deeper if erosion had not created an outlet to Elson Lagoon farther to the northeast. Most of the spit surrounding the lagoon from NARL to Point Barrow and Plover Point was flooded. "Areas near the village [of Barrow] were not flooded as badly because the beaches there are backed by higher tundra."[66]

Normal weather reports from Soviet Siberia did not reach Barrow because of poor radio propagation during the several days prior to the storm's arrival, according to the *Tundra Times*. The editor, Howard Rock, reported that "[t]he great storm that staggered Barrow on October 3 happened so suddenly that even the weather bureau did not have an idea it was coming. There was no warning, which meant, the people of Barrow had to do

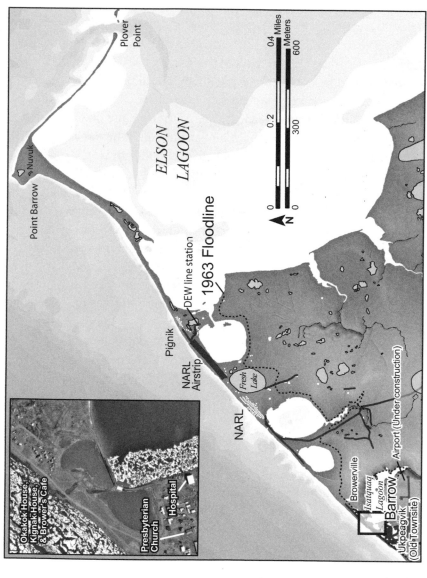

Fig. 3.2. Places impacted by the Great Storm in and around Barrow. Source: Leanne Lestak, William Manley, and Eric Parrish.[67]

some quick thinking to save themselves during the onslaught." Remarkably, no deaths and only one serious injury were reported. The injury occurred when "[t]he wind hit a pile of sheet metal and they took off like kites. One of them slammed into Lawrence Ahmouak [now spelled Ahmaogak] and knocked him unconscious. . . . He was the only casualty among some 1350 people living in Barrow."[68] Nevertheless, several close calls were reported by *Tundra Times* correspondent Guy Okakok. "If Leo Kaleak had waited two more minutes when he tried to reach the house which stood on the highest ground, he said he would have been washed out to sea. Visibility was very bad in the storm and he almost got lost. Also Claire Okpeaha, 70 years old, was knocked down by a wave and he was rolled over and over in the water. His grandson saw him and grabbed him just in time as he was being carried out to sea. He then dragged his grandfather into the house."[69]

Correspondent Okakok and his family also had close calls. Like the Weather Bureau, he did not expect the storm. He reported, "Early that morning, October 3, when we woke up around 6:30 A.M., the weather wasn't too bad. After we had our breakfast, my wife and I went across the lagoon fighting through and against the wind" to the local hospital where he worked.[70] They left several children at home. But the storm had picked up by 9:00 A.M. when "a neighbor noticed the door of the house opening, then shutting quickly as high seas pounded against the house. The neighbor, realizing somebody must be in the house, waited for a break in the storm, then ran over and found the children . . . wanting to come out, but afraid to leave."[71] Okakok reported that while he was at work his wife "came running to tell me that they could see big waves splashing over buildings. I didn't say a word to anyone. I went out and started to run toward my home. My children were in the house."[72]

After crossing back over frozen Isatquaq Lagoon, which was already flooded, he asked about his children. They were safe at Betty Kignak's house nearby. Okakok struggled over to his house, where big waves had been splashing against the north wall, to sack up the children's clothes. "While I was collecting the clothes a wave came through the door. All at once the water was up to my knees. As soon as the water dropped I opened the inside door. . . . Before I could let go my hands from our door knob, [a] second strong wave came in and pushed me to the stoves. The house was now half full of water. I couldn't wait any longer and I walked through the water and went out. I couldn't do anything so I watched. While I was watching a great strong wave came over and knocked down my walls. Stoves, fuel, food, and clothes scattered all over. What I could save, I saved but lost the children's

Fig. 3.3. View from hospital across flooded Isatquaq Lagoon to Browerville during the Great Storm. Photo: Grace Redding.

clothes, food, range and heater."[73] Okakok was given shelter, hot coffee, and dry clothes at Betty Kignak's house next door. There he heard by radio that a Caterpillar and another heavy vehicle would be sent over to rescue people from Browerville and take them across the outlet of Isatquaq Lagoon to higher ground in the main part of Barrow. After the Okakok family was rescued, they were given supper at the school, a place to stay overnight at the Presbyterian church, and clothes. "I will never forget those people who had done so much for us," Okakok reported.[74] Other storm victims also were treated kindly. Howard Rock later observed, "When the storm abated, about 200 Eskimos were left homeless. Except for three families that were housed in a church, all of the storm victims were quickly taken in by relatives and friends."[75] Here is another echo of the sharing tradition.

Less was recorded about immediate threats to life and limb northeast of Barrow at NARL and the construction camp where flooding was more extensive (Fig. 3.2). The laboratory's monthly progress report for October, an eyewitness account quoted by Hume and Schalk, noted that electric power was shut off to prevent fire before the storm peaked in midafternoon on October 3. During the peak, evacuation of the women and children to Barrow had been blocked after a timber bridge washed out and additional parts of

Fig. 3.4. House in Barrow destroyed during the Great Storm. Photo: Grace Redding.

the coastal road were badly eroded. "All women and children were evacuated from the Camp to the DEW line site. Most of the damage in the area occurred at this time. The force of the current through the camp was so strong that only Cats could safely be driven through the streets."[76] Even so, a weasel and a D-4 Caterpillar were sunk trying to rescue a wolf, two wolverines, and three foxes, all of whom drowned early in the peak period.

More property was exposed to flooding at the camp than in the village of Barrow, which was still in the early stages of development as a modern community in 1963. Consequently, property damages to government installations at the camp were greater, although estimates varied in the immediate aftermath of the storm. As repairs were being made, Hume and Schalk reported that actual costs were close to estimates of $3 million in damages to government installations and $250,000 in damages to the village. In addition to coastal-road damage, they listed as "major damage" to the camp "contamination of the water supply [in Fresh Lake northeast of the Camp], destruction of 70 per cent of the airstrip [at NARL], and loss of 6 buildings, two with scientific equipment. Supplies and stores were floated away and damaged by salt water. . . . In addition, the foundations of almost all the buildings were eroded, a process which usually resulted in structural damage. Three buildings, one a large quonset, were actually floated away. In all likelihood, more buildings would have been rafted if the water level had remained high for a greater period of time."[77]

The main losses in the village according to Hume and Schalk were "32 homes, 15 of which were totally destroyed; 250,000 gallons of fuel; and three small airplanes"[78] (Fig. 3.4). One of the earliest accounts in the *Tundra Times* reported "Total damage . . . over $1 million, with most of this occurring to private persons, the native Co-op Store, Central Construction Co., Wien Alaska Airlines, Golden Valley Electric, and federal property."[79] A few weeks later another account claimed "[m]ore than $600,000 in private property was lost at Barrow" during the storm.[80] Various accounts filled in specific details. The storm damaged or destroyed numerous skin and wood boats, and three generators recently landed from the Bureau of Indian Affairs supply ship *North Star*. Wien Alaska Airline's 200-foot radio tower for transpolar flights was toppled when two floating houses crashed into the guy wires supporting it. "About 1000 barrels [of fuel oil] were lost when the waves crashed into piles of drums and scattered them helter skelter. Some of them washed out to sea. . . . Huge oil tanks were ruptured by the force of the waves and they spewed their contents into the streets and into a fresh water lake used by Barrow for water."[81] One of the primary sources of fresh water was contaminated by salt water and sewage in addition to fuel oil.

Citizens and officials alike improvised emergency responses, focusing first on taking care of storm victims. But they were hampered by fire danger, damaged infrastructure, and the blizzard that followed the storm. While Guy Okakok waited in Betty Kignak's house, he heard by radio that "they needed able-bodied men to watch over people. They were urged not to smoke, even near the lagoon. The man on the radio said that big tanks had broken and were pouring oil all over the place."[82] Flooding restricted evacuation to heavy vehicles between Browerville and the main part of town; the only evacuation route from NARL to town, the coastal road, was impassable. The airfield under construction at the southern edge of Barrow was pressed into emergency service after the airfield at NARL was shut down and then became unusable. Another early report affirmed that "[t]he immediate work of caring for displaced persons is the first job being handled by officials. Even this work is being hampered by continued winds gusting to 40, with snow. Visibility in the area is poor, and no air traffic could get into, or leave Barrow on Sunday."[83]

However, one plane did reach Barrow on Friday, the day after the storm. It was a single-engine plane piloted by Wien Alaska Airlines's Ed Parsons on a mission to restore communications for transpolar flights. Repairmen jury-rigged an antenna, allowing flights between Alaska and Europe to resume on Friday evening. On Saturday, an Air Force cargo plane brought the first relief—blankets for several homeless families. A Wien DC-3 also brought freight. Relief officials arriving by air that day represented the Red Cross, state civil defense, the Bureau of Indian Affairs, the Army Corps of Engineers, and Golden Valley Electric Co-op of Fairbanks. The co-op sought to restore electric power to the village. The state and the airline sought to expedite compaction of the new airfield south of town to fly in needed freight. (Barrow cannot be reached by road, and shipping is blocked by sea ice most of the year.) Within a few weeks a representative of the American Red Cross announced hopes to have 26 destroyed or damaged homes in Barrow rebuilt and repaired before December 1. The Red Cross expected to prefabricate some replacement homes and package them for air delivery, at a total cost of $100,000. At about the same time, the Small Business Administration announced that it had declared Barrow a disaster area, making low-interest loans available to restore homes and businesses damaged or destroyed in the storm.[84]

The available reports on the restoration of utilities are relatively scant. Just after the storm, it was reported that the contaminated source of fresh water would be out of commission for several years. (It was eventually restored by pumping contaminated water from the lake, allowing it to refill with fresh water, and then repeating both operations.) Meanwhile, people would be able to get freshwater by melting ice from a nearby uncontaminated lake expected to freeze to the bottom soon. The hospital melted pieces of ice from large blocks washed ashore during the storm to help get through an outbreak of whooping cough later that fall. Within two months of the storm, Golden Valley Electric restored electric power using several generators, including one from the Native co-op store. The new distribution system for electric power, part of modernizing the town site, was all but completed by then. Surviving fuel oil reserves in Barrow were estimated to be enough to last four or five months. Fuel oil losses prompted the Navy to expedite a project, already approved by Congress, to drill a natural gas well for Barrow village nearby in the National Petroleum Reserve in Alaska. The first deliveries of natural gas from the well were made in December 1964.[85] The overall impression from published accounts and interviews is that the people of Barrow improvised and took the storm, the destruction, and the restoration of the community in stride. Even two non-Natives who lived through the great storm, Grace Redding at the hospital and her husband Bob at the camp, did not recall any great hardship associated with it.[86] But as modernization continued to accelerate, the vulnerability to disruption increased.

The significance of the great storm goes beyond the storm itself to how it was experienced by the Okakoks, the Reddings, and others, and how it has been contextualized. In an essay published a few weeks after the great storm, Howard Rock considered the great storm in the historical context of Eskimo survival in the unforgiving Arctic: "It was amazing and miraculous that no lives were lost in a storm of that magnitude at Barrow. It brought to mind the fact that Eskimos have long faced many and terrible dangers during the span of time in their survival in the Arctic."[87] His survey of adaptations began with the "main tool of combat against the Arctic dangers," which was "meeting them without panic." Another survival aid was avoiding unnecessary risks, which included listening faithfully to the elders. "A nervy man was much admired by the Eskimos but if he used it foolishly, he became something of a man marked with doom." He concluded the survey with Intensive Studies (his label) of the Arctic environment. Among other things, the Eskimos "had found that the only way to live with the Arctic was to know it from every angle—its behavior under certain conditions. In doing so they learned to cheat the many perils it dealt out from its many-sided and dangerous nature." But even with vast knowledge garnered from their experience in the Arctic, "the people still had to deal with one of its dangerous facets—its unpredictability." The great storm was an example. Rock concluded that "[t]he people of Barrow, both white and its predominantly Eskimo population, should be commended for their heroic behavior under the deadly onslaught of the vicious winter storm. It was a classic example of survival of man in the Arctic."

In an article published in 1967, Hume and Schalk considered the great storm in the context of a global warming future: "If, as has been suggested, the climate is becoming warmer as a result of the addition of carbon dioxide to the atmosphere, the likelihood of an open ocean and strong winds coinciding to produce such a storm is constantly increasing. Another such storm can be expected. . . ." They also drew some policy implications, recommending that "care should be exercised in the selection of building sites and construction methods." More specifically, "The buildings should be built on the highest points available, away from areas of rapid erosion, and should be erected on piles to put them above the reach of the water. Finally, borrowing from the beaches should be kept to a minimum. The best protection that such an installation can have is a naturally high, coarse beach. Building of groins, breakwaters, and other structures will never be an economical substitute where strong ice action is found."[88] As is often the case, advice on policy that is sound in retrospect often made little difference at the time. The story of Barrow's policy responses will resume later in this chapter. As noted there, the great storm of October 1963 still serves as the major historical baseline for understanding and reducing losses from big storms; the story as told here also helped us engage the interest and cooperation of Barrow residents for that purpose.

Vulnerabilities

As the story of the great storm suggests, both Barrow's past damages from big storms and present vulnerabilities depend on the conjunction of many human and natural factors. We begin this review of Barrow's present vulnerabilities by focusing first on the history of human factors, then on the variability of natural factors, and finally on specific storms in which these factors came together in unique ways.

Largely as a result of their own efforts, the Iñupiat empowered themselves in established decision-making structures to protect their lands and livelihoods from increasing outside encroachments. The most successful effort generated revenue from oil-industry developments at Prudhoe Bay, located on the North Slope coast 200 miles east of Barrow, which supported and accelerated modernization in the region. Modernization in turn exposed more people and property in Barrow to storm damage but also increased the community's capacity to reduce storm damage and to defend and advance its many other interests in the modern world.

Human Factors

With some help from outside, the Iñupiat began to empower themselves in the 1960s, not long after other American minorities had taken action to secure their rights. Here we rely primarily on *The Story of Self-Determination in the Arctic*, the subtitle of a North Slope Borough publication by Bill Hess titled *Taking Control*.[89] From the Iñupiat perspective, as described in *Taking Control*, "Government leaders and lawyers from outside looked at the Far North, and did not see legal papers nor deeds of property. . . . The Arctic Slope, they reasoned among themselves, was an empty, unpopulated region;

a place with no government. . . . [They] believed they could come to North Slope and do whatever they pleased. They felt no need to seek any kind of permission from the people already living here."[90] Examples of increasing encroachments echoing the colonial tradition are not difficult to find.

In July 1958 officials of the Atomic Energy Commission (AEC) initiated Project Chariot, which planned to detonate a 2.4-megaton atomic device to create a harbor near Cape Thompson in 1962. The site was about 30 miles southeast of Point Hope, in an area "where the people of Point Hope went to gather murre eggs, to hunt caribou, birds, and marine mammals, and to fish. . . . Yet, when these officials decided . . . they did not even bother to ask the people of Point Hope what they thought about the idea."[91] After AEC officials briefed the people of Point Hope for the first time in March 1960, the village council quickly and emphatically rejected AEC assurances that the blast would be no hazard to their way of life.[92] In 1961 the Secretary of the Interior decided to enforce for the first time in Alaska a 1916 treaty that banned hunting of migratory birds from March to September. Thus, "With no input from the Iñupiat, people from far away passed laws and signed treaties which forbade the Iñupiat to hunt ducks. Yet, here, ducks were hunted as a vital spring food source. These same laws allowed sports hunters living in California to shoot the ducks just for fun."[93] The ban triggered a mass act of civil disobedience resulting in the arrest of 139 duck hunters in Barrow. Officials later dropped the changes and quietly abandoned enforcement of the treaty. Hunters faced other hardships in the 1960s. "Oil exploration rigs ran across the land of the Iñupiat, conducting seismic tests. . . . Although the Iñupiat had been camping, hunting, and fishing on the involved lands for untold centuries, their permission was not sought nor were they consulted in any way."[94] Meanwhile, the state of Alaska began selecting lands under the Alaska Statehood Act of 1958. That act, "while acknowledging the right of Natives to lands they used and occupied, nevertheless authorized the new state government to select and obtain title to 103 million acres from the territory's public domain."[95]

In response, Natives scattered across Alaska began to organize from the bottom up. "In 1963 alone," according to Norman Chance, "a thousand Natives from 24 villages petitioned the Interior Secretary Stewart Udall to install a 'land freeze' on all land transfers from the federal government to the state government until Native rights had been clarified."[96] In Barrow in 1965, Charlie "Etok" Edwardsen, Jr., Guy Okakok, the *Tundra Times* correspondent, and the Rev. Samuel Simmonds of the local Presbyterian church began to organize the Arctic Slope Native Association (ASNA). In January

1966 ASNA filed with the Department of the Interior a claim based on aboriginal Native rights for nearly all land north of the Brooks Range, a total of 58 million acres.[97] Later that year, "more than 250 leaders from seventeen regional and local associations came together in the first statewide meeting of Alaska Natives." Overcoming their differences, they "unanimously recommended that a freeze be imposed on all federal lands until Native claims were resolved; that Congress enact legislation settling the claims; and that there be consultation with Natives at all levels prior to any congressional action."[98] Late in 1966 Sec. Udall, recognizing Native claims under the Statehood Act and the Organic Act of 1884, ordered a freeze on all federal land transfers until Congress settled the issues through legislation.[99] Governor Hickel of Alaska filed suit to force the transfer of federal lands to the state, but the U.S. Supreme Court eventually upheld the freeze.[100] Pressures on the state of Alaska to settle included the growing political power of statewide Native leaders, who organized the Alaska Federation of Natives (AFN) in 1967, and the state's financial interest in the sale of oil leases on federal lands. Pressures increased in 1968 after oil company explorations confirmed billions of barrels of recoverable oil at Prudhoe Bay on the coast of the North Slope.

Pressures on ASNA may be inferred from priorities recalled by Joseph Upicksoun of Point Lay, who joined ASNA in 1968 and became its president in 1969: "At ASNA, our mission, and our instructions, came from our people. ... One: we were to protect our land. Two: we were to bring in quality education. Three: we sought a basic improvement in housing. Four: we needed an improved health care delivery system for the people of the villages."[101] These priorities were underscored by widespread poverty confirmed in the federal government's Field Committee Report in the late 1960s. As *Taking Control* described it, the report "revealed that, along with other rural regions of Alaska, the economy of the North Slope was the poorest in the nation. It was no better than that of poverty-stricken nations in the Third World."[102] Focusing on securing the Iñupiat claim to 58 million acres, ASNA leaders found it necessary to extend the tradition of exploring new territory to Washington, D.C. "During repeated lobbying trips to Washington, the leaders of ASNA began to understand politics, and the workings of government." Concurrent intelligence from Washington indicated that the situation "was not good. Congress leaned toward a settlement which would drop some token cash on the Native communities, but which would leave them with only a sliver of their ancient lands." Apparently recognizing that "adaptability had always been key to survival in the Arctic, and would be now,"[103] ASNA pursued in parallel at least three courses of action to secure Iñupiat lands and livelihoods on the North Slope.

With Edwardsen once again taking the lead, the villages of the North Slope overwhelmingly voted to organize a tribal government under the Indian Reorganization Act of 1934. This established the Iñupiat Community of the Arctic Slope (ICAS) in 1971, the first regional tribal government; it federated Native village governments on the North Slope including the Native Village of Barrow, which was established in March 1940. ICAS was "envisioned to be the entity that would take control of settlement lands and the monies that came with them. ICAS would serve as the tribal government, with powers recognized under federal law" but not state law. *Taking Control* suggests that the role envisioned for ICAS was preempted by Congress, which sought to settle Native claims by conferring land and monies on Native-owned corporations rather than tribal governments.[104] According to Chance, ICAS "was only minimally active until 1975" when new legislation allowed the Bureau of Indian Affairs to fund ICAS "to establish welfare assistance programs and expand health services."[105] The amounts were small compared to the funds obtained by North Slope Borough after it was incorporated in 1972.

Meanwhile, ASNA also was involved in the politics that led to the Alaska Native Claims Settlement Act of 1971, which sought to assimilate Alaska Natives into mainstream American life by making them shareholders in corporations. Under pressure to compromise, AFN supported the act; in Chance's interpretation, they "saw the corporate solution both as a way to remove themselves from the bureaucratic yoke of the Bureau of Indian Affairs, and as a new tool in the fight to maintain their culture."[106] Within AFN only the ASNA opposed the act because of "Congress' decision to use population as the key criterion for the allocation of land and money rather than the size and value of the land."[107] Less than a tenth of 80,000 Alaska Natives enrolled under the act lived on the North Slope, but the North Slope was the second-largest region and included an estimated 9.5 billion barrels of recoverable oil. The act extinguished aboriginal Native claims in exchange for $962.5 million and 44 million acres of land, less than a tenth of all land in Alaska. These assets were conveyed to 12 regional and 200 village corporations chartered under state laws, not to tribal governments. Each person with at least one-quarter Native ancestry living in the villages (but significantly, not their unborn children) received 100 shares in a village and in a regional corporation. Natives were prohibited from selling their shares to non-Natives until 1991, a period of 20 years.

In July 1983 Thomas Berger, a Canadian, was appointed by the Inuit Circumpolar Conference to review the 1971 Settlement Act. The review in the mid-1980s took him "to Native villages all over Alaska to hear the evidence of Alaska Natives—Eskimos, Indians, and Aleuts. . . . It has not been easy for the people of village Alaska to be heard. For many years, they have been caught up in the cultural uncertainties of assimilationist policies."[108] He "found that, except for a few, the settlement had not worked for the benefit of Alaska Natives. The village corporations were many of them dormant or insolvent, and few Alaska Natives were employed by them. The same was true of the regional corporations. The $962.5 million that the regional and village corporations had received had declined in value, and the land that was to be theirs under the settlement . . . was in jeopardy."[109] The lands were in jeopardy because Natives could lose control by selling their shares to non-Natives beginning in 1991. Amendments to the act passed in 1988 extended with some exceptions the prohibition on selling shares, but did not correct what Berger considered the fundamental flaw: The prohibition suppressed the value of the shares, but without the prohibition "ownership will tend inevitably to be concentrated in the hands of non-Native interests."[110] However, from a business standpoint at least, the Arctic Slope Regional Corporation (ASRC) and the Ukpeaġvik Iñupiat Corporation (UIC) in Barrow are exceptions among the Native corporations; they have survived and thrived. ASRC is reportedly the largest corporation chartered under Alaska laws, with worldwide operations and annual revenues that exceed $1 billion.

As the Settlement Act was being worked out, ASNA faced the loss of almost 90% of the 58 million acres it had claimed in 1967. ASNA president Upicksoun saw a third course of action: incorporation of a borough under state law. As he recalled it, incorporation would "allow us to have control over all our homeland. We could do this through the use of two powers of local government—taxation and zoning. We could tax the oil industry at Prudhoe Bay to provide a base for a borough government. This would provide the revenue we needed to accomplish the four goals we had set at ASNA. Zoning: we would have the power to zone the industry—not to zone the industry out; the industry is our tax base but to zone for subsistence, to protect those places our people depend upon for their food and living."[111] At the request of Presbyterians in ASNA, the Synod of the Presbyterian Church granted ASNA $85,000 "to aid in their land claim efforts."[112] Upicksoun sent a petition for incorporation of a borough to the state Local Affairs Agency, triggering vigorous opposition. The oil industry, seeking to avoid taxation and regulation, argued that creation of a borough required an act of the state legislature. In an understatement, *Taking Control* observed that the industry's attorney "did not say that many in the Alaska State legislature and the administration of Governor Egan did not want to see the North Slope Borough created. They feared the Borough would claim revenues that otherwise would go to the State. Industry lobbyists had great influence inside the Alaska Legislature."[113] After hearings in February 1972, the Local Affairs Agency granted the petition over opposition from attorneys representing the oil industry and the state. Oil companies later filed suit to set aside the ruling in favor of ASNA and the borough, and to stay North Slope Borough elections already scheduled. "Despite the court case hanging over them, North Slope voters poured in to the polls on June 20. The vote in favor of the Borough was overwhelming."[114] Six days later a justice of the state Supreme Court heard the case and ruled in favor of the borough. The borough's logo includes the date of its incorporation as a first-class municipality under state law: July 2, 1972.

After its incorporation, North Slope Borough had only a $25,000 organizational grant from the state of Alaska to get started. It also had difficulty selling its revenue participation bonds. Reverend Charles White, then the Presbyterian minister in Barrow, recalled that "the oil companies had said to all the banks in Alaska 'if you in any way do anything to help the North Slope Borough, we will withdraw from all of our banking activities in Alaska.'"[115] However, the first mayor of North Slope Borough, Eben Hopson, succeeded in selling $150,000 of the borough's bonds to the Presbyterian Church; this encouraged other investors to buy the bonds. A special session of the state legislature in 1973 failed to prevent the borough from taxing oil industry property, but succeeded in capping the property tax rate at 20 mils, or $20 per $1,000 of assessed property value. The property tax revenues could be used only for operations, not capital improvements. However, Mayor Hopson and his advisors discovered that state law allowed the borough to use the property tax revenues to pay off general obligation bonds that could be sold to finance capital improvements. This opened the door to modern improvements in the quality of life on the North Slope.[116] *Taking Control* includes a fitting conclusion to this part of the story. In 1974 a delegation from the North Slope traveled to Kentucky for the general assembly of the Presbyterian Church. "Joe Upicksoun addressed the assembly on behalf of ASNA. 'During our time of need,' Upicksoun stated, 'you gave us $85,000.' He then presented them with a check for the same amount. 'You used your money for us,' Upicksoun explained. 'Now use it for somebody else that needs help.' It was the first time a grant had ever been paid back to the Presbyterian

Church." At the same time Oliver Leavitt on behalf of the borough paid off with interest the church's $150,000 investment in the borough's bonds.[117]

The North Slope's modern economy and the borough's revenues still depend primarily on property tax revenues from the oil industry. The borough's comprehensive financial report for the fiscal year ending June 30, 2006, acknowledged that "[s]ince 1968, oil and gas exploration and development on Alaska's North Slope has become the principal industry in Barrow and the employer of the bulk of the Borough's workforce. The other service providers, including the government sector, exist primarily due to the presence of the oil and gas industry."[118] Figure 3.5 summarizes trends in the borough's revenues and expenditures for fiscal years ending in 1980 through 2005. Property tax revenues increased sharply from $53.5 million in 1980 to a peak of $249.3 million in 1987. Eugene Brower, North Slope Borough's mayor in the early 1980s, recalled that "the Borough faced much jealousy elsewhere in the state, because it had a tax base."[119] But as the oil industry reduced its exposure, property tax revenues declined slowly by about 20.8% to $197.5 million in 2005. From 1987 to 2005, property tax revenues as a percentage of total revenues increased from 72% to 87% of total revenues, indicating more dependence on the oil industry. Most of the sharp drop in total revenues since 2000 can be traced to a decline in transfers from the state government and in miscellaneous revenues. Thus what was easily affordable in the 1980s and 1990s for various purposes, including storm damage control, became less affordable.

North Slope Borough's total expenditures increased sharply from $65.8 million in 1980 to a peak of $336.2 million in 1995. (The anomalies around 1990 reflect technical adjustments in debt service.)[120] During this period, "The administration and the assembly had agreed upon a six year plan to spend up to $150 million a year on capital improvement projects (CIP). Construction surged on all the kinds of facilities that people in Anchorage took for granted, but that the North had never seen. . . . Jobs were plentiful and high paying. . . . Anyone who desired could find work."[121] Among other things, in the outlying villages the borough constructed modern housing and upgraded airstrips, including navigation aids to help pilots locate them in poor weather. In Barrow the borough constructed the "utilidor"—a heavily insulated utility corridor buried deep in the permafrost—which cost about $270 million when it began providing water, sewer, and other utility services early in 1984. Later extensions of the utilidor into Browerville used a less expensive "direct bury" technology that also was used later in the villages.[122] Although hook-up costs kept some households from attaching right away,

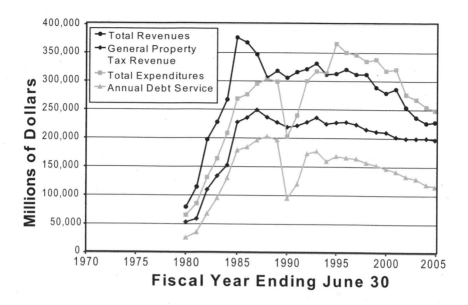

Fig. 3.5. North Slope Borough revenues and expenditures, 1980–2005. Source: *Comprehensive Annual Fiscal Report of the North Slope Borough, Alaska.*

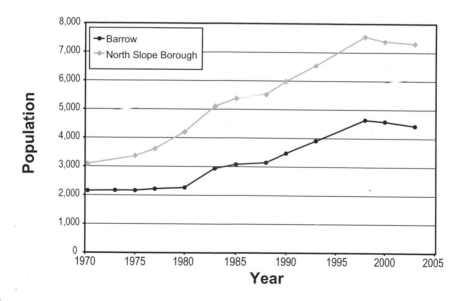

Fig. 3.6. Population of Barrow and the North Slope, permanent residents, 1970–2003. Source: Bob Harcharek, *North Slope Borough: 2003 Economic Profile and Census Report,* Vol. IX.

most homes in the Barrow area no longer require holding tanks for water or sewage or a honey bucket. After the peak in 1995, total expenditures began to decline in anticipation of further declines in property tax revenues. Expenditures for municipal services nevertheless increased erratically while expenditures for public safety (police and fire protection) held rather steady. But the two largest categories declined: Expenditures for general government (including jobs) dropped sharply in 1997 and continued down; expenditures for education dropped sharply in 2002. Expenditures for housing, a relatively small category, were zeroed out in 2005.

These trends are related to population trends in Barrow and the North Slope. As shown in Figure 3.6, the population of permanent Barrow residents held steady at about 2,200 in the 1970s, then increased rather sharply from 1980 until it peaked at 4,641 in 1998. Since 2003 the population has declined by about 150 persons per year according to informal estimates.[123] The population of permanent North Slope Borough residents followed a similar trajectory, peaking at 7,555 in 1998. In addition, at any one time about 5,000 temporary workers have been employed, mostly in the oil industry, on the North Slope during the last few decades.

Figure 3.7 compares Barrow including Browerville in 1964 and 1997, providing a visual image of the location and extent of modern development. Figure 3.8 focuses on development in the area flooded by the great storm in 1963, based on an analysis of eight aerial photographs of Barrow between 1948 and 2002. Our colleagues Leanne Lestak and Cove Sturtevant looked for structures added or rebuilt from one photo to the next. On average, just over one house per year was added or rebuilt in the 1963 flood zone between 1948 and 1962. After 1964, more than two houses per year were built or rebuilt in this zone. The most active period was between 1979 and 1984, when almost three buildings were added or rebuilt each year. In total, between 1964 and 2002, at least 65 new buildings were constructed and 12 were rebuilt or renovated within the zone, including the new Veterinarian Clinic. The clinic was completed in 1998 in the Browerville section of Barrow. Over the same period of 1964 to 2002, the road network, which hardly existed in the early 1960s, spread out to a total of 29 miles. Only 22 buildings were removed from within the historical maximum flood-prone area during this time, and 436 new buildings were constructed within 820 feet of this area.

Fig. 3.7. Map, Barrow and Browerville, development 1964 and 1997. Development of Barrow including Browerville, 1964 and 1997, compared. Source: Leanne Lestak and Eric Parrish.[124]

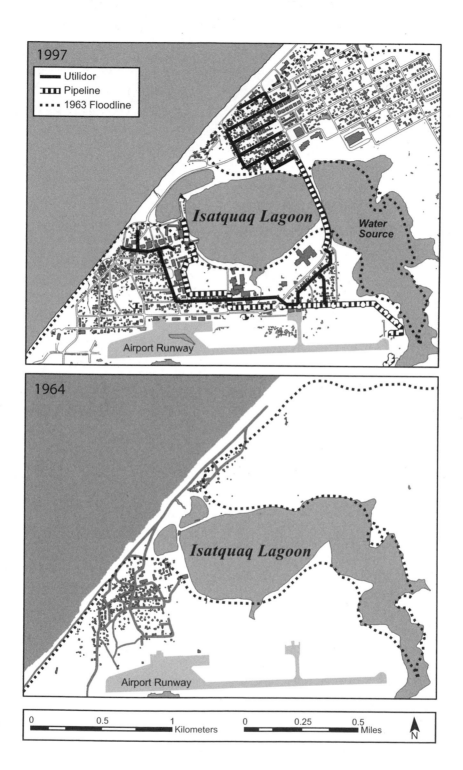

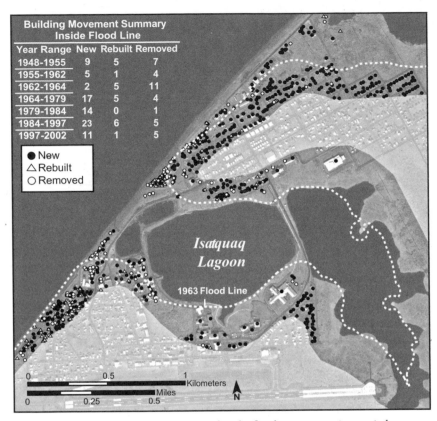

Building Movement Summary Inside Flood Line			
Year Range	New	Rebuilt	Removed
1948-1955	9	5	7
1955-1962	5	1	4
1962-1964	2	5	11
1964-1979	17	5	4
1979-1984	14	0	1
1984-1997	23	6	5
1997-2002	11	1	5

● New
△ Rebuilt
○ Removed

Isatquaq Lagoon

1963 Flood Line

0   0.5   1
Kilometers
N
0   0.25   0.5
Miles

Fig. 3.8. Buildings new, rebuilt, and removed in the flood zone, seven time periods, 1948–2002. Source: Cove Sturtevant and Eric Parrish.

Thus, as an unintended consequence, modernization has left more people and property vulnerable to big storms in Barrow, but the numbers do not tell the whole story. There are also changes in resilience. For example, buildings that were rather easily moved according to the new town site plan in 1964 would be more difficult to move now because they are tied into a modern water and sewer system, the utilidor. Similarly, in the early 1960s Barrow still relied on sleds and sled dogs that could be replaced with local resources. The snowmobiles that replaced most of the sleds in a single year in the mid-1960s could only be repaired or replaced with cash in the modern economy. But snowmobiles and other modern means of transportation—motor boats, navigation aids, and search-and-rescue helicopters—also support subsistence hunting over an area that in earlier times could only be covered from smaller dispersed settlements. The remaining subsistence economy provides some resilience to cutbacks in the modern economy. According to a survey

Fig. 3.9. Present vulnerabilities to big storms in Barrow. Source: Leanne Lestak and Eric Parrish.

in Barrow in 2003, "Eighty-nine percent of the unemployed and 76% of part-time workers said that half or more of their food came from subsistence hunting and fishing." The same survey found that "[o]ver 91% of the Iñupiat households that were interviewed participate in the local subsistence economy. . . . 66% said that half or more of their diet consisted of local subsistence resources."[125] Even among the roughly 40% of non-Iñupiat households in Barrow, there is significant use of subsistence resources.[126]

Based on a QuickBird satellite image from 2002, Figure 3.9 identifies the location of Barrow's major vulnerabilities to damage from big storms. (It also locates some policy responses intended to reduce future damage; we consider these in the next section.) At the southwest corner is the Old Barrow town site, Ukpeaġvik, where cultural artifacts of great value to the Iñupiat are buried (Fig. 3.10). Those artifacts could be washed out to sea and lost by further erosion of the bluffs. Moving up the coast are two of the utilidor's seven pump stations, No. 4 (Fig. 3.11) near the center of town and No. 3 in Browerville, which are exposed to flooding and erosion. Sewage flows by gravity to a holding tank in each pump station, from which it is pumped to the next station, and eventually from pump station No. 7 to the South and Middle Salt sewage lagoons. The pumps also circulate potable water heated to about 40°F to prevent freezing. Seawater flowing down through pump station hatches, or through manhole covers nearby, could short out electric pumps and controls, stop the circulation of fluids, and subject them to freezing in miles of pipes. That would be a major disruption in community life, and costly to repair.

The spillway across Isatquaq Lagoon supports a road and above-ground telephone, fiber optics and power lines, utilidor water and sewer lines, and the single line that supplies natural gas to town. All these lines are exposed to waves and flooding. A break in the natural gas line would shut down the central power plant; the utilidor has a diesel-power generator as a back-up for that contingency. The reservoir east of the spillway, the town's main water source, is vulnerable to flooding and contamination. Among the structures exposed in low-lying areas in Browerville are fuel tanks at the gas station, the Veterinarian Clinic across the street to the northeast, and farther in that direction, public housing units and heavy-equipment shops of the Department of Municipal Services. Still farther up the coast are the landfill, where the military buried unknown toxic material years ago, and South and Middle Salt Lagoons holding sewage. If breached by erosion or flooding, these would release hazardous materials into the sea, impacting fish, marine mammals, and other living things, including the people who depend on them for subsistence. The culvert from Middle Salt Lagoon to the sea has been repeatedly washed out by big storms. The bridge over the culvert is part of Stevenson Street, the coastal road that also has been washed out in various places, blocking evacuation from NARL and putting people who live or work or study there at risk.

Behind the modern vulnerabilities of Barrow lies an empowerment process that left the Iñupiat in a better position to protect what they valued from big storms and from further encroachments from the modern world.

Fig. 3.10. Old Barrow townsite (Ukpeaġvik) with remains of a whalebone house in foreground and Barrow in the background. Photo: Dora Nelson.

Fig. 3.11. Utilidor Pump Station No. 4 protected by temporary berms from the Arctic Ocean in the background. Photo: Ron Brunner.

In the empowerment process the role of the most powerful whaling captains (the *umialit*) was partially reconstructed after being largely destroyed in the colonial era. Nearly all of the leaders in the incorporation of North Slope Borough, all of its mayors since incorporation, and most of the leaders of UIC and ASRC since the Settlement Act have been *umialit*.[127] Similarly, the extensive trading system created prior to contact to cope with local material shortages has been partially reconstructed through the success of UIC, ASRC, and the borough. At least they function in similar ways: Natives explore a larger territory by leaving Barrow and the North Slope for offices in Anchorage, Washington, and around the world; increasing numbers also are returning home, bringing new knowledge and skills to help cope with the modern world.[128] The process of empowerment extended beyond national boundaries in 1977 when Mayor Eben Hopson took the lead in organizing the Inuit Circumpolar Conference (ICC) with representatives from Canada and Greenland to deal with transnational environmental concerns arising from expansion of the oil and gas industry in the circumpolar region. The ICC was accepted as a "Non-Governmental Organization" by the United Nations in February 1983.[129] In these various examples, the people of Barrow and the North Slope provided insight into how other communities might empower themselves in established decision-making structures.

Another initiative shows how differences between top-down and bottom-up perspectives can be resolved on an empirical basis. In 1977 the Iñupiat received what *Taking Control* described as a "terrible shock": "An organization from far away called the International Whaling Commission suddenly announced that there were so few bowheads left in the world that the Iñupiat hunters must stop whaling."[130] The U.S. government did not object to the moratorium; among other reasons, only 600 to 1,800 bowheads remained according to estimates by the National Marine Fisheries Service. But Iñupiat whalers relying on their local knowledge claimed at least 4,000 to 5,000 bowheads existed in 1977. "Backed by the Borough and ASRC, angry whalers living all along the arctic coast from Kaktovik to St. Lawrence Island gathered in Barrow on August 29. To spearhead the fight to protect their ancient subsistence rights, the hunters organized the Alaska Eskimo Whaling Commission [AEWC]."[131] During the next several months and years the AEWC and their allies lobbied the government, filed lawsuits, and sought to make their case in IWC meetings around the world. But they also understood, in the words of Ben Nageak, that their own knowledge of the bowhead population "doesn't mean anything to the outside world unless they can read it in a scientific report."[132] So Iñupiat hunters cooperated with scientists, including

Dr. Tom Albert of the NSB Department of Wildlife Management, to set up a system of four directional hydrophones to identify and count individual bowheads by sound during their migration. The system picked up bowheads scattered under the ice that had been missed by visual counts of bowheads in the leads. The results corroborated local knowledge and were a critical resource that made it possible for Iñupiat subsistence hunters to replace the moratorium with a series of larger quotas. When *Taking Control* was published in 1993, the quota reached 47 bowheads landed, or 57 struck, per year. "In the spring and fall seasons of 1992, Barrow hunters landed 22 whales, which were shared with villages landing no whales."[133]

Natural Factors

Understanding Barrow's past losses to flooding and erosion and present vulnerabilities also depends on observed trends and variability in the range of natural factors that influence the occurrence of flooding and erosion events. These factors include changes in the atmospheric circulation, temperature and snow and their influence on permafrost, and sea-ice retreat. None of these factors alone can lead to a flood event or an erosion event, but in their unique combinations they produce the broad range of events observed over the past decades.

At Barrow, the primary cause of flooding and erosion events is sustained strong winds from the west or southwest, which push water onshore in a process known as Ekman transport. The climate of the North Slope of Alaska is controlled in large part by the broad, persistent high-pressure area that is centered over the East Siberian Sea or Chukchi Sea in winter and the Beaufort Sea in summer. Because of these high-pressure conditions to the north, much of the time Barrow is subject to prevailing easterly winds. Because the town is situated on the western side of the peninsula, these prevailing easterlies, however strong, cannot lead to erosion or flooding events in Barrow. The North Slope region is also influenced by the North Pacific storm tracks,[134] although the Brooks Range to the south can isolate the North Slope from some of these effects.[135] These storm tracks show a tendency toward developing deep lows in the vicinity of the Aleutian Islands, but occasionally a cyclone will slip northward through the Bering Strait to cause strong southerlies or westerlies in Barrow. Also relatively infrequently cyclones track eastward from the East Siberian Sea to the Beaufort/Chukchi region. Because these cyclones typically cause westerly winds in Barrow, they are often a source of flooding and erosion events. The great storm of October 1963 was one such cyclone.

It has been demonstrated in many different ways[136] that average sea level pressures over the Arctic Ocean have decreased over the past several decades and that concomitantly cyclones have increased in both frequency and intensity. An increase in these cyclones suggests a shift from the prevailing easterly winds to a more frequent occurrence of westerly winds at Barrow, as well as an increase in the intensity of the wind events. Notwithstanding, the linear trends in the magnitude and direction of average winds and the trends in the magnitude of wind speeds during high-wind events at Barrow are insignificant. Most important for big storms is the record of extreme wind speeds. Unlike average wind speeds, the frequency of extreme wind speeds affecting Barrow is changing with time, but in a nonlinear way.[137] In fact, the record of these high-wind events (Fig. 3.12) shows a minimum during the 1970s and early 1980s, a period of intense development in the community. Both prior and subsequent to that period, the incidence of high-wind events has been greater. This is also true for both westerlies and easterlies separately. What is not clear is whether the recent increase in frequency of extreme wind events will continue to increase. The shortness of the record, the high nonlinear variability, and the shortcomings in current understanding of the physical processes influencing future changes means that the likelihood of a big storm in Barrow cannot be reasonably projected into the future with any confidence. Major uncertainties are inevitable.

In contrast, a strong trend in factors such as temperature, the length of the snow season and the length of the sea-ice season is readily apparent, consistent with an amplification of warming in the Arctic. Many researchers have documented increases in the annual mean and seasonal air temperature in Alaska over various periods between 1921 and 2003.[138] Barrow itself has experienced increasing air temperatures over the last 83 years. While this warming is not uniform or significant for all seasons, nor uniform over the entire period, an average increasing trend of around 2.9°F per decade is apparent in winter and spring since 1921. No evidence of an urban heat island effect has been found. In winter, postulating a simple linear trend differs substantially from postulating a polynomial trend (Fig. 3.13), and the difference between these two statistical models of the trend is significant in winter. There is no easy way to choose between alternative statistical analyses of this sort. It seems important, however, to recognize that the finding of a continuing warming trend of large magnitude in the late winter/early spring average temperatures in Barrow is not insensitive to different approaches.

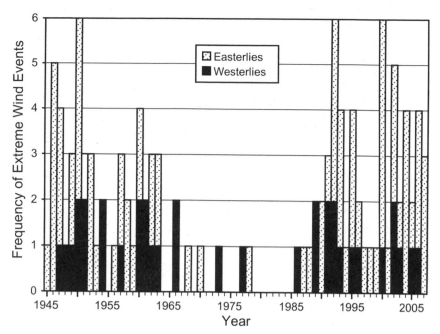

Fig. 3.12. High-wind events in Barrow, frequency of easterlies and westerlies, 1945–2008. Source: National Weather Service data.

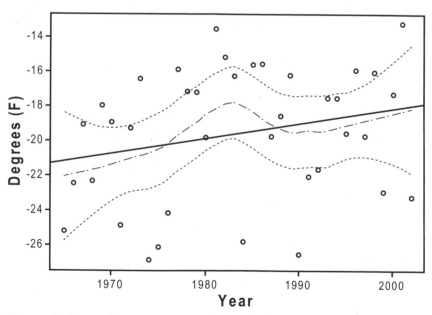

Fig. 3.13. Minimum winter temperatures in Barrow data. Source: Claudia Tebaldi and Eric Parrish.

That said, there are other measures of a warming climate much more robust to different statistical model choices. For example, despite some generally cooler conditions in the first five years of this century, the frequency of extremely cold days and the duration of cold snaps continue to decrease. This is true for all seasons and for all choices of low temperature thresholds and lengths of cold snap.

Increasing air temperatures are implicated in a decrease in the extent, in area and time, of winter snow cover in Barrow and its environs. In addition, winter total precipitation has reduced significantly. Snow cover onset shows very little change, but the record shows a significant trend toward earlier spring melt—almost one month earlier over the last 50 years. A decrease in snow cover exposes permafrost to greater potential for thaw, thereby destabilizing coastal sediments and increasing the risk of erosion. In addition, the date of snowmelt onset is increasingly variable over that same period, consistent with the increasing variability of spring air temperatures in recent years. This variability of snowmelt reflects temperature changes quite important to the community. For example, it complicates planning for return from winter hunting camps using snowmobiles.

Ocean surface temperatures along the coast near Barrow have also increased, by almost 4°F during the period from 1982 to 2002. The fairly localized increases in temperature may reflect the longer duration of ice-free conditions described below, or could be due to increased southerly currents or upwelling in the area. These ocean temperature increases reflect the general warming in spring, but there is a slight nearshore cooling in January and February in the Barrow vicinity.

The distance to the main ice pack from the Barrow nearshore zone (the "fetch") is important because a longer fetch allows for the build-up of larger storms surge and waves; conversely, shorefast ice prevents them. The size of the fetch at Barrow is strongly related to overall ice conditions in the Arctic Basin, which are in turn affected by the winds, the air temperatures, and the ocean temperatures. In 2007, James Maslanik and colleagues noted that in the Arctic "[t]he oldest ice types have essentially disappeared, and 58% of the multiyear[139] ice now consists of relatively young 2- and 3-year-old ice compared to 35% in the mid-1980s. Ice coverage in summer 2007 reached a record minimum, with ice extent declining by 42% compared to conditions in the 1980s."[140] Multiyear floes provide an "anchor point" for shorefast ice when the thick floes become grounded near shore. Typically, during recent years, the southward extent of multiyear ice in spring corresponds to the location of the overall ice pack edge at the end of the preceding summer.

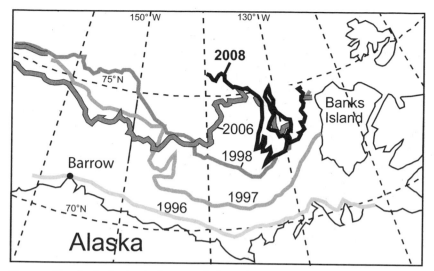

Fig. 3.14. Ice edge in the Arctic Ocean, end of March in 1996, 1997, 1998, 2006, and 2008. Sources: James Maslanik, National Ice Center data.

"The trend in the September ice extent for the period 1979–2004 is −7.7% per decade . . . a value twice as large as that reported for . . . 1979–95" and the trends are greatest along the Beaufort Sea coast.[141] These reductions have been associated with the persistence of open water offshore longer into the autumn season than was characteristic of the earlier record (Fig. 3.14). The average winter freeze-up date at Barrow in the last 30 years has been around October 17th and the trend in that date is around 15 days later each decade. This represents 15 additional days during which Barrow is at risk for exposure to the impacts of a severe storm.

Extreme Events

The individual records of climatic variables can provide important information, but as separate factors they do not tell the whole story. Each extreme event that has the potential to cause losses in the community arises through the coincidence of several variables in the context of a single event.

In the Barrow area, erosion is primarily determined by the coincidence of an extreme storm event when the ice is out. Along the North Slope coast, shoreline losses average nearly a meter a year, and in some places exceed 16 feet per year.[142] In general, erosion is more rapid along the ice-rich peaty soils of the Beaufort coast to the east of Barrow, with the high bluffs and

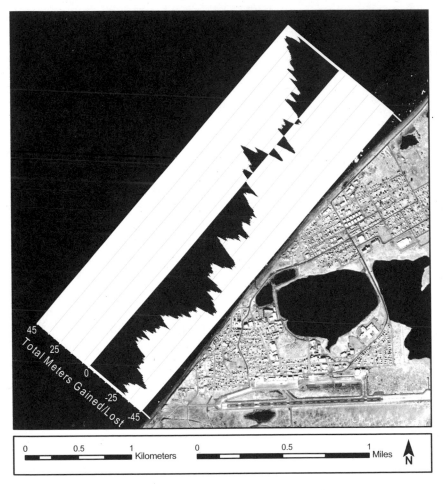

Fig. 3.15. Coastal erosion in Barrow, 1948–1997. Source: Leanne Lestak.

coarser beach sediment of the Chukchi shoreline to the west providing some protection. The contribution of undercutting and bluff collapse cannot be quantified but is known to be significant.[143] In contrast, the bluffs directly adjoining Barrow experience relatively little erosion activity, and most of that occurs during big storms.[144]

James Hume and Marshall Schalk estimated that the alongshore transport from the Barrow area between 1962 and 1964, attributed to the October 1963 storm, was 4 million cubic feet, which is equivalent to 20 years' worth of normal transport.[145] Our colleague Leanne Lestak obtained aerial photographs of the Barrow-area coastline from August 1962 and July 1964. She found that during this period erosion occurred along the bluffs, ranging from low values in the southwest far from town to values as high as 33 feet toward the northeastern portion of the bluffs. The highest levels of erosion occurred near runoff outlets and the old town site of Ukpeaġvik. The average erosion along the bluffs in this period was 12 feet—that is, 10 years' worth of average erosion—with a very high degree of spatial variation, consistent with the chaotic nature of wave action during a storm (Fig. 3.15). Lestak also observed rounding and slumping of the bluff edge in the 1962 coastline. The 1964 aerial photographs show much less slumping near Ukpeaġvik and also show a much more clearly defined bluff edge. It may be that the severity of the 1963 storm cleared away all of the slowly eroding material, leaving a sharper bluff edge of freshly exposed permafrost.

At the water's edge, Lestak found a mixture of erosion and aggradation between 1962 and 1964. Again, the results were highly spatially variable with an average of 10.5 feet of aggradation. In some places, nearly 46 feet were added to the beach. The water line varies from day to day depending on tide and atmospheric pressure, and this may have played a part in the difference in water line seen between the 1962 and 1964 coastlines. We wonder, though, if there may have been other processes at work on or near the beach during this period to produce the large beach growth in a small area.

Thus, one consequence of the reduction in sea-ice extent is increased potential for coastal erosion from greater wave exposure.[146] Reduction in the ice season is already contributing to Alaskan and Siberian coastline retreat at rates of feet per year.[147] The lengthening of the ice-free season suggests a potential for increased erosion in Barrow. But the episodic rather than continuous nature of erosion processes, combined with the immaturity of efforts to model transport in these ice-rich sediments, contributes to the difficulty in assessing future rates of change with any confidence.

There is no fixed or standard set of factors and linkages that explains every extreme event. Ice retreat in conjunction with a storm can lead to a very different outcome, depending on the specific nature of the event. The reduction in multiyear ice and the increase in the ice-free season lead to the presence of less stable shorefast sea ice in winter and spring. Under such conditions, a storm can drive shorefast ice onto the shore in an ice push or "ivu" event, damaging infrastructure in the way. One catastrophic example of this process, which was "apparently unprecedented in size, severity, and extent,"[148] occurred in May 1957, on a clear, calm day, when whaling crews were out on the ice edge. Eyewitness Wesley Aiken, Sr., said, "We didn't have any weather forecasts at that time. No radios. No radio station." The wind began to increase to a maximum of around 25 mph from the southwest, with an accompanying storm surge. As ice floes started to move toward the shorefast ice on which the whaling crews were standing, they tried to pack their gear and retreat to safety. But when the ice floes smashed into the shorefast ice, it shattered all the way to shore, something the hunters had not seen before. Usually, the crews could rely on the long experience of their whaling captains, but the speed and unexpected force of the event took them all by surprise. Arnold Brower, Sr., recalled the confusion: "This pressure ridge. I didn't expect it to break up. That's where I waited for the rest of the boats. And I got caught in between two. . . . It is impossible to remember which way the ice was moving, because there was something in every direction. As that ice was moving, some of it would push the other ice and swirl it around. It's unusual. It's moving in every direction." The orderly retreat turned into a flight. Harry Brower, Sr., had just started his own crew, and they were camped far out on an ice peninsula where the whales might pass directly underneath, taking shortcuts along the ice edge. He said, "We just threw away all the gear, 'cause we weren't going to be able to save it anyway. We were right in the middle of that moving ice. Most of the dogs were all caught in the ice, too. They were let loose to run around and try to make it through on their own, but we still lost quite a bit of our dogs."

Remarkably, no lives were lost. "We came in with our lives, nothing else," said Warren Matumeak. "The whole town was watching us trying to stay alive. . . . Lots of people were running around checking to see who made it." According to Matumeak, around half of the 10 crews that were out that day lost their equipment. Harry Brower, Sr., said it took him until 1971 to restore his equipment so that he could put out a whaling crew once again.

In the absence of ice close to shore, other combinations of factors can converge instead to result in damaging flood. Because central pressures in Arctic storms are generally not particularly low in comparison to, say, hurricanes, the effect of atmospheric pressure on sea level is usually small.[149] Rather, storm surges are caused primarily by sustained strong winds. Waves make an additional contribution to flooding. Based on an analysis of the most significant 21 storms of the last 50 years, we have found that forecast westerly winds of greater than around 30 mph blowing for at least 20 hours over relatively ice-free seas will most likely lead to a significant flooding event.[150] As noted earlier, winds from the west are associated with cyclones to the north of Barrow, in the Beaufort and Chukchi Seas. Such cyclones are relatively infrequent, compared to cyclones affecting the Aleutians, for example, and very intense cyclones are rare.

The seasons of greatest potential danger from flooding are late summer and early autumn, when the sea-ice extent is likely to be a minimum and the open water fetch is the greatest. Barrow residents historically have feared autumn storms the most and have prepared for them in recent decades. The ice-free season is increasing dramatically; hence, regardless of any uncertainty in future projection of the wind record, the danger represented by the decrease in ice protection is unequivocal.

Consider the potential consequences of this extended ice-free season. In the record of the past 50 years, one storm has demonstrated the potential for a more severe flood than the October 1963 storm; this occurred in November 1966. The high winds from this storm occurred in Barrow on November 17, 1966. The average daily wind speed for this day was 34 mph with a one-minute sustained wind of 54 mph. The prevailing wind direction was from the southwest, due to a low-pressure system that had slipped north from the Aleutian Islands through the Bering Strait, to the north of Barrow. The sea ice was fast in to the coast during this storm, and hence protected the community from storm surge inundation. However, as we have described, sea-ice setup has trended rapidly later in recent years, thus this storm represents an interesting hypothetical case. What if the ice had been out when this storm occurred? When applied to the forecast system we have developed, the November 1966 storm leads to a much more severe flood than the October 1963 storm (Fig. 3.16).

Conjunctions

Each of the factors contributing to Barrow's vulnerability combines in unique ways in dozens of extreme weather events that have hit Barrow in the last half-century. In the examples that follow, we demonstrate research that is centered on a series of extreme flooding events of different character, comprehensively accounting for the major natural and human factors contributing to losses, to reach an understanding of the convergence of factors that shape outcomes.

An important event in North Slope Borough's history of erosion and flooding occurred in September 1986, which was actually two storms that impacted Barrow in short succession, September 12 and 20. The fetch from Barrow to the ice edge was approximately 550 km. The average wind speed was 22 mph and the peak winds reached 56 mph on September 12. The

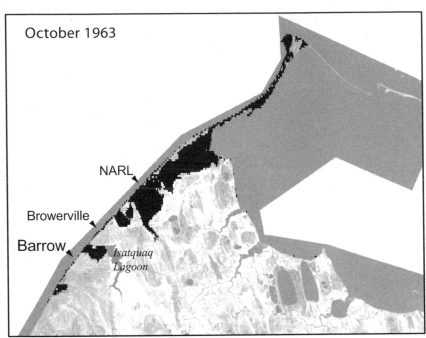

October 1963

NARL

Browerville

Barrow

Isatquaq Lagoon

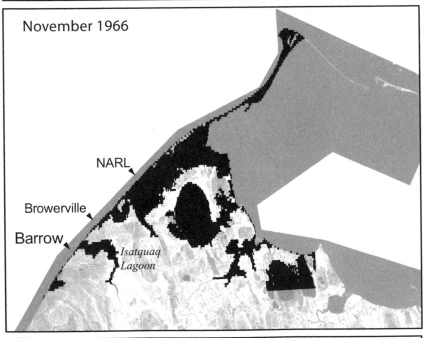

November 1966

NARL

Browerville

Barrow

Isatquaq Lagoon

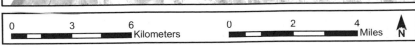

0   3   6
Kilometers

0   2   4
Miles

N

prevailing wind direction was from the southwest due to the passage of a low-pressure system to the west and north of Barrow. As in 1966, the low-pressure system moved up from the Aleutian Islands through the Bering Strait, passing to the north of Barrow.

The heavy rains just prior to the first stormy period exposed an archaeological site known as Mound 59 in Ukpeaġvik, which contained the remains of an ancestor. Only a leg clad in a mukluk could be seen from the top of the bluff. On September 20 the winds peaked again, reaching an average of 31 mph, but gusting as high as 65 mph. Before the community could make plans to properly secure and excavate the site, the second storm washed "Uncle Foot" out to sea. In addition to this loss, the second storm eroded a cliff under George Leavitt's house, which left it perched perilously 35 feet over the ocean (Fig. 3.17). The house was saved. Estimated damage to roads and structures in Barrow and in Wainwright exceeded $7.5 million.[151] These storms prompted the community to begin to explore options for protecting the town from flooding and erosion.

On August 10, 2000, the strongest summer storm in memory hit Barrow. A low-pressure system moved toward Barrow from Siberia, taking a very similar path as the October 1963 storm. The fetch to the ice edge was approximately 500 km. The prevailing wind direction was from the west, and the wind that day averaged 37 mph, with gusts at least as high as 64 mph. The National Weather Service anemometer may have "pegged out"[152]—a maximum wind of 74 mph may have been observed at the CMDL[153] weather station. Emergency management teams had insufficient notice to mobilize heavy equipment to build up temporary berms for protection along the coast. This was fortunate, because the extent of flooding was much less than in the 1963 storm, for example, due to the shorter duration of strong westerlies and the high surface pressure of this storm. Nevertheless, damages were reported all along the North Slope coastline. In Prudhoe Bay, a tide gauge reported a storm surge of 1.46 m, within 6 cm of the 100-year event.[154] In Barrow, there was wind, wave, and erosion damage. The major loss was the dredge *Qayuuttaq* valued at $7 million (Fig. 3.18). The storm ripped it from its anchors and washed it ashore, damaging the starboard hull and flooding it.[155] Roofs were removed from 40 buildings, the beach was eroded to within 100 m of a main pumping station for the utilidor, and a boat ramp was washed out.

Fig. 3.16. Big storm of November 1966, actual and hypothetical flooding. Sources: Leanne Lestak, Eric Parrish, and Amanda Lynch et al.[156]

Fig. 3.17. George Leavitt's house after the September 1986 big storms. Photo: *Open Lead*, a publication of North Slope Borough.

Fig. 3.18. Dredge *Qayuuttaq* beached after the August 2000 big storm. Source: Alaska Department of Community Services.[157]

About 35 private homes and 4 public housing units sustained roof and siding damage. Waves washed out a gravel seawall and a culvert crossing at Middle Salt Lagoon, and damaged roads along the coast once again. Storm costs to the Department of Municipal Services, which maintains and repairs roads and beaches among other things, was $829,772. The initial aggregate damage estimate was about $7.7 million, renewing interest in protective measures.[158] Unfortunately, in this and other cases, damage estimates reported in dollars are not necessarily complete, comparable, or otherwise reliable.

The strongest storm in recent years occurred over a three-day period from October 7–9, 2006. The winds peaked around 5 P.M. (local time) on the third day at 55 mph with gusts up to 67 mph. The winds started picking up the afternoon of October 3, a Tuesday, and continued to build through the week. The atmospheric pressure at Barrow during the storm was measured to reach a minimum of 985 mb, which indicates a very strong cyclone. Fortunately, the winds were predominately easterly until the frontal passage at around 8 A.M. on October 10, when they switched around to the west-southwest and started to weaken as the pressure began to rise. The National Weather Service official forecast for Barrow issued at 7:40 A.M. on October 9 stated, "The high wind warning is now in effect until 1 AM ADT Tuesday. . . . Roof damage is now being reported on several structures in Barrow." Dave Anderson at the Barrow National Weather Service said, "The saving grace with this storm was that the winds switched from east to south to west as the storm passed. That kept the winds offshore for the event. . . . We had rain, freezing rain, snow, blowing snow, blowing sand (gravel, actually), and a little fog on the 9th with the high winds. Our ice observations show that the sea was ice free for almost all of October 2006. No ice was reported during the storm."[159] Local archaeologist Anne Jensen noted significant erosion, with some artifact losses, at Plover Point to the east of Barrow and had heard reports of the loss of scientific equipment.

While these extreme events share some characteristics, each event is unique under a comprehensive and integrated assessment. What matters in understanding past damage and reducing vulnerabilities in the future is the coincidence of natural factors including ice retreat, strong westerly winds of long duration, and permafrost thaw, in conjunction with human factors including lives, property, and cultural artifacts/items of value. The remaining human factors—policy responses intended to reduce losses from big storms—are considered in the next section. Thus, there can be no standard model. In the face of profound uncertainties, members of the community often take the October 1963 storm, the most damaging to date, as the scenario for planning purposes. Remarkably, there have been no reports of fatalities or serious injuries attributed to any of them, but there have been close calls. This is partly good luck.

Policy Responses

Barrow's major response to storm damage from erosion and flooding so far has been the Barrow and Wainwright Beach Nourishment Program, which was motivated by damaging storms on September 12 and 20, 1986. These storms occurred near the end of a decade and a half with relatively few severe storms and a decade of rapid infrastructure development, including the first phase of the utilidor completed in 1984. The program *as planned* effectively relied on beach nourishment technology as a single technical solution in the tradition of scientific management. So did its successor, a joint feasibility study to reduce storm damage in Barrow undertaken by the U.S. Army Corps of Engineers with participation by the borough. Both had to be modified considerably in light of unanticipated problems. In contrast, Barrow's history of distributed policies and engaging the world at large is more consistent with adaptive governance.

Technical Rationality

In a special meeting on September 22, only two days after the second storm, the North Slope Borough Assembly passed a resolution declaring a disaster and state of emergency. It appropriated $500,000 for capital projects "for the prevention and minimization of injury and damage" caused by big storms.[160] In that same month the borough hired Tekmarine "to inspect the damages caused by the storm(s) and make initial mitigation recommendations," according to Frank Brown's chronology of the program.[161] (Brown was then director of Capital Improvements Projects Management [CIPM] for the borough.) In July 1987 the borough selected Barrow Technical Services (BTS), a subsidiary of UIC, the village corporation, in a joint venture with LCMF in Anchorage to collect material samples offshore. This initiated a continuing series of searches for material suitable for beach nourishment, each premised on the need for more or better material. These and other planning studies culminated in a report by BTS/LCMF, *Mitigation Alternatives for Coastal Erosion at Barrow and Wainwright*, dated April 1, 1989. The report first acknowledged and then quickly absorbed uncertainties about weather and climate: "The timing [of another big storm] is unknown, nor do we believe that the probability of such a occurrence can be forecast with much reliability . . . however, we feel that it can be stated that the probability of a storm with a 9 to 10 foot surge within the next 30 years is better than 50%."[162] Similarly, the report adopted the rate of erosion assumed by Tekmarine, 4 feet per year, while noting that reported erosion rates varied by authors and areas of the beach.[163] The report also concluded that "an offshore material source of sufficient volume and character to provide material for nourishment" had been identified. However, questions could have been raised about the suitability of offshore material samples for beach nourishment.[164]

The Executive Summary listed the active and passive alternatives available: "Active consideration of erosion control included 1.) beach armor in various configurations, 2.) sheet piling seawalls, 3.) utilidor section seawalls and 4.) beach nourishment by dredging or conventional material placement. Passive or non-active considerations include insuring capital investments against flood damage and a 'do nothing' concept."[165] BTS/LCMF rejected the beach armor and sheet piling seawalls recommended by Tekmarine because of cost and technical uncertainties. It also recommended against further expenditures for utilidor-section seawalls until the experimental seawall protecting Barrow's sewage lagoon was evaluated during and after storm conditions.[166] Finally, the report set aside the passive alternatives, including relocation as well as insurance and "do nothing," in the last 4 of 50 pages of text, not counting pages of engineering drawings and appendixes on the beach nourishment program. As Brown summarized it, the report "recommends that if the Borough is to seek *active* mitigation measures, that beach nourishment by dredging is the recommended course of action. Other *non-active* alternatives are discussed, but background information is limited."[167] *Mitigation Alternatives for Coastal Erosion at Barrow and Wainwright* became the seminal plan for further studies, public hearings and workshops, and a policy decision to proceed. The schedule, productivity, and costs of the beach nourishment program as planned in 1989 are reviewed below and compared with the actual results realized a decade later.

In July 1992 the North Slope Borough Assembly effectively decided to proceed, appropriating $16 million for the Barrow and Wainwright Beach Nourishment Program "based on civil equipment designs to date which incorporate the use of existing NSB owned machinery."[168] Undocumented reports claim that the assembly merely formalized a decision made in an older tradition by two of the most powerful whaling captains. In any case, the details of the program as prescribed—objectives, assumptions, capabilities, schedules, costs, and so forth—are not clear in the documents available. However, a third-party project and plan review by Ogden Beeman & Associates described the basic concept in September 1992: "Present plans are for the construction of a self-propelled, barge mounted dredge with a mechanical (clam shell bucket) excavator feeding a hopper. Material is fluidized in the hopper and pumped to shore via a submerged pipeline. The dredge is to be moored by means of two mooring anchors offshore and the submerged pipeline connection to the beach. The dredge should be attended by a crew boat and beach operations would be carried out by tractor dozers as standard in the dredging industry."[169] The equipment design changes recommended by Ogden & Beeman were incorporated into the project. Brown summarized the change and their significance: "the addition of a tug boat to support the program, several changes to the concept of the dredge such as anchoring systems and engine changes, and a 'backup' ladder style digging system." These changes "significantly added to the project cost. . . ." The project submitted an updated budget outlining the capital cost increases in September 1992, but the assembly did not make a supplemental appropriation.[170]

Because of unexpected problems in bidding and fabricating equipment, the Borough did not take delivery of the custom-designed dredge *Qayuuttaq,* a shore barge, and a dredge tender until three years later, in July 1995, in Wainwright. That one season of dredging for Wainwright lasted from the end of July to mid-September 1995. (Dredging was seasonal because of Arctic weather and sea-ice conditions.) Planning and review after that season determined that about six weeks would be needed for major equipment modifications before the next season. By late 1995, the program's prime contractor, BTS/LCMF, reported that the actual cost of capital equipment to date exceeded the budgeted cost by about $4.1 million and that operations for the 1996 season would cost about $3.4 million more than was available to the program.[171] The 1996 season nevertheless began with equipment modifications and preseason maintenance and preparations. These were completed in Wainwright in May and in Barrow in July, with excavation of a channel in the Barrow gravel pit to provide a safe harbor for the dredge. In August the equipment was moved to Barrow, where beach nourishment began on August 15 and continued through September 12. In this 29-day season, 17.5 days were operating days and 11.5 days were "down" days because of adverse weather.[172] The postseason evaluation and planning meeting in October 1996 concluded that "the material encountered offshore at Barrow contains a significantly higher percentage of clays and large rock than that at Wainwright.

This material is suitable for placement on the beach but caused considerable problems with production. .... Equipment modifications are planned to enhance production with this material."[173] The program suspended operations in the 1997 and 1998 seasons for financial and other reasons.

After further modifications of equipment and operations, 1999 was the first and only full season of operations off Barrow. Operations occurred in the period from July 18 to September 3, nourishing the beach from Ukpeaġvik just south of town and continuing northeast 1,800 feet to a point near the intersection of Aivik and Stevenson Streets in Browerville. Compare the 1989 planning estimates in *Mitigation Alternatives for Coastal Erosion at Barrow and Wainwright* with the actual results above and in the 1999 postseason report:

- Schedule: The 1989 plan called for "dredging to begin in Wainwright in 1990 and move to Barrow in 1991, with the Barrow Nourishment completed in 1996."[174] But dredging did not begin in Wainwright until 1996 and moved to Barrow in 1997. The program was terminated before the Barrow segment was completed.
- Productivity: The 1989 plan estimated a total of 800,000 cubic yards of beach nourishment material in place in Barrow at a unit cost of $15.27 per cubic yard in 1989.[175] The first full season of operations produced 64,000 cubic yards of material in place at a unit cost of $80.00 per cubic yard in 1999.[176]
- Total cost: The 1989 plan "estimated capital and operating expenditures" by year from 1989 through 1996 that added up to $14.4 million, but then described the sum as "Total Capital Expenditure." Thus the costs included were not clear.[177] The lowest estimate was that the program spent $27 million of the $38 million appropriated for it. The highest estimate was $100 million, including program costs charged to other accounts.[178]

Clearly, there is a lot of ambiguity in estimates of total costs and measures of productivity. Nevertheless, the differences between what was projected in 1989 and actual results are rough indicators of uncertainties absorbed up front. The differences arose from important factors misconstrued, discounted, or ignored in the planning process.

The program was terminated in 2001 after the big storm of August 10, 2000, damaged and sunk the dredge *Qayuuttaq*. It was towed to Seattle for repairs, but the assembly decided to accept an insurance settlement instead of making the repairs. This reflected a general consensus that the program was too expensive to continue. One informed citizen likened it to standing on the beach and throwing buckets of gold coins into the Arctic Ocean. An informed source at CIPM claimed there was no survey to determine how much material was washed away because the program "was a complete failure" and "no one is interested in proving their mistakes." Perhaps the closest approximation to a formal appraisal of the program's effectiveness came in an assembly meeting a month after the storm. Asked how much beach nourishment material washed away in the storm, CIPM Director Frank Brown replied that "we are having it surveyed now[.] [R]emember that this project is [a] sacrificial program to begin with. The beach nourishment program would have to be a constantly maintained program in order to keep from eroding. We think that most of what we had there washed away but the good news is that if we hadn't done it that we would have lost probably [a] considerable amount of stuff. So the program actually succeeded in that respect."[179] According to the Army Corps of Engineers, the material dredged in the 1999 season washed away because it was "about 70% silt and 30% sands with little if any gravel." That was not coarse enough. "Any beach nourishment program will be ineffective unless better material sources for nourishment are found."[180]

The program was not simply a matter of reducing damage from big storms; other interests figured in its design, approval, and operations. The third-party review by Ogden & Beeman in September 1993 noted that the borough's preference for the beach nourishment alternative is "based on economic and non-economic criteria including the preservation of use of the beach and employment of local citizens."[181] Local employment figured prominently in questions by assembly members before the decision to proceed with the program. (The program employed 45 crew members in two shifts when operations resumed in the 1999 season.)[182] The assembly at that time also waived competitive bidding to award contracts to local companies where possible. The Iñupiaq tradition of sharing surfaced in these practices. Other interests surfaced as technical issues as early as August 1992 when the borough commissioned the third-party review by Ogden & Beeman, and again in April 1993 when two prominent citizens wrote to the Planning Commission requesting further review. Dr. Tom Albert, senior scientist for the NSB Department of Wildlife Management, noted the scale of the project and concluded that "[u]nfortunately, the 'bottom line' is that there are serious risks associated with a major action such as is being proposed." Richard Glenn, then of BTS, acknowledged "the need . . . to reduce Barrow's vulnerability to coastal erosion." However, he continued, "I have serious doubts about this project. The project is important enough, both in cost and potential risk, that a review of the project by the North Slope Borough Science Advisory Committee should be conducted. I believe that the Planning Department's review of options was not a sufficient analysis of the project or of possible alternatives." The opening sentence of the committee's report in November 1994 acknowledged that "a beach nourishment project will be carried out, as opposed to other hard structural methods." Hence, it sought to improve the project, evidently because it was too late to explore other means to reduce storm damage from coastal erosion and flooding.[183]

In retrospect, the program was conceived by technical experts as a single comprehensive policy to mitigate coastal erosion and flooding in Barrow through beach nourishment technology, but it provided at best only temporary protection at an unsustainable cost. Uncertainties absorbed in the 1989 plan but operant in the context forced a series of modifications in the program that added to overall costs and produced material rather easily washed from the beach. The structure of the seminal plan as a cost-benefit analysis with metrics for some important variables left omissions and misconstructions of the data that later turned out to be important. So did various reviews commissioned by the borough. Meanwhile, passive alternatives mentioned in the seminal plan were not sufficiently well developed or publicized to gauge their political support. In balancing multiple local interests, the policy was not above politics. If technical rationality was an aspiration of the Barrow and Wainwright Beach Nourishment Program, it was not an achievement.

While the August 10, 2000, storm led to termination of the program, it also prompted the borough to participate in a joint feasibility study with the U.S. Army Corps of Engineers. As Mayor George N. Ahmaogak, Sr., explained in May 2001, "This study is a critical step in determining the protection of enormous capital infrastructure essential to the health and safety of the residents of Barrow. The recent storm last August attests to the urgency to provide coast storm damage protection for our community."[184] The Army Corps's 2001 proposal for a feasibility study focused on two active engineering alternatives, called "solutions," quite similar to those Tekmarine proposed in 1987:

■ Solution One would widen the beach by 100 feet from south of Barrow to northeast of the landfill, a total of 25,000 lineal feet, requiring about 2 million cubic yards of material, and elevate the coastal road along the same part of the beach from an average of about 9.5 to 16 feet above mean sea level, requiring another 500,000 cubic yards of material.

- Solution Two would widen the same beach by 50 feet with 1 million cubic yards of material; elevate the coastal road to 16 feet with 500,000 cubic yards of additional material; and add a concrete mattress revetment constructed over a filter cloth on the seaward side of the bluffs and the elevated roadway, a total of about 25,000 lineal feet.

The Army Corps's proposal assumed that suitable material (sand and gravel) could be taken from Elson Lagoon, leaving a navigation channel and harbor. Those navigation benefits could be combined with either Solution One or Two without significant additional cost. "The overall combined project cost would be about $80 million to $82.5 million. The average annual costs would range from about $5.6 to 5.8 million."[185]

In 2002 the Project Management Plan added a "No Action" alternative and allowed for "any other reasonable alternatives that develop during the study." The schedule of major milestones included the Feasibility Scoping Meeting in September 2005, the Alternative Formulation Briefing in January 2007, and the Draft and Final Report/Environmental Impact Statement (EIS) for public review in May 2007 and March 2008, respectively. The feasibility study would end with transmittal of a report to Congress at the end of October 2008. Assuming everything worked out as expected, the Project Management Plan projected that storm damage reduction measures could be in place in Barrow as early as 2012. The study was "joint" in the sense that the estimated cost, $7.2 million, was to be shared equally between the federal government and the borough, and that the Army Corps's technical expertise would be applied to reduce storm damage in Barrow. Only 1 of the 15 persons listed on the Project Delivery Team represented the borough; the other 14 were employees of the Army Corps in Anchorage. The study was subject to national technical requirements listed in 34 primary references in the Project Management Plan plus "other appropriate Corps documents, such as Policy Guidance Letters."[186]

After approvals were secured at higher levels in the Army Corps, the study began in February 2003. The Feasibility Scoping Meeting took place by teleconference in December 2005, with the Final Feasibility Scoping Meeting Guidance for Phase 2 approved in February 2006. The scoping meeting and guidance were based on documentation revised in September 2005. According to that documentation, the Army Corps dropped from further consideration a breakwater to reduce wave impacts at the base of the bluff; a groin to retain sediment transported alongshore; and a seawall to withstand flooding. It retained five alternatives for further study in Phase 2 and keyed them to five contiguous reaches: 1) from the gravel pit southwest of town to Barrow, 2) Barrow, 3) Isatquaq Lagoon, 4) Browerville, and 5) South Salt Lagoon including the landfill. The five alternatives were as follows:

- Alternative 1—*No Action Plan*
- Alternative 2—*Beach Nourishment* for all five reaches, but Elson Lagoon was rejected as a source of sand and gravel; other possible sources would be assessed
- Alternative 3—*Revetment* for reaches 1 and 2, using rock barged from Nome
- Alternative 4—*Dike with Road* for reaches 3 to 5
- Alternative 5—*Non-Structural Measures* for all reaches

Included in Non-Structural Measures were "opportunities to remove or re-locate facilities, structures, and other damageable materials" and "changes in policies, procedures, regulations which the local government(s) could implement. . . ." They were described as potentially "useful supplements to other measures" and given only minimal consideration: 17 lines out of 64 single-spaced pages in the documentation.[187] For evaluation of the alternatives retained, the Army Corps dropped cost-benefit analysis in favor of scenario analysis, citing climatic uncertainty that could have been based on our project's climate research: "Because of this uncertainty, setting a single most likely future climate for Barrow (which would drive the predicted storms and flood damages) over the study's 50-year analysis period is difficult. Scenario planning is generally used in situations in which there are a large number of uncertainties. Barrow is a fine example of this. . . ."[188] The first evaluation criterion for Phase 2 alternatives was "Completeness: Is it a complete solution to the identified problem(s)?" The next two criteria were effectiveness and efficiency; the final three were acceptability, Native benefit, and stability.[189]

On the evening of August 23, 2006, the Army Corps provided an update on the project at a public meeting in Barrow attended by 50 to 60 local people, including an Iñupiaq interpreter.[190] Project Formulator and Corps Planner Forest Brooks announced that "large beach nourishment has been dropped from active consideration" because beach erosion was no longer a big problem. The beach was stable during the last 50 years, according to their general results, and "most of the loss occurred [between 1954 and 1974] when material was removed from the beach to support construction of the airport runways."[191] But additional reasons for this change may be inferred.

According to the study's hydraulic engineer, "When we were looking at beach nourishment, one biggest stumbling block is getting material we could use."[192] Without beach nourishment, costs declined and gravel requirements were reduced from at least 2 million to 0.5 million cubic yards or less; at that level, "there should be enough gravel material available in existing commercial pits."[193] That eliminated the search for suitable beach nourishment material and avoided the environmental impacts of opening a new pit for gravel. The most significant impacts would have affected the Steller's eider, a threatened species of sea duck that nests in the Barrow area.[194]

With beach erosion set aside, the remaining problems were erosion of the bluffs and flooding. The proposed solutions were a revetment to protect the bluff and a dike along the beach to prevent flooding. Each was to be constructed with a core of HESCO Concertainers at mean sea level or slightly above it. Each Concertainer is a wire-basket cube containing a geotextile fabric sack holding a cubic yard of sand or other material; the cubes are connected in the field with spiral wires and steel rods. For the revetment, "Rocks will be placed over the core to provide protection from the waves. Backfill will be placed along the face of the bluff to reshape the bluff. . . . The surface of the bluff will be covered [to the top] with supersacks to take ocean spray and rainfall and runoff."[195] The design of the dike for flood control was similar but did not include the gravel backfill and supersacks; they were unnecessary because land behind the rocks and Concertainer core was lower in elevation. The dike would replace the temporary berms constructed by the borough with a permanent structure at about the same height, 16 feet above mean sea level. To provide access to the sea, the Army Corps included four breaks in the dikes to be filled with temporary material in advance of a big storm.[196] Once again, nonstructural alternatives were deferred. Brooks advised, "We *will* be working with the North Slope Borough on these options. As you see, the cost of structural projects is very big, [it's] going to be very difficult to justify under COE policy."[197] Brooks reported that "[e]ach piece, bluff protection and flood protection, looks like it will cost about 30 million dollars. The total project ranges from 50–70 million dollars. The price range depends on how high we build the rock . . . it looks like the federal share is about 60%, and local share will be about 40% of the construction cost."[198] The borough would be responsible for maintenance costs.

The public meeting included several confrontations between differently informed perspectives, from the top down and bottom up. Consider the following example. During Brooks's presentation of the design of the revetment, audience members became concerned about the plan to place large rocks on top of smaller rocks on the seaward side; one member raised the issue by citing a heavy concrete boat ramp on the beach in Barrow that sank in the sand and was almost swallowed up years ago. From the Army Corps's perspective, the proposed design was justified as standard practice: "This is the way we build our breakwaters. Nome and Homer are built like this. We always build revetments coarser as we go up," according to the Army Corps's hydraulic engineer. From the audience perspective, circumstances in the Arctic were different: The Army Corps's design was an attempt to prevent waves from overtopping the revetment and eroding it from behind, but that was not sufficient in the Arctic where sea ice along the beach (an *ivu*) shears from the bottom of the structure and lifts up the rock, while water under the ice removes material supporting the rocks from below. The hydraulic engineer deferred to modeling to resolve the issue: "Once we come up with a final design, we'll construct a little model and have the Cold Regions Lab [in New Hampshire] run ice up on it and see how it performs."[199] In general, the audience referred to detailed observations of local and Arctic conditions, and made inferences from them, to make their points; in other words, they appealed to their local knowledge. Members of the Army Corps's team appealed to standard practices and numerical models as surrogates for Arctic systems and as substitutes for learning by doing. In doing so, each continued his or her own distinctive tradition.

Whatever the influence of bottom-up perspectives from Barrow, it is already clear that nature and the Independent Technical Review mandated in Army Corps procedures have impacted the outcomes of the evolving feasibility study. The Army Corps was reconsidering use of Concertainers in the Barrow project within months of the public meeting, it explained, because "seawalls built out of Concertainers during the summer of 2006 at Kivalina and Wainwright were severely damaged" in storms that fall.[200] Sometime after July 2007, when the Independent Technical Review was approved, the Army Corps's review team of specialists from the Alaska district and other Army Corps districts sent the project back to the drawing board for major changes.[201] The review concluded that the study's beach and economic models had overestimated the damages averted by the alternatives selected and therefore overestimated the benefits of the project. It mandated scaling back the alternatives to make the project more economical, including consideration of nonstructural alternatives. Consequently, by January 2008, public review of the Draft Feasibility Report/Environmental Assessment (EA) was rescheduled for January 2009. Federal funding to implement any program in the Final Draft Feasibility/EA is contingent on approval by the borough and by the Alaska district's superiors to the top of the hierarchy in Washington, and on Congressional authorization and appropriations.

In retrospect, the feasibility study was conceived by technical experts as a search for a single comprehensive policy to reduce storm damage in Barrow through structural engineering. Thus, its origins were similar to those of the seminal plan for the Barrow and Wainwright Beach Nourishment Program, and so were some consequences: unanticipated changes in the Project Management Plan underscoring uncertainty in comprehensive planning, and diversion of attention from distributed alternatives to address separately such local priorities as the utilidor or the landfill. But unlike the seminal plan, the Army Corps's study was subject to national technical requirements. Substantively, the requirements prescribed explicit consideration of the net national interest in the project and a broad range of local interests, including environmental, archaeological, and cultural. Procedurally, they exposed the study's interim plans to public and technical reviews. This openness improved the dependability of the study and fostered creative comparison of initial and newer alternatives, leading to elimination of beach nourishment and more serious consideration of nonstructural alternatives. On the other hand, implementation of those technical requirements has cost time and other resources, and in the end could suppresses Barrow's distinctive needs and opportunities for storm damage reduction on behalf of technical compliance. The study team and its superiors in the Army Corps are subject to national technical requirements and effectively control the study's work on Barrow storm damage reduction. It should not be assumed that local needs and opportunities are automatically compatible with national technical requirements.[202] Differences are best resolved case by case rather than once and for all, because context matters; and they are best resolved on the basis of experience wherever possible.

Distributed Policies

Early in our field work in Barrow we met people who appeared to be well positioned and predisposed to act on separate parts of the overall problem of reducing storm damage from erosion and flooding. At the same time, we were aware of the community's preoccupation with active or structural alternatives developed by Tekmarine and BTS/LCMF in the late 1980s and proposed by the Army Corps in 2001. But prospects for timely progress from the joint feasibility study were uncertain, and the beach nourishment program had not succeeded according to local appraisals. Under these circumstances, and as part of our effort to help the community adapt to climate change, it made sense to expand the range of informed alternatives for those in Barrow who were willing and able to act. We began with the least obtrusive approach that was also consistent with our areas of expertise: researching other less visible policy alternatives that had been considered in Barrow in recent decades and bringing them to the attention of community members to provide a more comprehensive picture of the alternatives available. We found that many distributed policies already had been considered to reduce storm damage in Barrow, and sometimes implemented to good effect. The following survey covers emergency management, archaeological investigations, relocation, planning and zoning, insurance, retrofits, and new construction. Even at the local level, progress need *not* depend exclusively on a single comprehensive solution.

First, consider emergency management. Since at least the mid-1980s, in response to warnings of big storms approaching Barrow, the borough has deployed bulldozers to construct temporary berms on the beach to take the brunt of the storm, thereby reducing erosion and flooding damage.[203] On at least one occasion the department attempted to stockpile material on the beach in advance of the storm season, but abandoned that practice along populated sections of the coast after complaints from residents who preferred unrestricted beach access and ocean views. So far, emergency action to build temporary berms, place sandbags, and tie tarps over five maintenance manhole covers has protected the utildor from flooding at its lowest point, pump station No. 4 near the Top of the World Hotel. (Next door to No. 4, the Osaka Restaurant reportedly has been flooded up to the foundation level on several occasions, but not up to the floor level.) Such emergency actions, however, can be dangerous. In September 1986, as the second big storm approached Barrow, heavy equipment operators had to be ordered off the beach before the berms were completed because working in the surf became too dangerous.[204] In 2001 the National Weather Service station in Barrow acquired its own shortwave radio transmitter to eliminate about 10 minutes lost in transmitting storm warnings through the NSB police department.[205] Adequate early warning time is also necessary to allow families including the elderly and children to assemble in a safe place, to secure sheet metal, plywood, and other materials launched by high winds, and to gather supplies. As a big storm approached Barrow on September 9, 2003, "The NSB . . . asked everyone to tie down loose stuff, and get three days supplies in hand."[206] The value of credible early warnings has been recognized in these ways.

To help personnel at the National Weather Service station in Barrow improve local forecasting skill, and to compensate for personnel turnover there, our colleague Elizabeth Cassano took the lead in developing an empirical classification of 55 years of atmospheric pressure maps in the western Arctic. The classification identifies storm patterns that have produced the most damage in Barrow in the past.[207] To account more specifically for the coincidence of natural factors, we also developed heuristics to help improve early warning of storms that could cause severe flooding in Barrow. Based on sensitivity analysis of numerical weather and storm surge inundation models applied to past storms, the study demonstrated that for Barrow, a forecast for winds exceeding 30 miles per hour for at least 20 hours is the most significant predictor of a severe flood, as long as there is relatively open water close to shore." Other factors were less important.[208] This heuristic "looks pretty good," according to Dave Anderson at the Barrow station. Given the much larger fetch experienced in recent years, he suggested lowering the wind speed threshold to 25 miles per hour in a meeting of the Local Emergency Management Committee in September 2007. This has been implemented: The National Weather Service in Barrow now recommends that any forecasts above 20 to 25 miles per hour when the ice is out should be considered dangerous.[209] Anderson also believes that better instrumentation could improve credible early warnings—particularly television cameras on automated weather stations along the coast to observe wave and ice conditions, and buoys at sea to observe air and sea surface temperatures, wind speed and direction, atmospheric pressure, and wave action.[210]

In June 2001, a contractor for the borough submitted a Comprehensive Emergency Management Plan.[211] For each generic management function to be activated and performed during emergencies, the contractor inserted the names of borough departments, agencies, and other organizations with primary and secondary responsibilities. It also urged users to take note of the responsibilities of their own units, but most users knew little or nothing about the CEMP. To educate and motivate borough personnel in Barrow in preparation for the autumn storm season, borough Assistant Risk Manager and Disaster Coordinator Rob Elkins took the initiative to design an emergency management exercise based on our account of the great storm of October 1963; he supervised the exercise in late August 2004.[212] In addition to training, the exercise stimulated investigation of reserves and redundancies—for example, whether there were enough honey buckets and sewage and water trucks in Barrow to provide emergency services and protect public health if the utilidor shut down.

At about the same time Elkins took the lead in developing the *North Slope Borough Local All Hazard Mitigation Plan* and equivalent plans for each of the eight villages, to qualify for federal disaster mitigation funds from FEMA under Sec. 322 of the Disaster Mitigation Act of 2000 (P.L. 106-390). This is the borough's master plan, which is a comprehensive framework rather than a single comprehensive solution. It notes that individual "projects will be prioritized and selected for implementation based on community goals, planning objectives, funding availability, environmental concerns and public support" in addition to benefit/cost considerations.[213] Moreover, it "will be evaluated and updated every five years as appropriate when a disaster occurs that significantly affects North Slope Borough. . . ." Thus, the form of the plan is distributed and procedurally rational. After the plan came back preapproved from the state of Alaska and FEMA in late November 2005, the NSB Assembly "listened long and hard" and finally approved the plan but not the action items.[214] Some of the action items are mentioned below—and available for submission to the assembly in the wake of damage from a future big storm.

Second, consider archaeological investigations, which might reduce losses like the remains of Uncle Foot in September 1986, even if they lie outside the Army Corps's authority. Borough Mayor Jeslie Kaleak, Sr., clarified the significance of such losses at a 1991 symposium on saving the past: "The modern world will not leave us soon. We must come to a living accommodation with it. But doing so we must never forget who we are, for that is our strength. For so long as we remain Iñpuiat, we will have the ability to deal with the changes. Self-knowledge is the greatest of all knowledge for it gives us the courage to face each day secure in ourselves and our abilities. Erosion of our beaches and disintegration of our historical sites takes that knowledge from us."[215] The 1991 symposium recommended "a year-by-year site monitoring program to keep track of archaeological site conditions on the North Slope, and to develop a realistic schedule for excavation of the more endangered sites, keeping in mind the availability of money and resources. The initial focus would be placed on sites that contain irreplaceable data and that are most significant in terms of unique qualities."[216] But prioritizing sites turned out to be difficult for technical and financial reasons; excavations are still primarily efforts to salvage cultural remains after they are exposed by big storms. A second recommendation of the 1991 symposium was "training local residents in the areas of archaeology and anthropology so that future excavations and the interpretation of the evidence uncovered from them would reflect local perspectives." Ilisaġvik College, the NSB community college located at

NARL, has graduated at least one student with a science technician degree. Laboratories for analysis of archaeological and paleontological finds were designed into the new Barrow Arctic Research Center. The 1991 symposium's third recommendation was realized by construction of the Iñupiat Heritage Center in Barrow in 1999.

Third, consider relocation or rollback of people and property from vulnerable shoreline areas. This alternative has been mentioned, and occasionally acted on, for some time in Barrow. At an elder's conference, Bertha Leavitt recalled that she "had her house moved from Browerville in 1960 and if she hadn't have done that it would have succumbed to the '63 storm."[217] In the first half of 1964, as previously noted, Barrow residents, with help from the federal government, relocated houses according to a new town site survey as an alternative to moving the whole town to a new location.[218] In 1970 local government tried to relocate 22 residences on the Barrow beachfront that were vulnerable to flooding. None was relocated, apparently because the lots were restricted by the Bureau of Indian Affairs and the occupants refused to take the offer.[219] This attempt followed the Army Corps's recommendation in 1969 "that Barrow adopt a number of non-structural measures to reduce damage from erosion (i.e., relocate houses, businesses, and utilities and develop/enforce erosion zone ordinances)."[220] After the second September 1986 storm left the seaward side of George Leavitt's house on the bluff suspended 35 feet above the Arctic Ocean, neighbors helped him move it inland. In 1989, the seminal plan for the beach nourishment program mentioned that relocation had occurred along with other preventive measures, but omitted details beyond these: "Previous Borough mitigation efforts regarding erosion have resulted in expenditures for reconstruction of facilities, *relocation of homes* and preventive restraints against further erosion. These efforts over the past 2 or 3 years have cost in excess of $4,000,000."[221]

Relocation has been mentioned as a possibility as early as 1987 in an appendix to the Tekmarine report. Jesse Walker, an expert in coastal studies, recommended "that an evaluation (financial, emotional . . .) be made of initiating a setback scheme (especially for the Barrow and SW Barrow reaches)." He added that the cost of protecting these priority reaches—$14,000,000 according to Tekmarine—"would buy a lot of property and soothe a lot of frustration."[222] Relocation came up in the 1989 seminal plan, in a reference to a Planning Department memo by Rick Sampson: "Mr. Sampson estimates the cost of relocation of all Barrow structures west of Stevenson Street is approximately 3 million dollars, [but] the basis for this estimate was not stated. Mr. Sampson feels relocation of structures should be explored as an alternative solution to mitigation of erosion." BTS/LCMF discouraged further consideration without a full analysis of costs it chose not to provide. Moreover, "most structures cannot be moved. Capital investments, such as roads, the land fill, the sewage lagoon, the lower and upper dams and most buildings larger than single family residences are impractical to relocate or move."[223] Nevertheless, relocation was included in discussion and promotion of the beach nourishment program. For example, in a special meeting and workshop on May 16, 1989, NSB official Herb Bartel assured the assembly that some first-phase work would "look at the costs associated with moving a lot of the structures and moving those to other pieces of property both in Barrow and in Wainwright, and the overall program may involve, most likely would involve, both the nourishment for the prevention of erosion and the moving of some structures onto safer properties."[224] An action item authorized but not implemented in the *Local All Hazards Mitigation Plan* calls for the borough to "acquire the services of a suitably qualified individual or firm to conduct a Project Analysis Report on a rollback/relocation of the community, to include all utilities."[225]

Barrow's history indicates that relocation is not just one policy alternative but a family of alternatives depending on what is moved, when, and how. Evaluation of relocation alternatives is less amenable to program-wide metrics, like cost per cubic yard of gravel in place on the beach, because costs and benefits vary structure by structure. For similar reasons it is rarely a single comprehensive solution. Perhaps the Army Corps deferred structure-by-structure evaluation of relocation in the joint feasibility study for that reason; in any case, the Army Corps received a mandate to evaluate relocation for the Draft Report/EA. Meanwhile, evidence from a series of eight aerial photos of Barrow and Browerville from 1948 to 2002 suggests that relocation has been a de facto policy in Barrow. Our colleagues looked for buildings within the area flooded by the October 1963 storm that appear in one photo but disappear in the next (Fig. 3.8). They counted 54 buildings removed from that area over that period, about one per year. Eleven buildings were removed from 1962 to 1964, which includes the great storm; another twelve were removed from 1984 to 1997, a period of growth in population and infrastructure. This timing suggests that informal policy allowed nature to dictate when and where relocation took place. This minimized the political burden of relocation for public officials. This is an adaptation to extreme events, so long as the structures were not replaced in equally vulnerable areas.

Fourth, consider planning and zoning to prevent building or rebuilding in vulnerable areas. This possibility was mentioned in 1992 as part of the rationale for the Barrow and Wainwright Beach Nourishment Program. Borough Attorney Alan Hartig observed that "the intent of the beach nourishment program is to . . . allow the North Slope Borough to recover their capital cost on major facilities that are built in that area . . . subject to erosion [and] not build them back in that same area. . . ."[226] In 1994 the Science Advisory Committee report on the program recommended that "for any new development or redevelopment in the community, serious consideration be given to adequate set-backs from the eroding shoreline or bluff. . . ."[227] In 1997 Frank Brown's report on the program claimed that one of its components "from its inception [was] to halt further development along the coastal areas . . . through the Planning and Zoning process. This process, along with the time frame afforded by the sacrificial beach, would allow existing structures to reach their useful life and be replaced, further inland, in a controlled manner."[228] Other sources recall that a planning and zoning component was only discussed, not formally approved, as part of the beach nourishment program. We have found no documentary evidence to resolve the issue. However, if it existed, a planning and zoning component was not consistently enforced.[229] The evidence includes construction of the borough's veterinary clinic in 1998 on land exposed to flooding in Browerville. The clinic is 1 of 11 new buildings constructed between 1997 and 2002 in the area flooded by the October 1963 storm, according to research on aerial photos by two of our colleagues.[230] An action item authorized but not implemented in the *Local All Hazards Mitigation Plan* calls for the Planning Commission to "enact and enforce a moratorium on all building within areas deemed to be threatened by erosion" but not flooding.[231]

Fifth, consider insurance, which has a planning and zoning component in so far as the National Flood Insurance Plan (NFIP) is concerned. In September 1998 the NSB Assembly tabled indefinitely a resolution on insurance. The first part would have resolved "[t]o assure the Federal Emergency Management Administration that [the borough] will enact as necessary and maintain in force for those areas having flood hazards, adequate land use and control measures with effective enforcement provisions consistent with the Criteria set forth in Section 60.3 of the National Flood Insurance Program Regulations."[232] Mayor Benjamin Nageak requested tabling the amendment when he introduced it. The brief discussion in the assembly minutes suggested uncertain financial impacts involving the Indian Housing Authority and possibly homeowners. In addition, the resolution as written would affect the entire borough, but the only North Slope community with an expressed interest in NFIP was Barrow. An action item in the *Local All*

Hazards Mitigation Plan notes, "Although [Barrow's] participation in the NFIP has previously been discussed and rejected by the North Slope Borough Assembly, a careful assessment of the benefits of said participation is worth a second look."[233]

Sixth, consider retrofits of facilities difficult to relocate, beginning with the landfill. No one really knows what was buried there by the military years ago; core sampling has been deterred by fears it might break a transformer, for example, and release PCBs and other hazardous materials into the environment. In 1999 an Army Corps study agreed that "there was an erosion problem occurring in front of Barrow and that the landfill and sewage lagoon were vulnerable to overtopping during a severe storm event. However the cost of potential complete solutions ($20–40 million) greatly exceeded the Federal participation limits ($1 million) of the Section 14 program."[234] Since then the landfill has been closed under a plan negotiated by the borough and the Native Village of Barrow with the U.S. Navy, Air Force, and Department of Justice. As described by the Army Corps, "The plan provides for the Department of Defense to provide a majority of the funding for the closure, with the proviso that no additional federal funds be provided to support the landfill. The landfill closure plan includes some minimal measures (such as jersey barriers along the road seaward of the landfill) to reduce flooding damages that might be experienced in the future by the landfill. However, these measures are limited and assume that the beach and the road will remain in place and will not be eroded and/or damaged in the future."[235]

Barriers to protect the landfill go back at least to 1987 when surplus wooden cross sections from the utilidor were used to construct a 600-foot seawall. This old seawall apparently contributed to beach erosion on its seaward side but remained in good condition until a big storm in September 2003 removed sand that had covered it. Some minor damage is now visible, but its long-term performance exposed to waves and sea ice remains to be seen. In August and September 2004 the borough extended the utilidor cross-section seawall 300 feet on each side using Concertainers. These Concertainers have not worked well; a flaw in installation of the top of the geotextile sack allowed water to pump material out of the sack. The Concertainer seawall constructed at the same time to protect part of the bluff has worked well; the landfill behind the wall has become permafrost, creating in effect a new bluff. But the seawalls constructed in 2004 have not yet been tested by a big storm.[236] The old landfill also is protected by a cap consisting of a protective fabric layer covered with 2 feet of tundra. There are concerns in Barrow that the design of the cap assumes the material buried below it will remain a "frozen popsicle" forever, held in place by permafrost. But the permafrost could melt if the warming trend on the North Slope continues, releasing material potentially hazardous to sea mammals and subsistence hunters. To monitor the permafrost, holes were drilled in front of the seawall protecting the landfill and sewage lagoon areas, and casings were inserted in the holes. But as the beach recedes in those areas, these holes also are increasingly vulnerable to erosion and flooding.

To reduce the vulnerability of the utilidor and other facilities, the borough stabilized the spillway across Isatquaq Lagoon in 2006 and 2007 by removing old concrete, driving 50-foot steel pilings, and placing large boulders to protect it from erosion.[237] The spillway supports a road and above-ground telephone, fiber optics and power lines, utilidor water and sewer lines, and the single line that supplies natural gas to town. These are still exposed to waves and flooding. To reduce the vulnerability of pump station No. 4 to flooding, veteran Plant Superintendent for the Barrow Utility System Tim Russell suggested extending upward the steel cylinder encasing the pump station, adding an external staircase to access a waterproof main hatch that would be installed on the roof. Similar retrofits on five utilidor maintenance manholes along the coast could help prevent leaks that would drain into pump station No. 4.[238] In addition, bulkheads and shut-off valves might make utilidor equipment and operations more resilient by limiting damage to one area while allowing other areas to function. (Bulkheads already segment the underground utility corridor to limit damage in case of fire.) Pump station No. 3 in Browerville is a system-wide bottleneck; that is, where all sewage lines from the other pump stations converge into a single line to the sewage lagoon.

Seventh, and finally, consider new construction projects that reduce future vulnerabilities. The first phase of the new $62 million Barrow Arctic Research Center (Fig. 3.19) has been constructed just east of NARL near Imikpuk Lake. Glenn Sheehan of the Barrow Arctic Science Consortium took the lead in having the facility designed on a pad and pilings high enough—15 feet above mean sea level—to protect the bottom floor from a flood equivalent to October 1963. A new sewage treatment plant has been constructed near the southeasternmost point of South Salt Lagoon; operations were expected to begin in June 2008. The site was selected in part to withstand a 100-year flood. Other sites closer to the beach and lower in the tundra were rejected in part because of their vulnerability to erosion and flooding.[239] Tom Brower III of the Native Village of Barrow and others have secured approvals and partial funding for a new road that will serve

Fig. 3.19. Barrow Arctic Research Center, phase 1, Febuary 2008. Photo: Ron Brunner.

as an inland evacuation route from NARL. This is an alternative to Stevenson Street, the coastal road often flooded by big storms and washed out in places, including the bridge over the culvert connecting Middle Salt Lagoon with the sea. The new road runs south from NARL, past the new research facility and sewage lagoons, and links up with Cakeeater Road, a total of 2.6 miles. Marie Adams Carroll took the lead in locating a new hospital outside the area flooded by the October 1963 storm, in an undeveloped corner of Browerville just northeast of Isatquaq Lagoon and reservoir. She used the map of the 1963 flood line we had reconstructed from the work Hume and Schalk previously cited (Fig. 3.20). The hospital is partially funded through the Indian Health Service; site preparation was expected to begin in 2008. However, the recent extension of the airport runway west toward the ocean has raised some concern in Barrow. Significant erosion has occurred in that area over the last century.

No survey of past and potential policy responses can be complete. But perhaps this survey is comprehensive enough to show how different people and organizations in Barrow have taken the initiative on distinguishable parts of the overall task of reducing storm damage. This distributed pattern of inquiry and action allows multiple initiatives to proceed in parallel rather than in series, thus multiplying opportunities for learning by doing. It allows

Fig. 3.20. Marie Adams Carroll and others viewing the 1963 flood map, August 2003. Photo: Christina McAlpin.

for specialized expertise and motivations to be brought to bear on each initiative. It also allows for coordination of responses as needed by the Disaster Coordinator's and Mayor's offices and the North Slope Borough Assembly. The borough's *Local All Hazard Mitigation Plan* goes a long way toward coordination. But no pattern, including this distributed pattern, is fool-proof. Alternatives can fall through the cracks; for example, community members still lack inquiry on relocation and planning and zoning that is detailed enough to gauge costs and benefits as well as political support. Related alternatives are not always compared; for example, community members lack comparison of the costs and benefits of the borough's emergency-management berms and the Army Corps's proposed replacement, a permanent dike. Moreover, differences in priorities are not always resolved. The Army Corps's joint feasibility study presumes a single comprehensive solution to reduce storm damage. The *Local All Hazard Mitigation Plan* prioritized a number of action items relevant to the top goals of protecting public utilities and reducing damage from erosion and flooding. One leader in Barrow told us that "[t]he utilidor and the landfill must be protected; everything else can go" as nature dictates in future big storms.

How can such shortfalls be remedied? Progress depends on sharing among policymakers a more comprehensive picture of vulnerabilities and alternatives that includes both the joint feasibility study and distributed initiatives. That need not preclude action on separate policies that make sense on their own, like designing new facilities to lessen their vulnerabilities. Progress also depends on more systematic learning by doing, emphasizing appraisals of any alternatives implemented. Tekmarine spelled out the logic in its final report in 1987. Citing the large size of the proposed investment and uncertainties about whether the proposed hard structure would modify waves and currents and thereby destabilize the structure itself, Tekmarine recommended a phased implementation policy: "As a first step of this policy, it is recommended to install test sections and perform a monitoring program in their vicinity for at least 2 years to study the performance of the protection work and obtain site-specific environmental information. The lessons learned from this program, as well as the enhanced data base on local natural processes, will aid in the design of future protection work with assured performance and improved economy."[240] Similarly, the *Local All Hazard Mitigation Plan* includes monitoring the Concertainer seawalls as an action item.[241]

It would be a mistake to assume that storm damage reduction or global warming is a high priority or that responses to them can be isolated from other challenges in the open-ended process of community development in Barrow and on the North Slope. Responses to all the challenges are connected through the borough's budget, if nothing else, and more pressing challenges include making a living and adapting Iñupiat culture in order to sustain it. In an interview on National Public Radio in September 2007, Richard Glenn of ASRC observed, "Our people depend now on resource development for schools, health clinics, fire halls, runways; everything that you would expect larger governments to do elsewhere in America happens here because there's a tax base."[242] As previously noted, that tax base depends heavily on oil-industry property at Prudhoe Bay. But the value of that tax base has been declining in recent years, forcing cutbacks in borough revenues and expenditures including employment in local government. Perhaps the leading economic priority is development of a new pipeline to transport 4 to 6 billion cubic feet of natural gas per day from the North Slope south, alongside the existing Trans-Alaska Oil Pipeline, to markets in the lower 48 states. When an interviewer noted Glenn's impatience with "all this gloom and doom about climate change," Glenn's response was adaptive, not predictive, and reminiscent of Levi-Strauss's science of the concrete: "Our job is to know, study changes, day-to-day changes, month-to-month changes. And so we don't say that, for example, on November 15th, the ice should be 10 inches thick. We say, we should know how [thick the ice is] regardless of the day, and it's our knowledge that keeps our people safe. . . . And so we live with change. There are other changes going on. Change is walking in two worlds . . . sometimes I think MTV has more effect on us than global climate change."

The two worlds also are in conflict over offshore oil exploration. That could alter bowhead whale migration routes and disrupt subsistence harvesting by Iñupiat whalers, but it could also eventually help support Iñupiat and other North Slope residents in the modern economy. A coalition of plaintiffs that includes North Slope Borough and the Alaska Eskimo Whaling Commission has taken legal action to block offshore drilling by Royal Dutch Shell. Since 2005 Shell has paid about $80 million for leases in the Beaufort Sea, and in February 2007 it obtained a drilling permit from the federal government's Minerals Mining Service. According to borough Mayor Edward Itta, "It's a hell of a dilemma. . . . Without a doubt, America's energy needs are way up, and something's going to happen up there. It's a way of life against an opposing value. This way of life has value; nobody can put it in dollars and cents."[243] More recently, the mayor affirmed that "opposition to offshore industrial activity is necessary for the protection of the bowhead whales and other marine mammals so vital to our culture. But opposition by itself is not enough. We need to couple it with engagement. By engaging with industry and with the federal agencies, we remain at the table. It keeps us in the discussion, allows us to educate the industry in all aspects of our concerns, and keeps us informed about their evolving technologies and involved in their mitigation strategies."[244] These are contemporary versions of pragmatic responses by earlier Iñupiat leaders to major historical challenges of the past, making a living and sustaining the culture in the modern world.[245]

Engaging the World

Engaging the world at large is also important for reducing net losses from climate change, one of the less immediate challenges faced by Barrow and other North Slope communities. Potentially this includes organizing networks of communities, for at least two purposes. One is to diffuse policy-relevant information, including case studies centered on policies that have worked in one local community for possible voluntary adaptation by other communities. Another is to clarify their shared needs to inform and influence central authorities from the bottom up. Central authorities often control some of the resources needed to support policies that have worked on the ground, but without information from the bottom up they cannot reliably know how to allocate resources accordingly. Steps toward organizing networks of communities have emerged more or less spontaneously as part of the effort to reduce storm damage in Barrow that is consistent with adaptive governance. An explicit strategy built on such initiatives has the potential to expand the range of informed alternatives for policy decisions in each local community and at higher levels.[246]

Examples of the diffusion of policy-relevant information across communities are not difficult to find in Barrow's past. Tekmarine's 1987 final report recommended two alternatives—an all-block scheme for the bluff area and a raised-road scheme for the lagoon area—based in part on its experience with hard structures at Prudhoe Bay.[247] Planning studies for the Barrow and Wainwright Beach Nourishment Program included two reports on beach nourishment projects elsewhere. In March 1989, LCMF hired Coastal Frontiers, Inc., to compile documentation on 18 beach nourishment projects in California, Louisiana, New Jersey, New York, and North Carolina, and overseas in Australia, Scotland, and an island in the North Sea. In February 1990, Ducote Engineering Associates, Inc., was hired to compile a review of existing beach nourishment projects outside Alaska including Miami and Duval County, FL; Grand Isle, LA; and Ocean City, MD. The Army Corps of Engineers reconsidered the use of Concertainers in Barrow after seawalls built from Concertainers failed partially or completely in Kivalina and Wainwright during storms in the fall of 2006. When Kaktovik lost electric power in the blizzard of January 2005, shutting down water, sewage, and other services as temperatures dropped to −20°F, the Disaster Coordinator's office used a dummy budget code to track the cost of emergency services and repairs. This mundane but important lesson learned from previous disasters elsewhere in the borough highlights its government's networking function.[248]

For the future, Barrow has much to learn about reducing storm damage from the experience of similar communities, and vice versa. Consider, for example, Shishmaref, home to more than 600 people exposed to big storms on a barrier island at the southern edge of the Chukchi Sea. According to the U.S. General Accounting Office (GAO), Shishmaref voted to relocate in 1973 "when it experienced two unusually severe fall storms that caused widespread damage and erosion."[249] The vote was reversed in August 1974 after the site selected for relocation proved unsuitable, and it was learned that federal agencies would withhold funding for a new school and runway if the town did relocate. Shishmaref's experience also includes selective relocation. After a storm in the fall of 1997, "FEMA and state matching funds were used to help move 14 homes along the coastal bluff to another part of the village, and in 2002, the Bering Straits Housing Authority relocated an additional 5 homes out of harm's way." After a storm in October 2002, cracks appeared in seaside bluffs, indicating damage to the permafrost holding the bluffs together. "Several homes . . . had to be relocated to prevent them from falling into the sea"; "In July 2002, the community again voted to relocate, and is currently working with NRCS [the Natural Resources Conservation Service] to select an appropriate site."[250] Shishmaref also has experience with hard structures. Since the 1970s, Shishmaref has implemented "sandbag and gabion seawalls (wire cages, or baskets, filled with rocks) and even a concrete block mat. Each project has required numerous repairs and has ultimately failed to provide long-term protection," according to the GAO.

Consider also the hamlet of Tuktoyaktuk, the northernmost community in mainland Canada located at the mouth of the Mackenzie River on the eastern shore of a peninsula in the Arctic Ocean. It has been monitored by the Geological Survey of Canada. In 1976 Tuktoyaktuk used Longard tubes to build bulkheads and groins to control erosion, but these failed within five years.[251] Following a study that recommended beach nourishment in 1986, "Sand was dredged from the nearshore and placed on the beach with a sandbag system. From 1987 to 1993, the sandbags provided protection of the cliff, and acted as a form of time-release beach nourishment. Since no protection was provided for the toe of the sandbags, storms would undermine the revetment causing breakage of the bags, removal of their contents and collapse of the bags higher on the slope."[252] A severe storm in September 1993 removed sandbags from more than 50% of the protected area. "In 1998, forty monolithic concrete slabs were installed over a gravel pad, which was overlain by non-woven geotextile. . . ."[253] This remained stable in the first several years after installation. However, "While the present storm protection measures are more robust than the sand bags, they are still subject to some of the same failure mechanisms."[254] (This experience has direct bearing on one confrontation between differently informed perspectives in the Army Corps's public meeting in Barrow in August 2006.) Tuktoyaktuk's mayor concluded that "even a reinforced shoreline will only delay the inevitable. 'Global warming's coming hard,' he said."[255]

As potential steps toward organizing a network of communities, consider Sen. Ted Stevens's initiatives on behalf of Alaska Native villages including Barrow. A GAO report in December 2003, addressed to Sen. Stevens as chairman of the Senate Appropriations Committee and other members of Congress, examined 184 coastal and inland Alaska Native villages that had experienced some coastal erosion and flooding, and focused on nine of them. It found that "Kivalina, Koyukuk, Newtok, and Shishmaref . . . are in imminent danger from flooding and erosion and are making plans to relocate. . . . The five villages not planning to relocate—Barrow, Bethel, Kaktovik, Point Hope, and Unalakleet—are in various stages of responding to their flooding and erosion problems." The GAO concluded, "With few exceptions, Alaska Native villages' requests for assistance under [the Continuous Authorities Program of the U.S. Army Corps of Engineers] are denied because the costs exceed the expected benefits. Even villages that meet the Corps' cost/benefit criteria may still fail to qualify if they cannot meet the cost-share requirements."[256] Senator Stevens followed up the GAO report with Senate field hearings in Anchorage in late June 2004 and in mid-October 2007.

Follow-ups also included Sec. 117 of P.L. 108-447, the Fiscal Year 2005 Consolidated Appropriations Act: "Notwithstanding any other provision of law, the Secretary of the Army is authorized to carry out, at full Federal expense, structural and non-structural projects for storm damage prevention and reduction, coastal erosion, and ice and glacial damage in Alaska, including relocation of affected communities and construction of replacement facilities." A Senate report on an appropriations bill in 2006 went further: "The Committee has provided $2,400,000 for Alaska Coastal Erosion. The following communities are eligible recipients of these funds: Kivalina, Newtok, Shishmaref, Koyukuk, Barrow, Kaktovik, Point Hope, Unalakleet, and Bethel."[257] In May 2006 the Army Corps of Engineers cited Sec. 117 and the Senate report as authorities for a study of erosion protection in Shishmaref. The Army Corps could invoke the same authorities if results of the Barrow storm damage reduction study do not qualify Barrow for assistance under existing technical requirements. Meanwhile, at the Senate field hearings in October 2007, Senator Stevens reportedly "battered representatives from the U.S. Army Corps of Engineers and the Federal Emergency Management Agency," saying "it was 'mind-boggling' that the Corps of Engineers hadn't requested emergency money—available under a law [Sec. 117] he shepherded through Congress two years ago. . . ."[258] Senator Stevens also sought waivers for Alaska Native villages to qualify for assistance from FEMA. Such waivers were granted to U.S. Gulf Coast cities after Hurricane Katrina.

The field hearings in Anchorage in June 2004 revealed a major problem in decision making—the maladaptation of state and federal agencies' programs to problems on the ground. After listing more than a dozen agencies contacted by the Shishmaref Erosion and Location Coalition, Luci Eningowuk, representing the coalition, testified that "our experience has shown that there is a lack of continuity between the various federal and State programs and agencies. . . . For the most part, we have found that none of the agencies have programs that cover the full range of our needs."[259] More specifically, with respect to Shishmaref's efforts to relocate, "There is currently no one agency stepping forward to take the lead. To be blunt, no agency's programs are designed for a project as complex as a full village relocation. Each agency has its realm of responsibility, and often there is a gap program to program."[260] This illustrates a pervasive problem in the tradition of scientific management, forcing emerging complex problems into bureaucratic pigeon-holes or stovepipes designed long ago for other purposes.

The field hearings in Anchorage in October 2007 revealed a major policy problem—difficulty in reconciling needs and resources across levels. Tony Weyiouanna, Shishmaref's transportation planner, testified that Shishmaref needed $61 million to ward off erosion and floods that year, and more the next year. According to a press report, "Stevens told him such money is unlikely to be found, especially since Shishmaref is not the only village where needs are urgent. 'Clearly the concept of each of the nine villages getting $70 million over the next year is next to impossible,' he said. 'Each village is proceeding on the basis that they're going to come first, and that's not possible.' Villages and the state need to work together to determine priorities and help decide how to best spend the available money, he said." When Sen. Mary Landrieu of Louisiana "asked if villages have previously met to discuss working together and prioritizing needs," one of the village representatives replied, "Never." (Sen. Landrieu, chairwoman of the Senate's Disaster Recovery Subcommittee, had visited Shishmaref to see storm damage firsthand.) "Stevens expressed frustration that no single agency—federal, state or local—has been designated to take the lead on erosion issues."[261]

A structural solution is not likely to be found in the creation of another stovepipe with a mandate, jurisdiction, and other resources sufficient to dominate competing programs and agencies. "One-off" policy solutions for single villages are not likely to be practical either, except for field testing prototypes or pilot programs; there are too many villages in need. However, networks of similar local communities could clarify their *shared* needs for funding and for adapting and supplementing existing programs, and lobby elected and appointed officials at higher levels accordingly. Local communities working together are more likely to sustain interests in resolving their own problems on the ground and to have the understanding and influence to pursue those interests effectively. At higher levels, elected officials at least have incentives under the Constitution to represent their constituent communities. Representing them is no less legitimate than representing national industrial, environmental, or other interest groups, despite the negative stereotype of pork barrel politics. And lobbying by place-based communities is no less legitimate than lobbying by conventional interest groups. In practice, the important question in any case is whether common interests are served.

Of course, it should not be assumed that policies advocated from the bottom up are *necessarily* valid and appropriate. A case in point is Shishmaref. At the Senate field hearings in June 2004 Luci Eningowuk testified that "it is merely a matter of time before we experience greater losses. We are quickly running out of space on our ever-shrinking island. . . . The no action option for Shishmaref is the annihilation of our community" by dispersion of the population to other places.[262] Thus, efforts to relocate the entire village are justified by rapid erosion from the bottom-up perspective of local residents. But Owen Mason, updating a book he coauthored on *Living with the Coast of Alaska*, claimed that "[a] different perspective arises from inspecting aerial photographs from 1948, 1986, and the present; these seem to indicate that erosion was not quite so severe until the present." He suggested that engineering has contributed to erosion around Shishmaref since 1983 when a concrete-block revetment was built. It failed quickly but was subsequently restored and supplemented by rocks placed on the shoreline. Mason contends that the crux of the problem has historical roots: "The ancestors of the Shishmaref people apparently recognized the dynamic nature of the island and situated their community on the most landward, safest part. . . . In the 1920s, to ease barge off-loading, Shishmaref shifted to its present location— on the most active eroding face of the island."[263] In a report for the state of Alaska in 1996, Mason concluded that the location of old Shishmaref "is accreting both vertically and horizontally and provides a comparatively stable location."[264] He suggested that moveable structures and dune-trapping devices such as plants, fences, and matting could allow current residents to remain on the island in harmony with nature.

Without attempting to evaluate the merits of different perspectives on Shishmaref's situation, it is clear that influence from the bottom up is not fool-proof, and neither is networking. The compilation of beach nourishment cases prior to approval of the Barrow and Wainwright Beach Nourishment Program apparently did more to promote this alternative than to inform program planning and decisions by NSB assembly members. On the other hand, the scientific management alternatives are not guaranteed either. Information, recommendations, and decisions that flow from the top down also are subject to omissions and distortions. And some of those decisions, not limited to the colonial past, have imposed hardships on the ground. Scientific management dominates in the climate change regime not because it works according to its own ultimate objective but because it was established long ago and continues to serve other interests.

There are good reasons for policymakers to support networks of communities and to help them work. In contrast to counterparts at higher levels, policymakers at the local level cannot easily evade de facto accountability for local policy decisions; they must live with the direct consequences of their policy decisions unless they migrate. Local policymakers understand much better the changing community values necessarily involved in policies to reduce storm damages, as well as the effects of relocating, nourishing the beach, and related alternatives on other needs competing for limited local resources in their communities. Among the values involved are ways of life, whether in Barrow or the North Slope, New Orleans, or elsewhere. As Mayor Itta insisted, nobody can measure a way of life in dollars and cents; it cannot be reduced to a formal cost-benefit analysis without major distortions and opposition. Local policymakers also understand through personal observations and experience the unique local environment, natural and human, that conditions the performance and promise of the policy alternatives available. The ability of higher-level officials to develop a satisfactory understanding of local conditions and values declines sharply with the number of different local communities under their jurisdiction. Human cognitive constraints alone are enough to force decentralization of decision making, if the priority is to improve the rationality of decisions. Differently informed perspectives can be reconciled point by point and case by case in a decentralized decision-making structure. Here is a constructive role for intensive research by scientists willing and able to rise to the challenge.

Barrow is unique, but it is also a microcosm of things to come at lower latitudes. In Barrow as in other local communities, past damage and future vulnerabilities to extreme weather events and climate change are contingent on interactions among large numbers of human and natural factors, no one of which can tell the story. Context matters, which implies the need for intensive research in enough selected communities (not all of them) to improve the empirical basis for adaptation. In Barrow as in other local communities, reducing storm damage is a complex policy problem. It involves multiple changing community values and other community needs. It also involves multiple changing circumstances, not limited to climate change, fraught with profound uncertainties. Hence, every action taken to reduce storm damage is a matter of trial and error in some substantial degree, even if it is not perceived as such. But perceiving it as such can accelerate learning by doing, and networking can expand the range of relevant experience available to each community. At higher levels in decision-making structures, Barrow and other local communities face an array of agencies with programs of assistance designed for rather narrow purposes often long ago. Effective coordination and integration of the available assistance depends on the local community. In turn, the adaptation of assistance programs to the common needs of communities depends on organizing them from the bottom up and supporting them from the top down. That support comes most easily from elected and appointed officials concerned about alleviating problems on the ground.

OPENING THE REGIME

4

This chapter elaborates the proposals in Box 1.1 for opening the climate change regime to adaptive governance. For that purpose we pull together the historical case materials in previous chapters and relevant theoretical material. Recall that Chapter 2 reviewed the evolution of scientific management in the climate change regime and exceptions that point toward adaptive governance. Chapter 3 reviewed Barrow as a microcosm of things to come as signs of climate change become more obvious at lower latitudes, including steps toward adaptive governance. Beyond these historical case materials, however, various aspects of the proposals for adaptive governance have been accepted or recommended in general literature on climate change, environmental hazards, and related policies, and in more theoretical literature on science, policy, and decision making. These convergent sources from different and larger bodies of experience add support to the proposals for adaptive governance in climate change. In particular, these convergent sources document a latent but coherent frame of reference in which the case materials become more than mere historical curiosities. They become foundations for an alternative frame to understand and reduce net losses from climate change. The established frame in the climate change regime is not the only construction of the relevant past and possible futures.

The three sections below consider, respectively, proposals for more intensive science, more procedurally rational policy, and more decentralized decision making; each begins with a review of the proposals in Box 1.1. It should be understood, however, that the proposals in each section are overlapping and mutually reinforcing aspects of adaptive governance, not discrete parts of it. Opening the established regime depends on *integrating* the proposals to the extent practical as niche opportunities arise in the regime's continuing evolution. This task can be informed by knowledge of science, policy, and decision making, but knowledge and information are not resources above politics. "Knowledge is a form of power, and most institutions exhibit an understandable reluctance to dissipate this power in the absence of compensating advantages."[1] In particular, broader dissemination of pertinent knowledge and information, local and scientific, can empower people on the ground to participate in climate change decisions directly affecting them in their own local communities as well as at higher levels. One compensating advantage for sharing power is advancing the common interest.

Intensive Inquiry

First, in terms of science, we propose more systematic empirical inquiry *centered* on (but not bounded by) efforts to reduce net losses from extreme events on the ground, in local communities initially considered as single case studies. Such inquiry strives to be *comprehensive*, covering all the major interacting factors, human and natural, shaping important outcomes in each case. Under a comprehensive description, each case is unique; this is both an empirical finding and a consequence of the conjunction rule of probability. Such inquiry is also *integrative*, construing the significance of relevant observations as contingent on a working "model" of the single case as a whole. This includes quantitative observations, but their significance is not limited to the operational definitions of variables, the requirements of general theory, or formal models or methods. Gaps in comprehensiveness and inconsistencies arising from the integration of additional observations prompt revisions in a working model of a single case. Thus, a model of the single case is both falsifiable and open to cumulative improvements with additional observations and integration. We call this approach in the historical sciences, including the policy sciences, *intensive inquiry*.[2]

In procedurally rational policy processes, intensive inquiry facilitates cooperation between scientists and policymakers with differently informed perspectives on the same context, to improve the working models of the context on which they act. In a decentralized decision-making structure, intensive inquiries harvest experience on policies successfully field tested in local communities working in parallel, for diffusion and possible voluntary adaptation by other communities organized as networks and to inform those central authorities at higher levels who are willing and able to support what works on the ground. Intensive inquiry is an adaptation to the point that context matters for both scientific and policy purposes.

Context Matters in Practice

This point is well documented in previous chapters. For example, as noted in Chapter 2, it was a premise of IPCC's Working Group III, also known as the Response Strategies Working Group, which formulated choices for negotiation of a framework convention on climate change in the First Assessment Report in 1990: "Any responses will have to take into account the great diversity of different countries' situations and responsibilities and the negative impacts on different countries, which consequently would require a wide variety of responses."[3] This point was a conclusion of the U.S. National Assessment Synthesis Team in 2000: "The nature and intensity of impacts will depend on the location, activity, time period, and geographic scale considered. For the nation as a whole, direct economic impacts are likely to be modest. However, the range of both beneficial and harmful impacts grows wider as the focus shifts to smaller regions, individual communities, and specific activities or resources."[4] That context matters also was a general conclusion of the IPCC's Working Group II on impacts, adaptation, and vulnerability in the Fourth Assessment Report in 2007: "Costs and benefits of climate change for industry, settlement, and society will vary widely by location and scale."[5] Although generally accepted in the regime, the point that context matters challenges the regime's reliance on extensive research. Taking local variations in costs and benefits into account implies factoring the global problem into diverse local problems; any generalizations about costs and benefits at larger scales have many exceptions. Ironically, perhaps the only generalization about climate change science and policy consistent with the whole picture, and without obvious exceptions, is the generalization that context matters.

The exceptions to scientific management in Chapter 2 illustrate intensive research adapted to particular contexts. In the Pacific islands case in advance of the 1997–1998 El Niño, the Pacific ENSO Applications Committee (PEAC) centered its drought forecasts on specific U.S.-affiliated islands by developing statistical models based on each island's own historical rainfall data; large-scale numerical models informed the statistical models but lacked requisite spatial resolution. PEAC personnel helped integrate these local drought forecasts into local policy processes through publications and personal briefings. Differences in local policy responses indicate that local policymakers took responsibility for integrating local knowledge into their decisions to secure water supplies. In the Melbourne case, an early goal was to dramatically reduce per capita water usage as an efficient means for coping with a reduced average rainfall. Scientific research to support this goal was centered on the geography, demographics, climate, and infrastructure of the city of Melbourne. Subsequent research became more comprehensive, taking into account the importance of climate variability, particularly periods of extreme rainfall deficit, and the possibility of augmenting supply as well as reducing demand, including contested plans for a desalination plant. In the west Greenland case, the research available focused on understanding past adaptations, not reducing future losses. However, it underscored the importance of comprehensive and integrative inquiry in understanding the main economic transitions in the twentieth century, which were shaped by interacting human and ecosystem factors as well as climate change. Differences in the human consequences of the cod to shrimp transition were not evident at the national level but were quite significant between Sisimiut and Paamuit. These local differences depended on interacting skills, social networks and cohesion, and investments as well as the available seals, cod, shrimp, and crab. In both of these cases, neither changes in climate nor changes in natural resources tell the story; the significance of each was contingent on human factors.

Chapter 3 illustrates intensive inquiry in more detail. Consider the great storm in Barrow on October 3, 1963, as a single case. Little damage would have been done if it had followed the typical track north from Siberia *or* if it had generated normal winds *or* if it had occurred when shorefast sea ice protected Barrow, as was usually the case in early October. As it happened, the storm tracked eastward, generating unusually strong WNW winds of long duration in Barrow, and the long fetch allowed the build-up of a storm surge and waves. Material damage from the storm surge and waves depended on local geology and geography; the location, design, and construction of human infrastructure; and human responses to the emergency. Losses of buildings, fuel oil, and other supplies and equipment were greatest in developed low-lying areas at the Naval Arctic Research Laboratory (NARL), parts of Browerville, and the section of Barrow adjacent to Isatquaq Lagoon. The bluffs to the south protected people located farther inland, but erosion of the bluffs during the storm increased their future vulnerability. An early warning of the approaching storm would have reduced the potential for loss of life and limb in Barrow. But fortunately, a host of quick-thinking improvisations helped prevent fatalities. Although the storm destroyed the landing strip at NARL, the airport under construction was mobilized to fly in emergency and longer-term relief. Problems of recovery were mitigated by local reserves of fuel oil, backup generators, and redundant sources of water (including sea ice washed ashore) to replace sources contaminated by sewage, fuel, and seawater. Thus, for Guy Okakok and his family and others in Barrow, the lived experience of the storm was not contained within the storm itself.[6] In this and other cases, many interacting factors shaped the losses incurred and avoided. The significance of each factor for understanding outcomes was contingent on other factors in the same context.

Like the other storms in Barrow, the great storm is unique when the relevant factors are considered comprehensively. Hence, it cannot serve as a reliable substitute for understanding other damaging storms in Barrow's experience before or since 1963. But it can serve as a baseline for comparison with other storms to understand the many contingent factors involved. For example, the August 10, 2000, storm was the one most similar to the great storm, but largely because of its shorter duration and a better forecast, significant damage was limited to a dredge sunk and a bridge and culvert washed out on the coastal road to NARL. Furthermore, consider Barrow as a single case in understanding and reducing losses from extreme events. Barrow's experience with damaging storms is unique when relevant factors are considered comprehensively. Hence, it cannot serve as a reliable substitute for understanding other communities vulnerable to big storms. However, it can serve as a baseline for comparison with communities that share a family resemblance with Barrow to inform understanding and action. For example, experience in Barrow can inform policy planning and promotion regarding beach nourishment in other low-lying Native coastal villages in or near the Arctic with similar silty soils affected by permafrost and similar atmospheric and oceanic conditions. Conversely, experience in Tuktoyaktuk and Shismaref can inform decisions in Barrow and elsewhere on the performance of seawalls, the complications involved in relocating communities in whole or in part, and other policies under consideration.

We adapted our case study to the requirements of intensive inquiry. First, the case was centered on Barrow and its recent coastal erosion and flooding problems, but not bounded by them. Chapter 3 includes relevant historical background on the present situation in Barrow; interactions with other Natives on the North Slope and beyond, the state of Alaska, the U.S. government, the oil industry, and the International Whaling Commission, among others; and connections with related problems such as sustaining both subsistence and modern economies that support the people of Barrow and affect their efforts to reduce storm damage. Thus, the case was constructed around a central focus, not inherently fuzzy boundaries, according to the information available and our purposes. Second, within these constraints, we used the policy sciences framework to construct a more comprehensive working model of the particular context. (A complete model is beyond the reach of those who are less than omniscient.) For example, one part of the framework directed our attention to the functions performed—however well or poorly, or the level of effort—in any policy decision process: intelligence, promotion, prescription, invocation, application, appraisal, and termination. The results can be seen in our descriptions of decision processes, most obviously of the beach nourishment program in Barrow, and in other cases including the climate change regime. The framework also directed our attention to serial, parallel, and hierarchical connections among the different decision processes involved in decentralized decision making, including networks.[7] Third, our working model takes the form of a story, a narrative, which allows for the integration of relevant details of many kinds—qualitative and quantitative, visual and textual, personal and impersonal.[8]

As explained below, the integration of comprehensive and detailed information on the particular context is the key to falsifying erroneous assumptions and improving a working model of the single case. For example, an early expectation of the participants of the Barrow project, both the scientists and the residents, was that big storms track the ice edge. This implied that a retreating ice pack under a warming climate would lead to fewer big storms in Barrow. But a detailed examination of the historical behavior of a series of big storms using observations and numerical models convinced us that these storms typically did not track the ice edge. Indeed, in some cases the storms redistributed the ice ahead of them as they traveled across the Arctic Ocean. Considered in context, the longer ice-free season increased in significance relative to other natural factors contributing to Barrow's vulnerability. Similarly, only rather late in the project, as we expanded the scope of our research into the precontact era, did we appreciate that era's significance for understanding recent adaptations to climate change on the North Slope coast. Thus, we updated our understanding of the case in a series of approximations as both the research and the context of the research evolved.

Because context matters, extensive research cannot provide a dependable understanding for reducing vulnerabilities in *all* local communities, even if it is billed as globally comprehensive. Too much of importance is left out. Moreover, extensive research cannot provide a dependable understanding for reducing vulnerabilities in *any* local community, even if general propositions abstracted from multiple local contexts are generally valid. For example, one might expect from extensive research that a systematic shift across the whole Arctic to more and stronger storms would have caused more losses in Barrow already.[9] Furthermore, one might expect from extensive research that the exposure of many more people and much more property in Barrow must have increased losses there since 1963, even without an increase in big storms.[10] However, despite changes in Arctic climate as a whole and significant community development in Barrow since incorporation of the borough, an increase in the frequency or severity of flooding events has yet to be detected in Barrow, and no lives have been lost to big storms in Barrow. Moreover, inflation-adjusted total dollar damages from the great storm of 1963 could still exceed the total damages from all subsequent storms, if all the damage estimates were consistent and comprehensive. These apparent inconsistencies in the story can be resolved to some extent by reference to Native survival skills, improvements in emergency training and management, including better forecasts and the construction of temporary berms as storms approach, and other factors. But Barrow also has been lucky so far to have been spared a similar conjunction of wind direction, intensity, and duration when the ice was out.

Despite the need for intensive inquiry to understand and reduce losses from extreme events on the ground, the established climate change regime has relied primarily on extensive research. Such research is exemplified by the initial research plans or assessments of three major initiatives and their successors reviewed in the first section of Chapter 2: Report No. 12 of the International Geosphere–Biosphere Programme, the Fiscal Year 1990 Research Plan for the U.S. Global Change Research Program, and the climate science and impacts in the IPCC's First Assessment Report, which provided context for issues considered by the Response Strategies Working Group. Extensive research in this tradition tends to reduce diverse systems to their stable and standard parts, to integrate these parts into numerical models for predictions intended to reduce uncertainty, and to generalize for results of broad national or international scope. In the process, differences and changes in context tend to be washed out and only partially restored when and if general models are applied to particular cases. Such generalizations can help direct the attention of researchers and policymakers to vulnerabilities on the ground, but they do not accurately predict what can be found there, or prescribe what should be done there; comprehensive and detailed information still has to be filled in for each case through local knowledge and, where practical, intensive inquiry. Among the systematic omissions in extensive research are the behaviors of human beings and other living forms, highly contingent on context, that resist standard and stable generalization. In the results of extensive research one does not find leaders like Rob Elkins in Barrow, John Thwaites in Melbourne, and Elias Kleist in Sisimiut, who made significant contributions toward adapting their communities. They are depersonalized in extensive research, where authors only retain their proper names. Ironically, recent climate change is attributed largely to human behavior, and changes in human behavior are necessary to reduce losses from climate change. But most research justified by climate change has focused on modeling the behavior of the total earth system with human behavior considered exogenous.

Finally, it should be emphasized that intensive inquiry for adaptive governance and extensive research in the scientific management tradition are different approaches to simplification of the total earth system, which includes its inhabitants. Intensive inquiry aims to clarify what is context dependent; extensive research aims to discover what is context independent. Simplification is necessary because no one has a complete or completely objective understanding of the much more complex reality in which we live; pretenses to the contrary notwithstanding, no one is omniscient. Both approaches to simplification are "scientific" in the sense that they are systematic and empirical, where "empirical" means that assumptions about complex reality are (or at least should be) falsifiable and falsified by observations. Extensive research is relevant and important, but it would be foolish to rely exclusively on it. Intensive inquiry is the appropriate means of simplification where differences and changes in context matter for understanding and reducing losses from climate change, and where expanding the range of informed choices for ameliorating losses on the ground is a purpose accepted in practice as well as in principle.

Moving beyond cases we find general climate-related literature that has converged independently on the need to open conventional climate change research to something like intensive inquiry. Convergence, however, is a matter of overlap rather than congruence.[11] In 1998 Steve Rayner and Elizabeth Malone underscored the importance of differences and changes in local context. "Empirical local-level studies reveal such complex mosaics of vulnerability as to cast doubt upon attempts to describe patterns and estimate trends at the global or even regional scales." Poverty was perhaps most prominent among recognized vulnerability factors, but they considered it contingent, neither necessary nor sufficient for vulnerability. In addition the very young and the elderly also were often but not always vulnerable. Rayner and Malone also recognized "differences in health, gender, ethnicity, education, and experience with the hazard in question" as relevant factors in these complex mosaics of vulnerability.[12] Similarly, in 1999 Neil Adger understood from his case study of coastal Vietnam "the complex nature of social vulnerability and the importance of the political economy context." He considered it "not meaningful . . . to generalize from the analysis" because "[d]ifferent societies face different threats of global climate change over the next century."[13] Sarah Burch and John Robinson of the University of British Columbia came to a similar conclusion in 2007, attempting to understand why effective action on climate change has lagged behind understanding of its impacts. They affirmed the need to "more effectively address the multitude of temporally and contextually specific intricacies of human behaviour in response to risks such as climate change."[14]

Aspects of intensive inquiry can be found in the "science of the integration of the parts," a "second stream" of science that ecologist C. S. Holling distinguished and recommended for adaptive management in 1995.[15] The first stream is a "science of the parts" that "emerges from traditions of experimental science, where a narrow enough focus is chosen to pose hypotheses, collect data, and design critical tests for the rejection of invalid hypotheses. . . . It is appropriately conservative and unambiguous, but it achieves this by being incomplete and fragmentary. It provides bricks for an edifice but not the architecture." It is also insufficient for adaptive management, which requires a second stream of science. The second stream is "fundamentally interdisciplinary and combines historical, comparative, and experimental approaches at scales appropriate to the issue." It is "fundamentally concerned with integrative modes of inquiry and multiple sources of evidence. This stream has the most natural connection to related ones in the social sciences that are historical, analytical and integrative. It is also the stream that is most relevant for the needs of policy and politics." One assumption underlying this proposal is that we "cannot experimentally manipulate lost pasts" of a system even though we can experimentally manipulate models of a system, as noted below. Another assumption is that "knowledge of the system we deal with is always incomplete. Surprise is inevitable. Not only is the science incomplete, but the system itself is a moving target, evolving because of the impact of management and the progressive expansion of the scale of human influences on the planet." In Holling's version of adaptive management based on these assumptions, policy interventions are understood as factors that interact with other parts of an open, evolving system. They are also understood as a series of approximations based on learning by doing, not the experimental ideal.[16]

Similarly, in 1993 social psychologist Paul Stern called for "a second environmental science—one focused on human-environment interactions—to complement the science of environmental processes" for policy purposes.[17] A basic rationale was his critique of conventional assumptions about human behavior: "Policy failures repeatedly result from faith in intuitively attractive but mistaken ideas about behavior: That people will accept experts' risk analyses at face value; that firms will accept and implement regulations; that consumers will act on relevant information; and that free market or quasi-market incentives will work in practice as they do in theory."[18] This critique sheds some light on the limited progress of the European Union's Emissions Trading System and on expert risk analyses demanding reductions in greenhouse gas emissions going back at least to the Toronto Conference Statement in 1988. As an alternative to such mistaken ideas about human behavior Stern discerned certain principles for a second environmental science emerging from research on energy conservation. "One [principle] is that consumer behavior needs to be analyzed in terms of limiting factors. Technology, attitudes, knowledge, money, convenience and trust are all needed for behavior change, and attempts to provide any of these will fall short to the extent the others are missing. . . . Limiting factors can vary with the consumer and the situation and so must be identified empirically. Another principle is that behavior must be understood from the consumer's perspective, a principle that implies involving consumers in some way in programs intended to change their behavior."[19] In short, Stern conceived consumers as participants, not merely stakeholders, in policy-relevant environmental research;

he understood their behavior as contingent on multiple factors in particular situations, any of which could be a limiting factor; and he found empirical research in those situations to be necessary for a second environmental science. Stern was well aware of institutional barriers in academia and government that stand in the way of a new science and have limited its progress.[20] Those barriers were overcome to some extent in the Melbourne case.

The Historical Sciences

At a deeper level, this climate-related literature and our proposal for more intensive research may be grounded in the epistemology and methods of the historical sciences, both natural and human. The evolutionary biologist Stephen Jay Gould was a prominent advocate of the historical sciences and a critic of the reduction of all knowledge to one kind of scientific knowledge. But he is not the only one.[21] Gould characterized "experiment, quantification, repetition, prediction, and restriction of complexity to a few variables that can be controlled and manipulated" as "the stereotype of scientific method." He contended that the stereotype is not sufficient to explain the unique outcomes of historical, including evolutionary, processes.[22]

> Historical explanations . . . account for uniqueness of detail that cannot, both by laws of probability and time's arrow of irreversibility, occur together again. We do not attempt to interpret the complex events of narrative by reducing them to simple consequences of natural law; historical events do not, of course, violate any general principles of matter and motion, but their occurrence lies in the realm of contingent detail. . . . And the issue of prediction, a central ingredient in the stereotype, does not enter into a historical narrative. We can explain an event after it occurs, but contingency precludes its repetition, even from an identical starting point.[23]

This view of the nature of history implies profound uncertainties that lie beyond the reach of scientific method, including the stereotype. Gould credits Charles Darwin with development of a different but equally rigorous methodology for testing historical explanations. It is based on William Whewell's concept of "consilience, meaning 'jumping together,' to designate the confidence gained when many independent sources 'conspire' to indicate a particular historical pattern." Thus, "We search for repeated pattern, shown by evidence so abundant and diverse that no other coordinating interpretation could stand, even though any item, taken separately, would not provide conclusive proof."[24] The preponderance of evidence is the basis for the modern understanding and acceptance of the theory of evolution, for example. The underlying assumption is that the outcomes of historical processes are "dependent, or contingent, upon everything that came before—the unerasable and determining signature of history." This is not about randomness but about contingency, "the central principle of all history."[25] In short, "Invariant laws of nature . . . set the channels in which organic design must evolve. But the channels are so broad relative to the details" of interest that "contingency dominates and the predictability of general form recedes to an irrelevant background."[26]

Gould's views on the uniqueness of historical explanations are not limited to evolutionary biology. In atmospheric science, the seminal work of the late Edward Lorenz published in 1963 demonstrated that both the evolution of states of the atmosphere and the rate of growth of uncertainties are critically contingent on initial conditions, or the initial state of a system.[27] Thus, in some simulations of a given system but not all, small differences in the initial state of the system can lead to large differences in subsequent states. This sets fundamental constraints on the accuracy of weather predictions, for example. Such behavior is characteristic of dynamical systems, including the atmosphere, that are deterministic, dissipative, and nonlinear. The behavior is termed "chaotic." This does not imply "disorder," but rather "aperiodic order" in which previous states of the system never recur exactly.

Similarly, in 1972 Philip Anderson, a solid state physicist and Nobel laureate, challenged the "constructivist" assumption that system behavior can be deduced from fundamental laws. Even in physics, he claimed, "The ability to reduce everything to simple fundamental laws does not imply the ability to start from those laws and reconstruct the universe."[28] In a hierarchy of sciences arrayed by increasing complexity, each science obeys the laws of more fundamental sciences. However, at each level "entirely new laws, concepts, and generalizations are necessary, requiring inspiration and creativity to just as great a degree as in the previous one." Thus, social science is not applied psychology, and "[p]sychology is not applied biology, nor is biology applied chemistry." Anderson summed up this insight on complexity in his article title "More Is Different." This "became a rallying cry for [research on] chaos and complexity" including complex adaptive systems.[29] Anderson explicitly shared Gould's view that "life is shaped less by deterministic laws than by contingent unpredictable circumstances."[30]

In the study of complex adaptive systems, numerical models based on many simulated "agents" corroborate the central role of contingency in the epistemology of Gould and others working in the historical sciences. Each agent in a model acts on a highly simplified "internal model" of its simulated environment. Each internal model consists of multiple rules of interaction that anticipate preferred outcomes and prescribe actions for various contingencies that may arise in the simulated environment. But these "if-then" rules are subject to change in response to the outcomes of simulated actions: For example, rules associated with positive outcomes are reinforced, and plausible new rules may be generated by recombining parts of previously tested rules. Thus, each agent learns and adapts as a small part of a much larger distributed system. Consider the behavior characteristic of such models as described by John Holland, a leader in the study of complex adaptive systems:

> Because the individual parts . . . are continually revising their ("conditioned") rules for interaction, each part is embedded in perpetually novel surroundings (the changing behavior of the other parts). As a result, the aggregate behavior of the system is usually far from optimal, if indeed optimality can even be defined for the system as a whole. For this reason, standard theories in physics, economics, and elsewhere, are of little help because they concentrate on optimal end-points, whereas complex adaptive systems "never get there." They continue to evolve, and they steadily exhibit new forms of emergent behavior. History and context play a critical role. . . .[31]

History and context play a critical role because as the system continues to evolve, the value of each variable eventually affects all the other variables in the system directly or indirectly and in turn is affected by them. Designation of any variable as dependent or independent for statistical purposes is a major simplification.

Like other numerical models, agent-based models are attractive as surrogates to be manipulated for controlled experiments—experiments that would be difficult, if not impossible to perform on any real-world social systems the models may seek to represent. Holland for example affirmed the use of such models as surrogates for the real world: "The broadest hope is that the theoretician, by testing deductions and inductions against the simulations, can reincarnate the cycle of theory and experiment so fruitful in physics."[32] But what can be known through this stereotype of scientific method, as

Gould called it, if contingency dominates and history and context matter? In their assumptions, agent-based models are relatively realistic representations of social systems, but for that reason their capacity to predict accurately and precisely the detailed behavior of any particular social system is virtually nil. The obstacles include accurate measurement of initial conditions in the internal models of many agents at the same time; the specification, appraisal, and parameterization of the multiple rules of interaction of each agent and the mechanisms by which those rules are adapted to outcomes; and the practical necessity of excluding most real people from representation as agents in the model.[33] However, what is excluded from the model nevertheless interacts in the real world with what is included to shape the behavior to be predicted. According to critics, such models are data free, producing behavior that is at best reminiscent of actual systems.[34] Such models are more useful for exploring the consequences of their assumptions (including surprises) than as experimental surrogates for real-world systems. Furthermore, even advocates of complex adaptive systems like Holland understood that the general principles sought through experiments on models would not provide solutions to problems. Those principles would only "point to ways of solving the attendant problems" such as trade balances, sustainability, and AIDS.[35] The results may serve as heuristics for intensive empirical inquiries, but not as a substitute for them.

An alternative approach to the simplified representation of social systems, and people in them, assumes that human behavior is determined by impersonal laws of nature. (Thus fixed variables and mechanisms are impersonal substitutes for agents that adapt their internal models to changing surroundings.) This assumption often surfaces in the language of mechanics applied to human behavior. One example refers to "the chain of causality that drives climate change. In this chain of cause and consequences, societal forces such as population, affluence, or technology drive the varied human activities that produce greenhouse gas (GHG) emissions."[36] If it were strictly true that behavior of a human being, like the behavior of a billiard ball, is uniquely determined by external forces applied to it, then the choices and decisions that have been made in response to climate change would be illusory, and there would be no justification for research to inform future choices and decisions. However, unlike inanimate objects, the behavior of human beings and other living forms is mediated by internal predispositions, genetic and acquired, that differentiate their responses to external stimuli, including climate change and extreme weather events; and the perspectives on which they act are subject to change through natural selection and learning. Sooner or later, the validity of theoretical generalizations about human behavior turns out to be context-dependent—that is, restricted in space, time, or culture. Statistical aggregates also are limited because the probability that a damaging extreme weather event will be replicated in all its particulars is zero. Intensive inquiries can identify when and where statistical aggregates and other generalizations are invalid and shed light on how to improve them.

Another approach is rational choice theory, which assumes that people behave *as if* they were objectively rational according to researchers' specifications of the context, including actors' goals. If this were strictly true, then people would have accepted expert risk analyses of climate change at face value and the concentration of greenhouse gas emissions in the atmosphere already would be under control. This assumption is unrealistic, but the realism of assumptions is irrelevant in theory testing, according to Milton Friedman in his famous essay on positive economics.[37] "Its performance is to be judged by the precision, scope, and conformity with experience of the predictions it yields. In short, positive economics is, or can be, an 'objective' science, in precisely the same sense as any of the physical sciences."[38] The claim of objectivity evidently refers to only one part of every model, the model itself, "which is abstract and complete," a matter of logic in which there is no place for "vagueness, maybe's, or approximations." The other part of every model according to Friedman consists of the "rules for using the model [that] cannot possibly be abstract and complete. They must be concrete and in consequence incomplete. . . ." Therefore, no matter how explicit the rules, "there inevitably will remain room for judgment in applying the rules. Each occurrence has some features peculiarly its own, not covered by the explicit rules."[39] The need for judgment qualifies claims of objectivity based on the first part of a model alone. The need for judgment also requires some familiarity with the context at hand for practical applications, theory-testing, and the construction of new hypotheses. The latter also is important for progress in positive economics. However, according to Friedman, "On this problem there is little to say on a formal level. The construction of hypotheses is a creative act of inspiration, intuition, invention; its essence is the vision of something new in familiar material. Thus, the process must be discussed in psychological, not logical categories; studied in autobiographies and biographies, not treatises on scientific method; and promoted by maxim and example, not syllogism and theorem."[40]

Our point is that the stereotype of scientific method is quite limited where history and context matter—especially if human beings and other living forms are involved. This includes climate change, where human behavior is a major factor explaining observed climate change, and changes in human behavior are necessary to reduce the adverse impacts. Those skeptical about the limits of conventional scientific methods for extensive research in climate change are invited to reconsider our appraisal of progress so far in addressing greenhouse gas emissions near the beginning of Chapter 1. Those skeptical on more general empirical grounds are referred to the very modest track record of policy-relevant predictions and forecasts.[41] They might also consider the judgment of Alice Rivlin, economist and practitioner, on the past and future of macroeconomic forecasting, where a wealth of forecasting experience has been harvested for some time: "The poor showing of the forecasters is not due to any lack of effort or ingenuity. . . . The real problem is that the economic system is extremely complicated, that our economy is battered by forces outside itself which are inherently unpredictable, such as the weather or foreign wars. I doubt we will ever improve the accuracy of forecasting very much, especially the forecasting of economic turning points. Instead, we will have to learn to live with the uncertainty."[42] As physicist Brian Greene put it in a different context, "Exploring the unknown requires tolerating uncertainty."[43]

Our point also is that intensive empirical inquiries can compensate for certain limitations of conventional scientific methods, bringing into the picture more of what is deleted through extensive research but nevertheless important for understanding and reducing net losses from climate change. Well before climate change became an issue, the general point was recognized in various ways by Harold Lasswell, Herbert Simon, and other social scientists. Like Gould in evolutionary biology, they understood the importance of contingency in the social and policy sciences, and they developed concepts, theories, and methods to support intensive empirical inquiry. Like Stern in social psychology, they understood the significance of the actor's own perspective in the historical sciences. The basic assumption is that people act from their own subjective perspectives—patterns of identifications, demands, and expectations—that individually fall far short of what would be required for objectively rational behavior and collectively are quite diverse and subject to change. The historian employs a similar assumption "to make intelligible the subjective element in his accounts" according to Henri Pirenne.[44] These perspectives are represented by the internal models of agents in computer models of complex adaptive systems. The perspectives of real people are major factors in the complex dynamics and outcomes observed in the adaptation and evolution of social systems on various time scales. This includes Iñupiat society on the North Slope over the centuries. Taking subjective perspectives into account improves the dependability of empirical inquiry for scientific or practical purposes.

The most prominent formulation of the basic assumption is Simon's principle of bounded rationality: "The capacity of the human mind for formulating and solving complex problems is very small compared with the size of the problems whose solution is required for objectively rational behavior in the real world—or even for a reasonable approximation to such objective rationality."[45] What are the bounds? Rationality is constrained by the limited knowledge and information available for choices and decisions in various contexts; this is the premise of all research justified for policy purposes, of course. But rationality also is constrained by the narrow span of attention—about seven plus or minus two chunks of information at any one time, the magical number in George Miller's seminal article in 1956.[46] According to Simon, "The narrowness of the span of attention accounts for a great deal of human unreason that considers only one facet of a multi-faceted matter before a decision is reached."[47] However, human personalities are not limited to a reasoning capacity alone. Simon recognized that uncertainties provide "an enhanced opportunity . . . for unconscious, or only partly conscious, drives and wishes to influence deliberation." He concluded that "[p]eople are, at best, rational, in terms of what they are aware of, and they can be aware of only tiny, disjointed facets of reality."[48]

Methodologically, this implies the need for empirical inquiry centered on particular contexts. As Simon put it (emphasis added) in an address on human nature in politics:

> My main conclusion is that the key premises in any theory that purports to explain the real phenomena of politics are the empirical assumptions about goals and, even more important, about the ways in which people characterize the choice situations that face them. *These goals and characterizations do not rest on immutable first principles, but are functions of time and place that can only be ascertained by empirical inquiry.* In this sense political science is necessarily a *historical science,* in the same way and for the same reason that astronomy is. What happens next is not independent of where the system is right now. And a description of where it is right now must include a description of the situation that informs the choices of the actors.[49]

Simon also observed that "a more careful look at the natural sciences would show us that they, too, get only a little mileage from their general laws. Those laws have to be fleshed out by a myriad of facts, all of which must be harvested by laborious empirical research."[50] Similarly, in 1952, Harold Lasswell and colleagues in the policy sciences emphasized that "evaluation of the role of any institutional practice calls for a vast labor of data gathering and theoretical analysis." This is a consequence of the principle of contextuality: "If modern historical and social scientific inquiry has underlined any lesson, it is that the *significance of any detail depends upon its linkages to the context of which it is a part*."[51] The context is constructed in the course of inquiry, not fixed or given or assumed at the outset.

To make the most of opportunities for intensive scientific inquiry, consider some context-sensitive methods that elaborate more or less independently Gould's notion of consilience. "Remember that consilience literally means the 'jumping together' of disparate observations under the only common explanation that could, in principle, render them all as results of a single process or theory—a good indication, though not a proof, of the theory's probable validity."[52] Recall also Holling's insistence on the integration of the known parts of a situation. Thus, for any working "model" centered on reducing losses from extreme events in a particular community, for example, it is important to ask, "If this understanding is correct, what else should we expect to observe in the same context?" The multiple implications of this understanding—in effect, a theory of a single case ($n = 1$)—are testable and tested through additional observations on the case. As the social psychologist D. T. Campbell explained in a seminal article on case study methods, the inquirer "does not retain the theory unless most of these [implications] are also confirmed. In some sense he has tested the theory with degrees of freedom coming from the multiple implications of any one theory" of the single case.[53]

Thus, testing relies on the integration of multiple independent streams of information on the same context. This is the test that practitioners typically use.[54] Some version of the integrative test also is used on a global scale in the established climate change regime. For example, Working Group I in the IPCC's Third Assessment Report rested its conclusions not on a single piece of evidence but on "an enormous body of observations" that gives "a collective picture of a warming world."[55] More recently, climate scientist Ignatius Rigor acknowledged that "we do not have all the pieces . . . but as the I.P.C.C. reports, the preponderance of evidence suggests that global warming is real."[56] However, the integrative test is *not* dependable without detailed and comprehensive observations. Rigor affirmed the importance of comprehensiveness when he criticized climate change skeptics. He alleged that they "typically take a few small pieces of the puzzle to debunk global warming and ignore the whole picture that the larger science community sees by looking at all the pieces. . . ."[57] Susanne Moser and Lisa Dilling noted that climate change skeptics and alarmists alike "use selective decontextualized scientific findings" among other political tactics. "Meanwhile," they conclude, "the public is more confused than ever, and a growing problem remains unaddressed."[58]

Tendencies to take a few small pieces of the puzzle and ignore the rest are not limited to climate change skeptics or alarmists. Within the discipline of Holling's first stream of science, or Gould's stereotype of scientific method, it is an integral part of mainstream extensive research to break down a whole system into separate parts relevant to a stable relationship or standard measure or method and to exclude details not relevant according to these criteria. It is also an integral part of political propaganda to focus attention on a few aspects of a controversy, often as sound bites, and to ignore details that complicate or contradict the position advocated. Machiavelli understood that "[m]en are apt to deceive themselves upon general matters, but not so much when they come to particulars. . . . The quickest way of opening the eyes of the people is to find the means of making them descend to particulars, seeing that to look at things only in a general way deceives them."[59] Walter Lippmann understood that "[w]ithout some form of censorship, propaganda in the strict sense of the word is impossible. In order to conduct a propaganda there must be some barrier between the public and the event."[60] The enormous barrier between the public and climate change as an irreducibly global problem facilitates competing propagandas by skeptics, alarmists, and others that absorb and suppress relevant details. The much smaller barrier between extreme weather events and the public impacted by them constrains but does not eliminate propaganda. For example, the seminal planning document on *Mitigation Alternatives for Coastal Erosion at Barrow and Wainwright* in 1989 ignored significant details on offshore material samples and passive alternatives in order to promote beach nourishment. It functioned as propaganda, even if not intended as such.

A comprehensive conceptual framework can support the search for more thorough and detailed observations and their integration, thereby contributing to a more dependable working model of the single case for scientific and practical policy purposes. (A complete model lies beyond the reach of boundedly rational people, including us.) For these purposes, as previously noted, we have used the policy sciences framework for heuristics to revise and improve our working models of the Barrow case, the global case, and cases in between. Among the other frameworks available, the Adaptation Policy Framework (APF) developed under auspices of the UN Development Programme offers a "flexible approach through which users can clarify their own priority issues and implement responsive adaptation strategies, policies and measures." The five components of the APF and the two cross-cutting processes overlap and are approximately equivalent to what policy scientists have conceived as the tasks of the problem orientation and the main functions in the decision process.[61] This illustrates the point that "[i]n principle all comprehensive conceptual maps of the social and policy process are equivalent to one another."[62] One implication is that a standard framework is unnecessary, even if it were feasible, because translations across comprehensive frameworks are always possible. Another implication is that such frameworks are best kept in the background in a group setting, used not to impose terms or meanings but to discover equivalencies.[63] The priority for climate change researchers is to use those comprehensive frameworks they consider satisfactory for new insights into contexts and problems they consider important. Creating new frameworks is a diversion from empirical inquiry supporting action to address what Al Gore called "the planetary emergency."

While the integrative test is available to improve a working model of the single case, the most definitive test is action on that model to fulfill one's purposes, including the reduction of net losses. For boundedly rational people, every action is an opportunity for field testing and falsifying what we think we know about the context, and as such an opportunity for learning by doing. The consequences of action bring into conscious awareness significant considerations overlooked or misconstrued when action was taken, but only if we are prepared to perceive them. Direct experience in the context at hand is the best preparation for making the obvious inescapable. Lee Cronbach's critique of the two disciplines of scientific psychology is relevant here: "There *are* more things in heaven and earth than are dreamt of in our hypotheses, and our observations should be open to them." In each situation, "Intensive local observation goes beyond [experimental or statistical] discipline to the wide-eyed, open-minded appreciation of the surprises nature deposits in the investigative net." "As [the scientific observer] goes from situation to situation, his first task is to describe and interpret the [expected] effect anew in each locale, perhaps taking into account factors unique to that locale or series of events."[64] Also important, in addition to field work or participant observation, are various intellectual tools employed as heuristics to open, not close, observations to what could be important. These include well-designed comprehensive conceptual frameworks, case studies, precedents or analogues, and theoretical generalizations.[65]

The test of action differs from an experiment because the consequences of an intervention in an open system cannot be compared with nonintervention in an equivalent system used as a control. The test of action also differs from the test of prediction, arguably the mainstay test of extensive science in the established climate change regime.[66] A precise and accurate prediction of a disaster is a success by the mainstay test but a failure by the test of action, if action was intended to prevent or reduce losses from the disaster. Moreover, the reduction of scientific uncertainty through better predictions and the improvement in the communication of scientific uncertainty are not the only means of reducing losses and vulnerability. Action on "no regrets" and other strategies considered in the next section are procedurally rational despite profound uncertainties that cannot be eliminated by scientific research. The test of action is the mainstay test in American pragmatism. As Abraham Kaplan summarized it, "Knowing is not one thing we do among others, but a quality of any of our doings. . . . To say that we know is to say that we do rightly by our purposes, and that our purposes are thereby fulfilled is not merely the sign of knowledge, but its very substance."[67] The importance of field testing what we think we know in an uncertain world is corroborated in the advice of a prominent business strategist: "Reward success and failure equally. Punish only inaction."[68] This advice makes sense only if action is modest enough to fail gracefully if it does fail. Like other generalizations, its applicability is contingent on the context.

Gould's criterion for science, including the historical sciences, is relevant here: "The firm requirement for all science—whether stereotypical or historical—lies in secure testability. . . . History's richness drives us to different methods of testing, but testability is our criterion as well."[69] By the test of action, the knowledge used so far in efforts to fulfill the ultimate objective of the UNFCCC in Article 2 does not fare well—nor does reliance on numerical models and aggregate scientific assessments for integration by the integrative test, which depends on more comprehensive and detailed information for understanding and reducing losses on the ground. In contrast, intensive case studies provide the empirical basis for the falsification of scientific expectations on which the established climate change regime is based, and for the discovery of what is important for understanding and reducing net losses from climate change previously marginalized or overlooked. Such case studies also can inform decision makers on the ground and at higher levels directly, without being reduced to generalizations.

Concerns that this might compromise objectivity in climate change science are unfounded. Subjective choices are already involved in nearly exclusive reliance on extensive research over intensive inquiry. They are also involved in applying scientific theories. In the 1970 postscript to his famous treatise *The Structure of Scientific Revolutions*, Thomas Kuhn affirmed Friedman's view on the unavoidability of judgment in applying scientific generalizations, using Newton's second law of motion as an example. According to Kuhn, "the fact that [scientists] accept it without question and use it as a point at which to introduce logical and mathematical manipulation does not of itself imply that they agree at all about such matters as meaning and application." Kuhn emphasized the heuristic function of Newton's second law as a law-sketch or law-schema. "The law-sketch, say $f = ma$, has functioned as a tool, informing the student what similarities to look for, signaling the gestalt in which the situation is to be seen. . . . [Different situations] are no longer the same situations he had encountered when his training began. He has meanwhile assimilated a time-tested and group-licensed way of seeing."[70] It is the time-tested and group-licensed way of seeing that provides the illusion or guise of objectivity, which survives even though applications of science are routinely contested—by scientists in scientific forums and by scientists and others in policy arenas where science is typically instrumental to other values at stake. Yet, "Some scientists are still scandalized by [Kuhn's] historical insight that science is not a process of discovering an objective mirror of nature, but of elaborating subjective paradigms subject to empirical constraints."[71]

To guide choices in climate change research justified as policy relevant, consider the policy sciences on the role of science *in human relations*. In that context, science as a value is insufficient to guide the selection and design of research projects to advance the common interest (see Chapter 1) or to evaluate the social consequences of research results. Of course science is valuable for the *execution* of a research project as objectively as possible. Beyond that, "It is insufficiently acknowledged that the role of scientific work in human relations is *freedom* rather than prediction." This means bringing into conscious awareness factors that once determined choices, freeing people to take them into account in making future choices and decisions. "Hence it is the growth of insight, not simply of the capacity of the observer to predict the future operation of an automatic compulsion, or of a non-personal factor, that represents the major contribution of the scientific study of interpersonal relations to policy." As insight exposes behaviors and behavioral relationships that have held in the past, it may also modify or destroy them—including behaviors based on ignorance or misunderstandings of impersonal factors such as the greenhouse effect. "[A]ll scientific propositions about character or society must be read 'subject to insight.'"[72] In other words, their validity is contingent on the context. They are "*hypotheses-schema*: statements which formulate hypotheses when specific indices relate them to the conditions of a given problem." They "serve the functions of directing the search for significant data, not of predicting what the data will be found to disclose."[73]

For additional recommendations and expectations in this approach, see Box 4.1. We consider the box useful, if nothing else, in stimulating readers to clarify their own perspectives on science and social responsibility. (That introspective exercise could begin by comparing the established perspectives summarized in Boxes 2.1, 2.2, and 2.3.) In Box 4.1, social responsibility entails taking the policy process and ideal aspirations of society into account and goes beyond predictions derived from numerical models. Contextual and problem-oriented inquiry integrates policy goals and action alternatives with the three tasks of science narrowly construed. This integration is an iterative process; "a fundamental principle is that postulated goals are to be held *tentatively* until they have been disciplined by exposure to the consideration of trends, conditions, projections, and alternatives."[74] The general expectations under the visibility and vulnerability of science are particularly relevant to climate science.[75] Climate science in the aggregate promised to reduce scientific uncertainty about the behavior of the total earth system as a prerequisite for rational policy decision and action in the early years of the regime. In so doing, it acquired a great deal of visibility and, to the extent it serves as a substitute for policy decisions and action, a great deal of vulnerability to outraged publics and public officials, especially if and when catastrophic climate change occurs. In the policy sciences there is nothing of that hubris Tennekes challenged in the formative years, as reported in Chapter 1. Scientists, too, are boundedly rational. As Lasswell put it, "To some extent we are all blind and no doubt will remain so. But there are degrees of impairment, and so far as decision outcomes are concerned, it is the responsibility of the policy scientist to assist in the reduction of impairment."[76]

To summarize this section, then, there are good reasons to consider understanding and reducing net losses from climate change and variability to be a quasi-evolutionary process of trial and error. It is evolutionary because contingencies dominate within the broad channels set by natural laws, but quasi-evolutionary because the most significant contingencies—human

Box 4.1. Science and Social Responsibility

Social Responsibility

It is our responsibility [as scientists] to flagellate our minds toward creativity, toward bringing into the stream of emerging events conceptions of future strategy that, if adopted, will increase the probability that ideal aspirations will be more approximately realized.

Contextual and Problem-Oriented Inquiry

A contextual map . . . is an indispensible preliminary to the examination of any particular problem. The map does not, however, supply the answers. It provides a guide to the explorations that are necessary if specific issues are to be creatively dealt with.

An adequate strategy of problem solving encompasses five intellectual tasks. Five terms carry the appropriate connotation, or can acquire them readily: goal, trend, condition, projection, alternative. (Many equivalent analyses are in current use; as usual, the important point is not choice of term but equivalency of concept.)

Goal clarification: What future states are to be realized as far as possible in the social process?

Trend description: To what extent have past and recent events approximated the preferred terminal states? What discrepancies are there? How great are they?

Analysis of conditioning factors: What factors have conditioned the direction and magnitude of the trends described?

Projection of developments: If current policies are continued, what is the probable future of goals realizations or discrepancies?

Invention, evaluation, and selection of alternatives: What intermediate objectives and strategies will optimize the realization of preferred goals?

Since apart from its context no detail can be adequately understood, the five questions furnish an agenda for allowing the context to emerge at the focus of individual or group attention. . . . The most productive procedure is to examine the whole problem by returning again and again to the separate tasks.

Visibility and Vulnerability

[S]cience has grown strong enough to acquire visibility, and therefore to become eligible as a potential scapegoat for whatever disenchantment there may be with the earlier promises of a science-based technology.

If the earlier promise was that knowledge would make men free, the contemporary reality seems to be that more men are manipulated without their consent for more purposes by more techniques by fewer men than at any time in history.

Sources: All text is selected and quoted from Harold D. Lasswell, The Political Science of Science, *American Political Science Review* L (December 1956), 966; Harold D. Lasswell, *A Pre-View of Policy Sciences* (New York: Elsevier, 1971), 39; and Harold D. Lasswell, Must Science Serve Political Power? *American Psychologist* 25 (1970), 119.

choices, decisions, and actions—are not random but at most intendedly rational. Among the important outcomes of this quasi-evolutionary process at any cross section in time are human communities that share a family resemblance: They are similar with respect to some vulnerability factors but different with respect to others, and some communities are further along than others in reducing their vulnerabilities. Therefore, one priority for intensive inquiry is expanding the range of informed policy alternatives available in selected communities and assisting in the evaluation of policies field tested by each of them. We consider these roles in the policy process in the next section. Another priority is harvesting experience from particular communities, making relatively successful policies anywhere available for diffusion and possible voluntary adaptation by similar communities elsewhere and available to higher-level officials motivated and able to support what works on the ground. These are functions of community networks that we consider in the third section. Intensive inquiry must be open to details that are possibly significant locally but typically lost in the absorption of uncertainty through statistical aggregates or general relationships. With due allowance for human insight, choice, and decision, evolution is a better template for policy-relevant climate science than mechanics.

Procedurally Rational Policy

Second, in terms of policy, we propose more emphasis on *appraisals* in policy processes, especially in the aftermath of climate-related disasters that reveal specific problems of vulnerability and open windows of opportunities for corrective actions. Corrective actions are constrained by the historical context of each community in the short run. They terminate failed policies and build on relatively successful ones in a series of approximations. To accommodate inevitable uncertainties, they also rely on portfolios of policies, reserves and redundancies, learning by doing, and the like. A disaster and major proposals for corrective action often activate diverse interests that must be balanced or integrated to advance the common interest of the community; politics are unavoidable under these circumstances. In these policy processes, scientists and policymakers can work collaboratively toward overlapping aims, sharing differently informed insights on the same context. But scientific research is not a substitute for local knowledge in the community, nor is scientific research a necessary condition for rational policy decisions by the community. Such processes can adapt policies to differences and changes in context, including surprises, and to uncertainties that inevitably expand as the time horizon for planning extends further into the future. This approach to policy is called "procedural rationality."

Procedural rationality is an adaptation to profound uncertainties arising from complexity in the world at large and human cognitive constraints, among other resource limitations. Cognitive constraints include limited time and attention, knowledge and information, and other factors subsumed under the principle of bounded rationality. They directly affect policy appraisal and intelligence (including plans) for corrective actions. (Policy intelligence "includes the gathering, processing, and dissemination for the use of all who participate in the decision process."[77]) They indirectly affect the allocation of funds, authorities, and other scarce resources through corrective actions.[78] Procedurally rational policies can be improved through intensive inquiry in each community. What works on the ground in each local community can inform policies elsewhere, in similar local communities and at higher levels, in decentralized decision making. All of this occurs within the broad channels set by invariant laws of nature where, as Gould put it, "contingency dominates and the predictability of general forms recedes to an irrelevant background."[79] Among the major contingencies in adapting to climate change are individual choices and collective decisions intended to realize preferred outcomes—values, in other words—amid a plethora of possible outcomes. Choices and decisions tend to reveal subjective values and identifications and their supporting expectations.

Rationality in Practice

Consider selected case material in Chapter 2 relevant to procedural rationality. In Nepal in 1998, the possible collapse of a moraine dam containing the rising waters of Lake Tsho Rolpa prompted the national government together with local and international participants to reconsider existing policies. To avert an impending disaster, the coalition redirected limited resources, including attention, from other concerns to corrective action to reduce local vulnerability. And they succeeded in advancing the common interest through action on a portfolio of policies: a gate to control release of water, an early warning system, and emergency management exercises. (Any politics involved were omitted from the IPCC's account.) These actions were considered the first but not the last in a series to avert catastrophic collapse of the dam as nearby glaciers continued to melt.

Variations on the pattern turned up in other cases as well. In the Pacific in 1997, PEAC's forecast of impending drought from the El Niño of the century prompted officials in the U.S.-affiliated Pacific islands to reconsider their respective water policies and divert attention and other resources to corrective actions. We lack information on any local politics involved and the details of subsequent corrective actions; we suspect that islanders are still building on what they learned in 1997–1998. We do know that PEAC capitalized on its success by making the transition from a pilot project to operational status around 2001. Similarly, droughts in Australia in 1994 and 2002 prompted officials in Melbourne to reconsider their water policies and take corrective action on a portfolio of policies that significantly reduced water demand. A third drought in 2006 prompted a second portfolio of policies to reduce water demand *and* increase water supplies, including controversial plans for a desalination plant. Similarly, action to reduce Shishmaref's vulnerability through relocation of the village has taken the form of a series of partial relocations and more ambitious plans prompted by appraisals of vulnerability following extreme weather events, including severe storms that damaged the community in 1973 and 2002. More generally, prompted by losses of seals and then cod in west Greenland waters, Sisimiut adapted relatively successfully to other resources through a series of local policy decisions sometimes assisted by Denmark; scientific research apparently was not a significant factor.

The Barrow case in Chapter 3 includes more detailed material relevant to procedural rationality. Recall that the great storm in October 1963 prompted multiple emergency responses and actions to reconstruct lost services and facilities. The big storms of September 1986 washed the remains of an ancestor out to sea and damaged other things of value, prompting the North Slope Borough Assembly to declare a disaster and state of emergency and to appropriate funds to prevent and minimize further losses. The big storm of August 2000 sank the dredge and caused other damage, leading to termination of the Barrow and Wainwright Beach Nourishment Program. The mayor cited that storm when he announced and justified a joint feasibility study with the Army Corps to reduce storm damage. Our reconstruction of the great storm of October 1963 drew attention to potential but not necessarily imminent vulnerabilities, informing distributed policies to reduce specific vulnerabilities. In these and other instances, when actual or projected disaster prompts corrective action, it typically does so by changing matter-of-fact expectations about threats to existing values, opportunities for realizing existing values, or both—not by significantly changing values themselves in the short term. The values activated by disaster-related changes in expectations in Barrow were not always transparent, but they were multiple and context specific, even if described in general terms. In addition to storm damage reduction, the existing values included feasibility and effectiveness, more local jobs in the modern economy, and sustained cultural heritage. The latter includes self-determination, artifacts, subsistence hunting rights and opportunities, access to the sea, and unrestricted views of the sea.

Storm damage protection was a generally accepted value, but major plans and other proposals expected by some to reduce storm damage turned out to be controversial. The major plans were the beach nourishment program and the Army Corps's proposed revetments and dike especially; less visible were proposals for federal flood insurance, planning and zoning restrictions on further building in vulnerable coastal areas, and relocation inland of structures damaged or destroyed in those areas. These proposals activated multiple community interests, initiating a process of giving and withholding support for them—in other words, a political process attempting to reconcile the diverse interests of community members self-identified, for example, by family, crew, and village. Evidently, community members expected such proposals to have major consequences for multiple community values in addition to storm damage reduction. (You cannot change just one thing, according to an old adage.) In contrast, relatively modest distributed policies to reduce storm damage were handled primarily as technical matters that attracted the attention and engaged the interests of relatively few community members. These included an emergency management exercise, the design and location of individual buildings and other infrastructure, the construction of the utilidor-section seawall to protect the landfill in 1987, and Concertainers to extend that seawall and protect part of the bluff in 2004. Both modest and major corrective actions were taken despite scientific uncertainty about the future of big storms and their impacts on Barrow.

Where corrective actions have succeeded in reducing vulnerability to storm damage, they have done so in large part because action was procedurally rational. Each new set of policy alternatives proposed in the ongoing joint feasibility study appears to be more promising as an adaptation of means to ends, and vice versa, than the previous one. Much of the improvement can be attributed to learning from the experience of the earlier beach nourishment program and from successive appraisals by the people of Barrow, the Army Corps, and others. Meanwhile, in the distributed effort in Barrow, multiple policies in parallel already have reduced vulnerability to storm damage regardless of how the joint feasibility study turns out. Reliance on portfolios of multiple relatively modest policies, in series and in parallel, makes overall progress on storm damage reduction less dependent on the success of any one of them. In the great storm of October 1963 Barrow capitalized on redundancies and reserves to reduce losses and expedite recovery. Although the airstrip at NARL was destroyed early in the storm, the airport under construction provided back-up to fly in emergency relief and other supplies. Much of Barrow's fuel oil supply, especially critical for heating in winter, was destroyed by the great storm, but enough reserves survived to supply the community for several months. These and other procedurally rational means of reducing losses, despite scientific uncertainty, were largely a matter of common sense in Barrow. But they also have a theoretical pedigree reviewed below, and more importantly, can become a matter of conscious strategy to improve future outcomes.

For the research project in Barrow we took a procedurally rational approach by design and practical necessity; we could not reliably predict research outcomes in detail. In 1999, in collaboration with a number of colleagues, we submitted a proposal for a Small Grant for Exploratory Research (SGER) to inform design of a larger project. The proposal stated, "The primary goal of the project is to help stakeholders clarify and secure their common interest by exchanging information and knowledge concerning climate variability on seasonal and decadal time scales . . . on the North Slope of Alaska, and particularly the vulnerable coastal region." To design a larger project with this primary goal, we proposed meetings with local groups in Barrow for answers to the question: "What are the understandings of past and future climate events among the people of the North Slope, and what policies and policy processes have they used to cope with those events?"[80] For a week in August 2000, we had meetings in Barrow to obtain answers that provided a historical baseline for designing a larger project. The proposal, submitted in September 2000, retained the primary goal and meetings with local groups in Barrow but elaborated additional means. These included the description of relevant trends, analysis of past climate variations and future scenarios using physical models, and various methods for an integrated assessment in support of local policy decisions. For the integrated assessment we proposed "initially to focus on coastal erosion and flooding in and around Barrow, and eventually to shift the focus as warranted by the progress of the research and the evolution of community concerns."[81] The proposal was supported by people in Barrow and by scientists in anonymous peer reviews, and it was funded on that basis by the National Science Foundation.[82]

Concurrent research on the experience of community-based initiatives elsewhere, part of what we now call "adaptive governance," informed our expectations about how to work collaboratively with people of the North Slope on adaptation to climate change. We understood that "[s]ome focus is necessary for the effective integration of scientific research and local knowledge on the causes and consequences of climate events, including policy responses to them."[83] The initial focus on coastal erosion and flooding in Barrow supported the quantitative description and analysis of relevant long-term trends, but it also led rather quickly to a specific focus on extreme weather events—including damages and policy responses associated with them—beginning with the great storm of October 1963. The common focus of attention engaged the diverse interests of policymakers in Barrow and scientists in the project; their participation was essential for the integration of differently informed perspectives and diverse capabilities. Annual group meetings in Barrow, usually in August, typically included technical presentations for self-selected officials and other community leaders in the morning, lunch and an open house for discussion in the afternoon, and a public presentation at the Iñupiat Heritage Center in the evening. These meetings were supplemented by intermittent contacts among individuals, presentations to high school science classes, and interviews on KBRW, the local public radio station. We helped sustain local participation in the project by sharing interim research payoffs in these contacts and meetings with people in Barrow, and by aligning further research with their evolving interests. In our judgment *what* we communicated—information within local interests and capabilities to act upon—was more important for working collaboratively than *how* we communicated it.[84]

The project's design also allowed flexibility to adjust lines of inquiry in response to what we learned each year and in accord with the primary goal, which was relatively fixed. Our initial interest in developing scenarios of future climate variations for the integrated assessment waned as research into natural factors, especially their conjunctions in big storms, revealed more complexity and uncertainty than predictability. Concurrently, our growing understanding of the human factors, quite limited at the outset, helped clarify our role. From inquiry into the human factors involved in big storms, the beach nourishment program and the joint feasibility study, and specific vulnerabilities amenable to distributed decisions, it became increasingly clear that many policy alternatives for reducing storm damage already had surfaced in Barrow's recent history, even though most of the community's attention and other planning resources had been invested rather narrowly in beach nourishment, revetments, and the like. Thus, we could help to expand the range of informed alternatives available to the community by reporting on Barrow's relevant past. Recommendations were unnecessary because people in Barrow proved to be quite capable of selecting whatever research results they found useful. Likewise, recommendations were unwarranted because we lacked confidence in our understanding of local interests and the possible consequences of recommendations for those interests. Avoiding recommendations was consistent with our preference for minimizing any influence in local politics.

As noted in Chapter 1, whether we helped the community advance its common interest is a judgment best left to the people of Barrow and the North Slope. However, the use of some of our results is an indication that they were policy relevant. Community members quite appropriately prescribed policies and assumed de facto responsibility for them—they had to live with the direct consequences. We took responsibility for the limitations of our results by making corrections and filling in gaps within our resource constraints. According to a senior planner for North Slope Borough, we made some progress both in pulling together more relevant knowledge and information on the great storm than was generally available and in drawing the policy implications for the community. In a signed review of an early manuscript focused on the great storm, he wrote, "There are very few, if any, individuals living on the North Slope or in Barrow who have even a fraction of the understanding of the events and policy implications" of the great storm as summarized in the manuscript.[85] The integration of different streams of historical and scientific information with local knowledge had the potential at least to fill gaps in the personal experience of the public and public officials in Barrow. Only the older members of the community had experienced the great storm directly, and none had experienced any of the big storms in so many relevant particulars. In any case, our experience working collaboratively with policymakers in Barrow represents a procedurally rational alternative to the linear model.

In contrast, under the linear model institutionalized in major climate science programs, the primary research goal is to reduce scientific uncertainty about the behavior of the total earth system through basic research unfettered by practical considerations. The assumption is that contacts with policymakers are unnecessary at best and a threat to objectivity at worst.

In Rep. George Brown's appraisal and in ours, this left politicians free to use scientific assessments as they saw fit, allowing scientific uncertainty to serve as an impediment to action in the United States at least. Meanwhile, the regime's quest at the international level for mandatory targets and time-tables for reductions in greenhouse emissions, legally binding in industrial nations, activated more powerful interests in opposition than in support. Because of politics and a lack of political will, among other limiting factors, the quest has failed so far to realize significant progress toward the target in Article 2 of the UNFCCC. In view of the resources invested in this approach relative to other approaches, the quest amounts to a single point failure, the equivalent of putting all the eggs in one basket. On the evidence available, the quest is hardly the "one best way" to reduce greenhouse gas emissions, despite the aspirations for technical rationality supporting it. Meanwhile, the quest neglects people on the ground and what they might do about climate change and its impacts within their own local interests and capabilities. "The danger," Al Gore recognized, "is that people will go from denial to despair without stopping in between to ask themselves what action they can take."[86]

Opening Policy Processes

Selected climate-related literature has recognized the limits of technical rationality, the importance of politics, and the value of collaboration between researchers and local policymakers. For example, at least three of Rayner and Malone's 10 suggestions for policymakers raised concerns about technical rationality as we understand it and moved toward procedural rationality including pluralistic politics:

- ■ *"Recognize the limits of rational planning"* largely because "[u]ncertainty is a pervasive condition of policy and decisionmaking." Uncertainty is perhaps "more properly characterized as indeterminacy. . . . We have inaccurate and conflicting theories about how and why people make choices, for themselves and in societies."[87]
- ■ *"Use a pluralistic approach to decisionmaking"* that "may appear to be irrational and conflictual, but the potential exists to make the most of diversity and the variety of decision strategies that diversity offers to decisionmakers." Policy must "accommodate different world views, institutional structures, levels and timescales" through linkages.[88]

■ *"View the issue of climate change holistically, not just as the problem of emissions reduction."* Rayner and Malone called into question the emphasis on emissions reductions under the UNFCCC, which in turn called into question "whether an exclusively rational technocratic approach to policy making is appropriate at all."[89]

According to Rayner and Malone, adaptation may be less amenable to technical rationality than emissions reductions but "more directly relevant to stakeholders. Adaptation is by nature a variegated response. . . . That is to say, adaptation is a bottom-up strategy that starts with changes and pressures experienced in people's daily lives."[90] This is consistent with the observation that "[p]eople will change when they have to, not when we tell them to."[91]

Rayner and Malone recognized two ways "to incorporate climate concerns into the everyday concerns of people at the local level and the big concerns of policymakers at the national level." In one way, "climate change . . . has to be shown to be a compelling threat that overshadows other policy demands. . . ." This is what damaging storms do by imposing changes and pressures on people's daily lives. In the other way, climate change "has to be integrated into the routinized decisionmaking frameworks of government organizations and agencies whose primary policy concerns (such as finance and energy) are widely recognized as compelling."[92] This is called "mainstreaming" in the *Summary for Policymakers* of scientific research on climate change adaptation by Working Group II in the IPCC's Fourth Assessment Report.[93] Examples include the distributed policies to reduce storm damage in Barrow, as distinguished from the joint feasibility study. Mainstreaming is implied in another of Rayner and Malone's 10 suggestions for policymakers: *"Incorporate climate change concerns into other, more immediate issues such as employment, defense, economic development and public health."*[94] According to contributors to one of the four volumes they edited, mainstreaming is important because "[c]limate policies per se are bound to be hard to implement meaningfully. This conclusion recasts the issue of compliance and implementation from the idea of a rational instrumental framework of evaluation, decision, and implementation to a continuous framework of interactive negotiation in which policy explicitly becomes the formalization of actions being undertaken by participating parties."[95]

Cash and Moser also suggested moving toward procedural rationality, urging assessors and decision makers "to employ adaptive assessment and management strategies—constructing long-term, iterative, experiment-based processes of integrated assessment and management." However, they acknowledged, "actual implementation of adaptive management regimes has been limited, and has exhibited varying success." In their diagnosis of these problems, two of the three "critical barriers to the implementation of adaptive management" have been value differences and politics—or in their words, "fundamental differences in how environmental resources are valued" and "threats to existing power structures and interests."[96] "Management" in both the scientific and adaptive management traditions often misleadingly connotes that the problem at hand is merely technical, that any issues arising from different interests already have been resolved. However, it is better to acknowledge such issues where they exist and to include the reconciliation of multiple interests as part of the problem to be solved in a procedurally rational policy process. This involves politics when a lot is at stake.

On the politics involved in climate policies, an essay by Michael Shellenberger and Ted Nordhaus stands out; it was front-page news in the *New York Times* early in 2005.[97] Their appraisal of the U.S. environmental movement focused on its preoccupation with technical rationality and neglect of the politics necessary to enlist nonenvironmental interests in support of action to mitigate climate change. They contended, for example, that in the 1990s, "the big environmental groups and funders put all of their global warming eggs in the Kyoto basket. The problem was that they had no well-designed political strategy to get the U.S. Senate to ratify the treaty. . . ."[98] In recent proposals by environmentalists to reframe global warming, the common element was "the shared assumption that a) the problem should be framed as 'environmental' and b) our legislative proposals should be technical. . . . The implication is that if only X group were involved in the global warming fight *then things would really start to happen.* The arrogance here is that environmentalists ask not what we can do for non-environmental constituencies but what non-environmental constituencies can do for environmentalists."[99] The authors contended that "the seeds of failure were planted" decades ago "at the height of the [environmental] movement's success. . . ."[100] However, "The entire landscape in which politics plays out has changed radically in the last 30 years, yet the environmental movement acts as though proposals based on 'sound science' will be sufficient to overcome ideological and industry opposition."[101]

Shellenberger and Nordhaus projected that environmentalism "will continue to be a special interest so long as it narrowly identifies the problem [of global warming] as 'environmental' and the solutions as technical."[102] As an alternative they helped organize an Apollo Alliance early in 2000. Initially, it "focused not on crafting legislative solutions but rather on building a coalition of environmental, labor, business, and community allies who share a common vision for the future and a common set of values."[103] They expected the vision to "set the context for a myriad of national and local Apollo proposals, all of which aim to treat labor unions, civil rights groups, and businesses not simply as means to an end but as true allies whose interests in economic development can be aligned with strong action on global warming."[104] In the Forward to the Shellenberger–Nordhaus essay, environmentalist Peter Teague drew attention to issues of responsibility for disappointing outcomes to date: "It would be dishonest to lay all the blame on the media, politicians or the oil industry for the public's disengagement from the issue that, more than any other, will drive our future. Those of us who call ourselves environmentalists have a responsibility to examine our role and close the gap between the problems we know and the solutions we propose."[105] (Here, introspection is not only an opportunity but a responsibility.) Similarly, columnist Thomas Friedman has envisioned the integration of climate change interests with priority interests in economic prosperity and national security through a green ideology.[106] And the private nonprofit Climate Group promotes "the development and sharing of expertise on how business and government can lead the way towards a low carbon economy whilst boosting profitability and competitiveness."[107] Such pragmatic politics recognize that environmental progress usually depends on accommodating multiple other interests impacted by environmental policy alternatives. Scientific assessments and, more generally, sound science are not the only means, nor often even the most effective means, for advancing the common interest in a democracy.

The politics of reconciling diverse community interests have been all but neglected in the scientific literature on climate change. As Rayner and Malone observed, in both the United States and on a global scale, "Political questions are often posed as technical questions that can be referred to experts without confronting the value differences that are the real origin of conflict."[108] This expectation was corroborated by our search of the Institute for Scientific Information (ISI) Web of Knowledge for every item published in *Nature* and *Science* that referred to "politics" or "political" in the context of "climate change" or "global change" or "global warming." For *Science* the search turned up seven articles, six news items, one editorial, one letter, and one interview, for a total of 16 items published from 1976 to 2007, inclusive. For *Nature* the search turned up a total of 15 items published from 1988 to May 2008, inclusive. By comparison, between 2001 and 2007, *Science* published on average 71 items *per year* referring to "climate change" or "global change" or "global warming"; *Nature* published on average 56 items per year. The trend in the number of such papers published annually is increasing sharply but cyclically; the number ramps up prior to deadlines imposed by an IPCC assessment, then drop off somewhat.[109] Despite the neglect of politics in climate science, they are a critical part of the total earth system and go a long way toward diagnosing two decades of disappointing outcomes documented in Chapter 1.

Other climate-related literature has recognized the importance of researchers working collaboratively with people on the ground. In a recent review of 40 years of hazards and disasters research, Dennis Wenger reaffirmed the "pioneering studies of Robert Yin and his colleagues [who] found that the adoption and utilization of research findings by practitioners was significantly improved when researchers and potential users seriously engaged in interaction and collaboration on all phases of the research process" beginning with definition of the problem. But "[u]nfortunately . . . this approach is rarely utilized within the social science research communities."[110] Similarly, an appraisal of the U.S. National Assessment supported stakeholder participation from the outset of a scientific assessment.[111] According to 10 researchers led by Granger Morgan, and informed by participants in their workshop, "Many believe that, in scoping out a policy-relevant assessment, one should begin by defining the bottom line from the target audience's (stakeholder) perspective. . . . [Workshop participants] suggested that more success could have been achieved through using a stakeholder definition of the scope and providing a consistent approach/framework for incorporation of these bottom lines."[112] An earlier modest proposal by one of us called for local, regional, and national decision makers to participate in the appraisal of global change research, not just the planning. In this proposal, competing teams of researchers present their results at periodic national conferences, and decision makers would select for further funding those teams that had best met decision makers' needs. If nothing else, this would motivate researchers to work collaboratively with decision makers in support of improvements in policy and policy making.[113]

Morgan and his collaborators also found in the U.S. National Assessment "a difference between the perspectives and information sources of global/national-scale analysts and regional/local-scale analysts" that did *not* reflect a difference in expertise. Both sets of differently informed viewpoints were considered valid: "It would be a mistake if [top-down] guidance material were to force regional people to ignore information that they have a valid reason to believe is better than what is being supplied."[114] The integration of relevant expertise should not be a monopoly of national or international experts. Similarly, Rayner and Malone concluded from an integrated assessment of social science research that "real people are not consistently experts or lay people. There are no universal experts and, in the civic arena, even the most modest lay person has some relevant expertise." They criticized the structuring of "communication as a unidirectional process in which expert knowledge is passed to the public either to alleviate its ignorance or redress its misperceptions." The unidirectional model tends to elicit political resistance from publics who believe that they or their local expertise has been ignored. Public information campaigns based on the unidirectional model "are bound to fail. Effective communication about climate change issues requires understanding of the frames of reference being used by all participants."[115]

An alternative to the unidirectional or "pipeline" model of communication in climate-related literature is based on "boundary organizations" at the interface among differently informed groups. It "suggests a more nuanced relationship between scientists and decision-makers, and proposes mechanisms that account for two-way interactions between science and decision-making and across scales."[116] Approximate equivalents are called "decision seminars" in the policy sciences and, with less emphasis on researchers and more on the representation of pluralistic interests, "community-based initiatives" in natural resource policy.[117] The common insight is that communication is not merely the transmission of words or word equivalents in one direction or another but knowing concurrence in their frames of reference. Establishing communication among people with differently informed perspectives requires interaction across a range of topics over a period of time. Only then can participants reliably anticipate how their comments will be understood—that is communication. Through direct contacts over several years, we and our colleagues made progress toward establishing communication with a few dozen community leaders in Barrow.[118]

The Behavioral Model

At a deeper level, this climate-related literature and proposals for more procedurally rational policies may be grounded in the behavioral model of bounded rationality. No one is more closely identified with the behavioral model than Herbert Simon; it was the foundation of his Nobel Prize in economics and his "sciences of the artificial" focused on design. For Simon,

"Design . . . is concerned with how things ought to be, with devising artifacts to attain goals."[119] Artifacts are human made, including the climate change policies of primary interest in this section and structures for decision making in the next section. Successful human artifacts adapt the "outer environment" of people to their goals, despite limited knowledge and information and cognitive capacity in the "inner environment" of people.[120] Of course, the point of inquiry and education is to relax those constraints, even though they cannot be eliminated. Design so construed is ubiquitous: "Everyone designs who devises courses of action aimed at changing existing situations into preferred ones. . . ." In addition, "Design so construed is the core of all professional training; it is the principal mark that distinguishes professions from the sciences."[121] In contrast to the sciences of the artificial concerned with what ought to be, "The natural sciences are concerned with how things are."[122] Simon's frame of reference clearly overlaps with Gould's as described in the previous section: "If natural phenomena have an air of 'necessity' about them in their subservience to natural law, artificial phenomena have an air of 'contingency' in their malleability by environment."[123]

The behavioral model describes how real people solve complex problems and explains their survival despite bounded rationality.[124] The most basic constraints on our cognitive capacity "arise from the very small capacity of the short-term memory structure (seven chunks) and from the relatively long time (eight seconds) required to transfer a chunk of information from short-term to long-term memory."[125] Nevertheless, a popular model of rationality in economics and related disciplines ignores these constraints and their consequences. Simon called it "the Olympian model," a reference to the omniscience of the gods of Mt. Olympus in ancient Greek tradition. Real people can apply the Olympian model only "to a highly simplified representation of a tiny fragment of the real-world situation"; the results depend less on the abstract model than on highly simplified empirical assumptions about the real-world situation.[126] "Within the behavioral model of bounded rationality," in contrast to the Olympian, "one doesn't have to make choices that are infinitely deep in time, that encompass the whole range of human values, and in which each problem is interconnected with all the other problems in the world."[127] As more realistic assumptions are taken into account, "the problem gradually changes from choosing the *right* course of action (substantive rationality) to finding a way of calculating, very approximately, where a *good* course of action lies (procedural rationality)" in an expanding maze of possible alternatives.[128] According to Simon, "The behavioral model gives up many of the beautiful formal properties of the Olympian model, but in return for giving them up provides a way of looking at rationality that explains how creatures with our mental capacities—or even with our mental capacities supplemented with all the computers in Silicon Valley—get along in a world that is much too complicated to be understood from the Olympian viewpoint. . . ."[129]

Within the constraints of bounded rationality, Simon observed, a person factors the environment into nearly independent problems. "Sometimes you're hungry, sometimes you're sleepy, sometimes you're cold. Fortunately, you're not often all three at the same time. Or if you are, all but one of these needs can be postponed until the most pressing is taken care of. You have lots of other needs, too, but these also do not impinge on you all at once."[130] The environment is amenable to factoring: "In actual fact, the environment in which we live, in which all creatures live, is an environment that is nearly factorable into separate problems."[131] This means that some problems or clusters of problems can be addressed satisfactorily and more or less independently of others, especially in the short run. Simon singled out three characteristics an organism needs to proceed rationally in such an environment:

- First, "it needs some way of focusing attention—of avoiding distraction (or at least too much distraction) and focusing on the things that need attention at a given time." The things that need attention are problems including missed opportunities.
- "Second, we need a mechanism capable of generating alternatives. A large part of our problem solving consists in the search for good alternatives, or for improvements in alternatives we already know."
- "Third, we need a capability for acquiring facts about the environment in which we find ourselves, and a modest capability for drawing inferences from these facts."[132] These capabilities are needed to focus on problems and to generate and test alternatives.

These characteristics interact in the boundedly or intendedly rational behavior of individuals, groups, and communities. Here we focus on procedural rationality in local communities factored out of larger-scale climate change problems; in the next section we consider the implications for decision making in larger structures.

The focus of attention in a community is structured by an institutionalized division of labor, the legacy of past decisions about clusters of problems—for example, in public safety, education, or finance—that are separable or nearly decomposable from other clusters. "Much of classical organization theory in fact was concerned precisely with this issue of alternative decompositions of a collection of interrelated tasks."[133] Each organizational unit, for example, attends to its own tasks in series, one at time, while the others proceed in parallel attending to their own tasks. The division of labor also stabilizes the search for better alternatives, ones that promise improvements over a historical baseline in the same unit or other norms inferred from the experience of comparable units. Baselines and other norms economize use of our limited cognitive capacity for problem solving; zero-based budgeting and related attempts to ignore history, and to start anew, do not. For example, taking the allocation of funds in last year's budget as a baseline simplifies preparation of next year's budget; taking the last great storm as a baseline simplifies preparation for the next one; and taking what comparable organizations have achieved as a norm simplifies any particular organization's estimates of what it can or should do. Thus, an institutional division of labor is an adaptation to cognitive constraints that attempts to make the most of the limited time and attention of officials and other community members while cultivating and using their specialized expertise through distributed processing. Taken together, "institutions provide a stable environment for us that makes at least a modicum of rationality possible."[134]

But the allocation of attention at any cross section in time is subject to disruption. Simon postulated for each dimension of community life that "expectations of the attainable define an aspiration level that is compared with the current level of achievement." The level of achievement can be forced down abruptly by exogenous events (e.g., a natural disaster), producing perhaps enough dissatisfaction to reallocate attention to the new problems created and initiate a search for alternatives. Simon uses the term "satisficing" to describe the best that real people can do. Optimizing is out of the question in part because "[t]here is no simple mechanism for comparison *between* dimensions." For example, not every level of aspiration can be reduced to a single metric to serve as a target; similarly, the net level of achievement or satisfaction must be assessed in multiple dimensions that are often incommensurable. It is difficult to compare gains in apples and losses in oranges, for example. Hence, "the system's net satisfactions are history-dependent"—a series of adjustments in multiple dimensions that strive for overall improvement.[135] The logic of satisficing is rather straightforward: "An alternative satisfices if it meets aspirations along all dimensions. If no such alternative is found, search is undertaken for new alternatives. Meanwhile, aspirations along one or more dimensions drift down gradually until a satisfactory new alternative is found or some existing alternative satisfices."[136] Thus by implication at least, rationality in the behavioral model is conceived as the mutual adaptation of aspirations and alternatives: in other words, the adaptation of means to ends *and* vice versa. Note that aspiration levels are preferred outcomes—values—that are not independent of achievement levels, which are empirical matters of expectations. Normative and empirical considerations are interdependent in behavior and in the behavioral model of bounded rationality.

The behavioral model emphasizes feedback over predictions. However, "In simple cases uncertainty arising from exogenous events can be handled by estimating the probabilities of these events, as insurance companies do— but usually at a cost in computational complexity and information gathering. An alternative is feedback to correct for unexpected or incorrectly predicted events."[137] Feedback is emphasized partly because "the record in forecasting even such 'simple' variables as population is dismal" and because "we need to know or guess about the future only enough to guide the commitments we must make today."[138] Moreover, in Simon's assessment, "Few of the adaptive systems that have been forged by evolution or shaped by man depend on prediction as their main means for coping with the future. Two complementary mechanisms for dealing with changes in the external environment are often far more effective than predictions: homeostatic mechanisms that make the system relatively insensitive to the environment and retrospective feedback adjustments to the environment's variation."[139] Homeostatic mechanisms increase resilience and reduce dependence on short-range predictions. Examples include reserves and redundancies like those that mitigated damage and facilitated recovery from the great storm in Barrow in 1963. "Feedback mechanisms, on the other hand, by continually responding to discrepancies between a system's actual and desired states, adapt it to long-range fluctuations in the environment without forecasting."[140] The discrepancies are problems turned up in the appraisal process of a system.[141] Each real person, like the simulated agents in Holland's models of complex adaptive systems, and each community functions as a feedback system with somewhat different problems.

The search for a satisfactory solution to a problem in a nearly decomposable system involves "first, the generation of alternatives and, then, the testing of these alternatives against a whole array of requirements and constraints. There need not be merely a single generator-test cycle, but there can be a whole nested series of such cycles."[142] The ability to search successfully "depends on building up associations, which may be simple or very complex, between particular changes in states of the world and particular actions that will (reliably or not) bring these changes about."[143] These associations are equivalent to policies connecting goals and alternatives, not to projections assuming present policies remain the same. The maze of possible alternatives to be searched may be enormous; it "arises out of the innumerable ways in which component actions, which need not be very numerous, can be combined into sequences."[144] Hence, the problem-solving process "ordinarily involves much trial and error. Various paths are tried; some are abandoned, others are pushed further. Before a solution is found, many paths of the maze may be explored. The more difficult and novel the problem, the greater is likely to be the amount of trial and error required to find a solution. At the same time the trial and error is not completely random nor blind; it is in fact rather highly selective. . . . Indications of progress spur further search in the same direction; lack of progress signals the abandonment of a line of search. Problem-solving requires *selective* trial and error."[145] The criterion for selection is rationality, the mutual adaptation of ends and means. According to Simon, "Analogous to the role played by natural selection in evolutionary biology is the role played by rationality in the sciences of human behavior."[146]

To make the most of our limited problem-solving capacity in climate change policy, and despite profound uncertainty at all levels, consider a few of the many implications of the behavioral model. Simon himself concluded that "[p]redicting the exact course of global warming is a thankless task. Much more feasible and useful is generating alternative policies which can be introduced at appropriate times for slowing the warming, mitigating its unfavorable effects and taking advantage of favorable effects."[147] In addition, the behavioral model corroborates the logic of harvesting experience to cut down the size of the search for possible solutions to a community's problem: "We see this particularly clearly when the problem to be solved is similar to one that has been solved before. Then, by simply trying again the paths that led to the earlier solution, or their analogues, trial-and-error search is greatly reduced or altogether eliminated."[148] The behavioral model corroborates the logic of distributed processing, portfolios of policies, and opening the climate change regime: "In carrying out . . . a search, it is often efficient to divide one's eggs among a number of baskets—that is, not to follow out one line until it succeeds or fails definitely, but to begin to explore several tentative paths, continuing to pursue a few that look most promising at a given moment."[149] Finally, the behavioral model corroborates the logic of factoring the global climate change problem into thousands of local problems, each nearly decomposable from others in the global system. We have more to say about this in connection with constitutive decisions below. But note here limits to reason based on logic alone. "Reason, taken by itself, is instrumental. It can't select our final goals, nor can it mediate for us in pure conflicts over what final goal to pursue—we have to settle these issues in some other way. All reason can do is help us reach agreed-upon goals more efficiently."[150] Agreed-upon goals are typically matters of politics.

The behavioral model provides insight into an important political problem that arises when the costs of collective action to individuals are disproportional to individual rewards. For example, many parties to the UNFCCC have an obligation to reduce their emissions as part of the international effort to mitigate global climate change, but some apparently have a stronger interest in avoiding the costs of mitigation by free riding on the efforts of other parties. The behavioral model recognizes that such collective action problems are overcome through organizational loyalty, which is "perhaps better labeled *identification*, for it is both motivational and cognitive. The motivational component is an attachment to group goals and a willingness to work for them even at the sacrifice of personal goals." The cognitive component fosters a shared outlook among members of an organization by surrounding them with "information, conceptions, and frames of reference quite different from those of people outside the organization or in a different organization."[151] Both the motivational and cognitive components are apparent in case studies of local communities that have reduced their emissions and realized related local goals at the same time, including Ashton Hayes and Samsø. Similarly, case studies of adaptation show how the collective action problem has been overcome by PEAC and the U.S.-affiliated Pacific islands, Melbourne, and other places where nature has helped motivate, if not forced collective action to reduce losses from climate-related disasters. To overcome the collective action problem at the global level, the climate change regime might build on these local models to cultivate identifications with complementary national and international efforts. But this would involve focusing on local activists in a position to organize their neighbors, providing models and other resources relevant to their distinctive interests and capabilities, and attending to feedback from the bottom up.[152] The mass marketing of messages about a planetary emergency will not be sufficient, even if repeated with more urgency at higher volume. Especially in the information age, "The real design problem is not to provide more information to people but to allocate the time they have available for receiving information so that they will get only the information that is most important and relevant to the decisions they will make."[153]

Aspects of the behavioral model and the practical insights associated with it have been more or less independently discovered by others with different perspectives using different vocabularies. In a classic critique of scientific management in 1969, Martin Landau drew attention to the Olympian model in public administration, the pretense that "certainty exists as to fact and value, instrumentation and outcome, means and ends. All that needs to be known is known and no ambiguities prevail."[154] However, "Conditions of certainty, or near certainty, appear to be rare facts in the life of a public agency, and when they exist, their scope is likely to be severely restricted." Indeed, uncertainty prevails. It is more rational "to construct organizations so that they can cope with uncertainty as to fact and disagreement over values. If the facts are in question, then we simply do not have knowledge of the appropriate means to use in seeking an outcome." This means that plans are only hypotheses, however elaborate and comprehensive they may be. "It is, therefore, an obvious and 'rational calculus' to employ a pragmatic and experimental procedure: that is, a policy of redundancy which permits several, and competing, strategies to be followed both simultaneously and separately. . . . [A]ny attempt to 'program solutions' prematurely is the height of folly." If there is disagreement over values, "and the parties involved value the existence of the organization, it makes good sense to compromise, to negotiate differences, to be 'political'. . . ." Negotiation "requires the redundancy of ambiguity, surplus meaning, for it is precisely such surplus that permits values to overlap the parties in dispute providing thereby some common ground for agreement." Landau concluded that redundancy "provides safety factors, permits flexible responses to anomalous situations and provides a creative potential for those who are able to see it." However, scientific management in various forms sought to eliminate redundancies, to streamline agencies on behalf of efficiency, and to control from the top down. "Time after time, control systems, imposed in the name of error prevention, result only in the elimination of search procedures, the curtailment of the freedom to analyze, and a general inability to detect and correct error."[155]

The social psychologist D. T. Campbell elaborated the notion that policies are only hypotheses to be evaluated through field testing. (This is the core idea of what later became known as adaptive management.) In 1969 Campbell proposed "Reforms as Experiments" to counter commitments to finding and imposing "the one best way," a characteristic of scientific management: "If the political and administrative system has committed itself in advance to the correctness and efficacy of its reforms, it cannot tolerate learning of failure. To be truly scientific we must be able to experiment. We must be able to advocate without that excess of commitment that blinds us to reality testing."[156] He suggested we advocate the seriousness of the problem for which alternative solutions were plausible: "By making explicit that a given problem solution was only one of several that the administrator or party could in good conscience advocate, and by having ready a plausible alternative, the administrator could afford honest evaluation of outcomes."[157] Campbell recognized that the ideal controlled experiment is rarely possible in field testing policy alternatives. He suggested "a self-critical use of quasi-experimental designs" and explained pragmatically that "[w]e must do the best we can with what is available to us."[158] This is an approach "in which we learn whether or not these programs are effective, and in which we retain, imitate, modify, or discard them on the basis of apparent effectiveness on the multiple imperfect criteria available."[159] Campbell also suggested staged innovation to capitalize on logistical necessity. "Even though by intent a new reform is to be put into effect in all units, the logistics of the situation usually dictate that simultaneous introduction is not possible."[160] Adapted to a world of contingency rather than necessity, staged innovation implies the harvesting of experience from each stage to improve the stock of field-tested alternatives available for subsequent stages. Campbell also emphasized the replication of social experiments. However, if the results of social experiments are contingent on differences and changes in context, it would be more rational to adapt policies rather than to replicate them.

A concern for improving procedural rationality in climate policy should not exclude nonrational factors in human choice and decision. Recall that Simon recognized the important role of emotional factors, particularly in focusing attention. Emotional factors are manifestations of unconscious struggles that are intensified for many people at the same time by widespread disturbances. As Lasswell once put it, "Famine, pestilence, unemployment, high living costs, and a catalogue of other disturbances may simultaneously produce adjustment problems for many people at the same time. One of the first effects is to release [emotional] affects from their previous objects, and to create a state of susceptibility to proposals. All sorts of symbols are ready, or readily invented, to refix the mobile affects. . . . The prescriptions are tied up with diagnoses, and the diagnoses in turn imply prescriptions."[161] Reason has a difficult time under these circumstances; the process easily can converge on emotionally satisfying solutions rather than rationally effective ones. Perhaps these dynamics shed some light on why several years of Congressional testimony warning about global warming did not gain traction until James Hansen's testimony during a widespread disturbance, the extreme heat wave and drought in eastern North America in June 1988. In any case, nonrational factors should be included in the picture to clarify the task. In general the task is "discovery of the *means by which all who participate in a policy-forming and policy-executing process can live up to their potential for sound judgment*."[162] Compare Rayner and Malone's recommendation with respect to climate change: "Instead of trying to make the world conform to the normative tenets of the rational choice model, we should attempt to understand how decisions really are made (outside of well-behaved markets) and shape information pertinent to decision makers and their available options."[163]

This recommendation is consistent with Saul Alinsky's decades of experience as a community organizer. Writing in 1971, Alinsky noted that "if people feel they don't have the power to change a bad situation, *then they do not think about it*."[164] This helps explain the remarkable passivity of the general public with respect to global climate change, which is a bad situation, if not a planetary emergency as described in messages repeated over and over again in the last two decades. However, Alinsky continued, "Once people are organized so that they have the power to make changes, then, when confronted with questions of change, they begin to think and to ask questions about how to make the changes."[165] This helps clarify the significance of local leaders in the models of mitigation and adaptation documented in Chapters 2 and 3: They organized their own communities to take on those nearly decomposable parts of the global climate change problem within their local interests and capabilities. "It is when people have a genuine opportunity to act and to change conditions that they begin to think their problems through—then they show their competence, raise the right questions, seek special professional counsel and look for the answers. Then you begin to realize that believing in people is not just a romantic myth."[166] Alinsky also anticipated much of what has been learned about communication in the climate-related literature, as reported above. Alinksy insisted, for example, that "[p]eople only understand things in terms of their experience, which means that you must get within their experience to communicate. Further, communication is a two-way process. If you try to get your ideas across to others without paying attention to what they have to say to you, you can forget about the whole thing."[167]

To make the most of our limited capacity for sound judgment in climate policy processes, consider some working criteria from the policy sciences. They assume a common-interest standpoint for each community at any level. Box 4.2 summarizes criteria only for the intelligence and appraisal functions, the two in which scientists, other experts, and reporters often are most influential. Readers might find the criteria for all seven functions (listed in Box 4.2) useful in clarifying their own preferences for policy processes in general and for reconsidering climate policy processes in that light. As an introduction to such introspection, consider the regime's quest for mandatory, legally binding targets and timetables in light of the first criterion under appraisal, 7.1 *Dependability and Rationality.* The regime's quest to stabilize concentrations of greenhouse gases in the atmosphere at safe levels is a policy and criterion agreed upon within the epistemic community but is controversial within and outside the Conference of the Parties to the UNFCCC, where economic growth and other criteria prevail in practice. Within the epistemic community, disappointing outcomes with respect to the stabilization criterion are acknowledged, with formal and effective responsibility often attributed to a lack of political will in general and to the political influence of the ExxonMobil's of the world in particular. (This was made explicit in Al Gore's *An Inconvenient Truth.*) Such appraisals tend to divert attention from introspection about the effective or formal responsibilities of the epistemic community and what the community might do differently and better. Readers identified with the epistemic community might use their own criteria or those under intelligence for evaluation of the regime's quest in comparison with exceptions to scientific management in the history of the regime and with the proposals in Box 1.1. The potential advantages of the proposals with respect to intelligence criteria—dependability, comprehensiveness, selectivity, creativity, and openness—stem primarily from factoring the global climate change problem into thousands of local ones.

In summary, the proposals accommodate our limited capacity for sound judgment to advance the common interest—whether taken as Article 2 of the UNFCCC or, as we prefer, reducing net losses to things we humans value. The climate change problem as defined in any local community is much more tractable for scientific, policy, and political purposes than the global problem. In that community, fewer vulnerability factors need to be taken into account, and local interests beyond reducing climate-related losses can be included. This makes it easier to find common ground on policy, enabling action and field testing of agreed-upon policies. Field testing provides rather direct and timely feedback to improve the policy process and outcomes in a series of approximations over time. Additional advantages can be expected from many diverse communities working in parallel on their own unique problems with their own unique resources. This transforms the diversity of interests and capabilities across communities from a liability into an asset

Box 4.2. Criteria for Ordinary Policy Processes

Intelligence and Appraisal

The conceptual model of ordinary policy decision processes in the policy sciences distinguishes seven functions and their outcomes: intelligence, promotion, prescription, invocation, application, termination, and appraisal. This box summarizes criteria for only two of them.

"The *intelligence* outcome includes the gathering, processing, and dissemination of information for the use of all who participate in the decision process." Examples include projections and plans, or other information reported (or not reported) to decision makers. Criteria:

1.1 *Dependability*. "Statements of fact that are made available to other members of the decision process are dependable; and if there is doubt, an indication is given of probable credibility."

1.2 *Comprehensiveness*. Outputs are inclusive of the goals sought; favorable and unfavorable trends; major conditioning factors, including controversial ones; projections relevant to goals and objectives; and the benefit, cost, and risk of each alternative.

1.3 *Selectivity*. "Outputs are related to perceived problems. . . . Priorities are indicated when a problem is imminent and important according to the values at stake."

1.4 *Creativity*. "New and realistic objectives and strategies are compared with older or less realistic alternatives."

1.5 *Openness*. Inputs pertinent to both immediate and long-range problems are obtained by consent from the public or particular groups or individuals. Outputs are broadly disseminated.

However, outputs are "closed to unauthorized participants for appropriate periods" and when "lawful goals would be compromised" by disclosure "or when avoidable deprivations are imposed on third parties."

"The *appraisal* outcomes characterize the aggregate flow of decision according to the policy objectives of the body politic, and identify those who are causally or formally responsible for successes or failures." Examples include policy evaluations and assessments. Criteria:

7.1 *Dependability and Rationality*. "The policies and the criteria are agreed upon. . . . The data are dependable. . . . The explanatory analyses are relevant and explicit. . . . The imputations of formal responsibility are explicit."

7.2 *Comprehensiveness and Selectivity*. "These are especially pertinent to the appraisal of total impact" on society of agreed-upon policies. As a guide to total impacts consider the values of power, enlightenment, wealth, well-being, skill, respect, rectitude, and affection.

7.3 *Independence*. "Appraisers are insulated from immediate pressures of threat or inducement, and either involve the entire context or representatives of third-party judgment. . . . Internal appraisers are supplemented by external appraisal."

7.4 *Continuity*. "Although intermittent appraisals remobilize needed attention and support, the effects are greatest when basically continuous."

Source: Selected and adapted from Harold D. Lasswell, *A Pre-View of Policy Sciences* (New York: Elsevier, 1971), 28–29, 85–97.

for understanding and action: Among other things, experience harvested anywhere can expand the range of informed choices available elsewhere. This also makes global progress independent of the success of any one policy initiative, and dependent only on enough successes to sustain the quasi-evolutionary process. Overall, the proposals would allow people on the ground to participate in reducing losses from climate change within their own interests and capabilities, to take responsibility for their decisions, and perhaps on that basis eventually to identify with the regime and accept its global objective as their own. Is this enough to meet what Al Gore called "the planetary emergency"? No one knows, or can know, the answer. The rational response is to consider the range of approaches available rather than to bet the planet on any one of them.

Decentralized Decision Making

Third, we propose opening centralized, top-down decision making to the *experience* of local communities, working in parallel, that have field tested climate policies in their own unique contexts. The diversity of contexts across communities as they evolve through time is an asset for generating creative policy alternatives; experience on the ground is essential for selecting what worked according to local policy goals. Case studies of what worked on the ground can be diffused directly from each local community for possible adaptation by similar local communities organized formally or informally as *networks* (see Box 1.1). Networks of similar local communities can also clarify what external resource needs they have in common, to advise central authorities who are willing and able to support what works on the ground but lack understanding of realities on the ground. As such, many communities proceeding from the bottom up can influence the resource allocation decisions of central authorities from the top down. Those decisions are matters of politics; they must reconcile shared interests in local communities and their networks with the interests of state, national, or international communities. This approach to constitutive policy is *decentralized*. It increases at the margin the power of local communities. It can also increase at the margin the power of those higher-level officials who choose to cooperate with local communities.[168]

Decentralized decision making is an adaptation to cognitive constraints on the capacity of central authorities to govern large systems. The significance of these constraints increases with the scope and domain of their jurisdictions. Decentralization encourages intensive inquiry and the adaptation of policies on a procedurally rational basis; it depends in turn on intensive inquiries and procedurally rational approaches. Decentralization can be initiated by relaxing central control over the attention frame, particularly in the intelligence function in the policy process. This expands the range of informed alternatives for people on the ground and at higher levels in the structure of decision making. The problem, as Rayner and Malone observed in 1998, is that "almost all of the climate change policy research and analysis is aimed at high-level policymakers."[169] That is still largely the case. Research and analysis for policy purposes is intelligence, one of many bases for power understood as participation in making important decisions. As noted in Chapter 2, only formal power is centralized under the UNFCCC. Effective power is dispersed among the parties to the UNFCCC, but it could be decentralized to include participation by local communities. What works in practice then could be authorized more explicitly.

Decentralization in Practice

Consider briefly selected case material in Chapter 2 relevant to decentralization. Through the U.S. National Assessment of the Potential Consequences of Climate Variability and Change, federally supported researchers directly engaged stakeholders in various locales and regions. This could have given voice and influence to people on the ground in federal planning processes, but the U.S. National Assessment was suppressed during the Bush Administration. Through the Regional Integrated Sciences and Assessments (RISA) program, NOAA officials and researchers sought to expand the range of informed alternatives for regional decisions makers, taking their contexts into account. This recognized the importance of regional decision makers and sought to support their decisions through research; however, evidence of RISA's influence on regional or federal policy decisions still appears to be anecdotal. In the Pacific, PEAC informed local decision makers about impending drought from the 1997–1998 El Niño and catalyzed Pacific islanders to reduce local impacts by taking control of their own local water demand and supply policies. PEAC also lobbied central authorities in the United States and Japan for resources to support local needs and policies in the participating islands, and succeeded in some instances. Similarly, in Nepal, the national and international effort to deal with the catastrophic threat of the rising waters of Lake Tsho Rolpa recruited local villagers into the design of the early warning system and the conduct of emergency management exercises. It is not clear whether local villagers participated in other decisions, including the gate constructed for controlled release of water from Tsho Rolpa. The important point is not the locus of leadership but the extent to which it serves the common interests of diverse communities.

In other cases, leadership came primarily from the bottom up. In west Greenland, Sismiut's relatively successful economic transitions were based on local entrepreneurs and self-reliance, as well as some success in lobbying the Danish government for support of the local fishing industry. In Melbourne, amid continuing drought, like-minded professionals in and outside Melbourne Water coordinated and cooperated through an informal network to influence scientific reports, green papers, white papers, and eventually major amendments to local water policies. The general public participated extensively, especially through submissions and meetings on the green paper in 2003. On the mitigation side, national officials sparked the interest of Garry Charnock in Ashton Hayes and Søren Hermansen on Samsø, but these local leaders of community-based initiatives had little guidance or support from national officials. Similarly, ICLEI was primarily an initiative of cities organized to reduce local greenhouse gas emissions and to influence higher-level decisions affecting sustainability. The expanding network grew from 13 North American and European cities in the Urban CO_2 Project in 1991 to 546 local governments worldwide in the Cities for Climate Change Protection program by 2006. ICLEI appears to be more successful in reducing local emissions than in influencing higher-level decisions. Meanwhile, a growing number of states led by California have developed or implemented mandatory emissions reduction programs in the United States. One of their interests is to catalyze federal legislation to limit greenhouse gas emissions.

The Barrow case in Chapter 3 illustrates the emergence of networking to expand the range of informed alternatives, but shows only initial steps toward organizing new networks to reduce storm damage reduction. In policy planning that followed the September 1986 storms, North Slope Borough contractors brought to Barrow practical experience with hard structures in Prudhoe Bay and with beach nourishment from around the world. Borough personnel in the Disaster Coordinator's office in Barrow were networking when they disseminated lessons from the Kaktovik blizzard of January 2005 for use in emergency management in other North Slope villages. The Army Corps of Engineers was networking when it used information on the performance of Concertainers in Kivalina and Wainwright to adapt its plans for storm damage reduction in Barrow after the briefing in August 2006.

Such efforts to prevent the replication of failed policy alternatives or to adapt more successful alternatives tend to arise more or less spontaneously and independently, even within a scientific management milieu. In general, those who believe they have successfully managed a problem are pleased to tell the story, and those who believe they face a similar problem are eager to listen. And if they have not been in direct contact, intermediaries typically have something to gain from bringing them together. Under these circumstances networking is a persuasive strategy, avoiding the passive political resistance, if not backlash often provoked by coercion. If networking were consciously conceived as a strategy, it would be easier to organize networks and realize more of the potential to expand the range of informed choices across communities. Barrow, for example, still has much to learn from experience in Shismaref on relocation and engagement with higher authorities and from experience in Tuktoyaktuk on the long-term performance of hard structures on the Arctic coast. Barrow in turn has experience to share with Shishmaref and Tuktoyaktuk, if there is interest there.

As a potential step toward organizing a network, recall Senator Ted Stevens's initiatives on behalf of his Native constituents beginning with the GAO report released in December 2003. The report surveyed the erosion and flooding problems of 184 coastal and inland Alaska Native villages and included brief case studies for nine of them. Senator Stevens followed up with field hearings in Anchorage in June 2004, a temporary boundary organization that brought representatives of the Native villages and state and federal agencies together as participants in the federal process. Senator Stevens sponsored subsequent legislation to correct the main problem identified in the GAO report, cost-benefit criteria and cost-share requirements that had disqualified the villages for federal assistance. The legislation authorized the Army Corps of Engineers to carry out relevant projects at full federal expense and appropriated modest funds for that purpose. At field hearings in October 2007, the senator vented his frustration that the Army Corps and FEMA had not implemented this legislation on behalf of Native villages and that no single agency had been designated to take the lead on the erosion issues. He advised his constituents to work together to prioritize their shared needs and thereby help determine how to spend what limited federal funds were available. But the villages had not met for that purpose according to testimony at the hearings, indicating a missed opportunity for self-empowerment through networking and networks to inform federal policy from the bottom up. This may be a more realistic and effective alternative than designating a lead agency. Without extraordinary overriding authorities and a blank check, even a lead agency would have to contend with other agencies having overlapping mandates and jurisdictions and other resources of their own.

The Barrow case in Chapter 3 also sheds light on decentralization through self-empowerment. Recall that during the colonial era, people on the North Slope could not participate in important decisions directly affecting them. By the 1960s, most of the important decisions were made by the Bureau of Indian Affairs, the Department of the Interior, the Atomic Energy Commission and other federal agencies, the state of Alaska, and international oil companies. Increasing encroachments and growing threats to aboriginal lands and rights provoked a political backlash that led to the organization of the Arctic Slope Native Association (ASNA) in 1965 and the statewide Alaska Federation of Natives (AFN) two years later. In one strategy employed for taking control of lands and monies expected from the settlement of Alaska Native claims, ASNA organized Native villages on the North Slope into the Iñupiat Community of the Arctic Slope (ICAS) in 1971. In another strategy, ASNA attempted to influence the Alaska Native Claims Settlement Act of 1971, but Congress in that Act allocated lands and monies to Native corporations rather than ICAS. ASNA succeeded in incorporating North Slope Borough as a first-class municipality in 1972. This strategy paid off in gaining legal authorities, property tax revenues, and other resources necessary for participating in decisions affecting Iñupiat culture and modernization, among other important values of people on the North Slope. Clearly, the distribution of power and other resources was not fixed.

Self-empowerment also occurred at the international level. The borough with the leadership of Mayor Hopson joined with communities in Canada and Greenland to establish the Inuit Circumpolar Conference (ICC) to protect and advance Inuit environmental interests as the oil and natural gas industry expanded in the circumpolar region. To restore subsistence bowhead whale hunting rights unilaterally withdrawn without prior notice by the International Whaling Commission (IWC), the borough in collaboration with local and regional Native corporations (UIC and ASRC, respectively) organized whaling captains from Arctic coast villages as the Alaska Eskimo Whaling Commission (AEWC). In collaboration with scientists and others, AEWC succeeded in persuading the IWC to replace the moratorium with a series of quotas that restored some aboriginal rights. Those inclined to dismiss decentralization in adaptive governance as politically impossible might reconsider in light of the evidence: With some help from the outside, the people of the North Slope—about 3,000, mostly Iñupiat, in 1970—empowered themselves through the organization of networks and coalitions. They gained a seat at the table on certain state, national, and international decisions, accepted compromises pragmatically when prudent or necessary, and sometimes prevailed despite vigorous opposition from powerful established interest groups. There is no single formula to replicate for empowerment, but there are models from the United States and elsewhere available for adaptation. In the North Slope model, the necessary conditions included losses and threats to basic community interests sufficient to motivate action and persevere against opposition; also included were leaders able to sustain Natives' identification with the collective effort. Together they were able to capitalize on latent opportunities for self-empowerment that are available in open, representative democracies.

Our project had no role in Barrow or on the North Slope until August 2000, and only a modest role in storm damage reduction after that. As previously noted, we sought to harvest experience from expert and local sources relevant to storm damage reduction in Barrow. We found that attention and other resources had been focused rather narrowly on hard structures and beach nourishment; then we sought to expand the range of informed alternatives for policymakers in Barrow. For that purpose we began preparation of a technical compendium of our results early in the summer of 2004, distributed an early version in connection with our fourth annual project meeting in Barrow that August, and distributed a revised edition in Barrow that fall. Concurrently, we were encouraged by Senator Stevens's initiatives to address erosion and flooding problems in Alaska Native villages, including the field hearings in late June 2004, as possible steps toward networking and informing and influencing federal allocation decisions from the bottom up. We began to consider a follow-up project based on networking that would expand our research beyond Barrow as envisioned in 2000, build on what we had learned in Barrow, and allow us to field test some possibilities for networking as another part of adaptive governance. To evaluate initial plans we invited representatives from four Native villages to participate in our fifth annual project meeting in Barrow in June 2005. Luci Eningowuk from Shishmaref, Earl Kignak from Point Hope, and Tom Brower III of Barrow participated in the meeting; Wainwright sent an observer. The three representatives agreed to participate in the project after we explained the basic design and rationale and reviewed some of our research in Barrow. After the meeting, all three inquired about our subsequent efforts, ultimately unsuccessful, to fund a networking project.

These cases indicate multiple parallel pathways to decentralization of decisions directly affecting local communities, while maintaining important roles for central authorities in guiding, coordinating, and supporting local policy decisions within enabling frameworks that already exist. There is no "one best way," no "best practice" that fits all circumstances, but there are improvements over historical baselines or other norms that fit particular circumstances—if they can be found. Empowerment of a local community can be initiated by researchers who engage people on the ground and expand the range of informed alternatives available to them. This is especially important where available alternatives have been restricted, unintentionally or not, through censorship and propaganda. And it is not necessary for every community to rediscover on its own what other communities have already learned about climate change adaptation or mitigation. Self-empowerment is initiated by a local community that takes the lead in integrating local and external resources to advance its common interest. Self-empowerment can be augmented if similar local communities organize as networks to share what they have learned and to clarify shared needs for resources from higher-level officials to support what works on the ground. Finally, empowerment of local communities can be initiated by higher-level officials, elected and appointed, who consider intelligence and support from the bottom up as resources for gaining power to pursue their legitimate interests and responsibilities in helping constituents adapt to climate change or reduce greenhouse gas emissions. The possibility of cross-level coalitions organized around such interests and responsibilities indicates that decentralization need not be a zero-sum power game. It can be a positive-sum game for finding common ground among participants across levels.

In contrast, without intelligence from the bottom up, officials near the top and center of a large decision-making structure rarely can have dependable pictures of realities on the ground. Because context matters, the complexity and diversity of information available on the ground are major cognitive challenges in themselves, challenges that grow in proportion to the scope and domain of jurisdictions. Moreover, information flowing from the bottom up through bureaucratic channels and into the center is subject to distortion through selection and aggregation, if not censorship and propaganda by subordinates who prefer to minimize interference from their superiors. Beyond the absorption of uncertainty, the limited time and attention of officials near the top may be preoccupied by bureaucratic or legislative rivals with overlapping interests and responsibilities, and by others to whom they are formally or effectively responsible. Officials unaware of the consequences of centralized decisions on the ground, or unconcerned about those consequences, sooner or later impose policies that generate passive resistance or political backlash from the bottom up. In doing so, they forego opportunities to gain additional intelligence and political support from the bottom up and turn politics across levels into a negative-sum game in which most players lose power.

Including Local Decisions

Some climate-related literature focuses on interactions across communities at the same level and across different levels in the international system. In 1998 Rayner and Malone drew attention to certain limitations of centralized decision making as formalized in the UNFCCC: "Diversity, complexity, and uncertainty will frustrate the search for top-down global policy making and implementation." The Response Strategies Working Group in the IPCC's First Assessment in 1990 also recognized the significance of diversity but focused on clarifying issues for an international agreement, which centralized authority in the UNFCCC. Rayner and Malone affirmed the need for high-level agreement on goals but suggested it was insufficient: "Social science research in all disciplines indicates that policy-makers should attempt to reach agreement on high-level environmental and associated social goals, then look for local and regional opportunities to use policy in various ways appropriate to the institutional arrangements, cultural values, economic and political conditions, and environmental changes." But people on the ground also had important roles as decision makers in their synthesis: "Whether or not humanity realizes the potential to get ahead and stay ahead of climate change impacts depends on what happens at the level of decision-making in households, firms, and communities."[170] They suggested decentralization, referring to an earlier chapter on "Institutions for Political Action" in their series. The chapter argued that "the bulk of climate change politics may have to devolve to the local level, if policies are to become effective in the informal institutional dynamics of individuals and households."[171] Decentralization is also a better way than centralization "to build responsive institutional arrangements to monitor change and maximize the flexibility of human populations to respond creatively and constructively to it." Flexibility was preferred because "decisionmakers cannot predict the unpredictable" as social change accelerates.[172]

In 2000 William Clark offered a compatible construction of past trends and future possibilities in global environmental governance, in which climate change was singled out as the "poster boy." Clark wrote that "a unique and possibly significant feature in the recent evolution of environmental governance may be the emergence of actor networks that cut across the conventional divisions of the Nye matrix." Joseph Nye had conceived divisions in the international system as a matrix of actors (private, public, and third sector) by levels (supranational, national, and subnational). From nation-states at the center of the matrix, the important decisions, the real "action," had been moving out to coalitions advocating particular programs. "These coalitions involve networks cutting across not only actors but also scales. The result is globe-spanning networks of knowledge and practice, connected to multiple local action coalitions that are individually attuned to the politics and ecology of particular places."[173] The significance of these and related trends may be decentralization: "[T]he real 'action' in environmental governance for the next several decades may occur more and more at regional rather than global or national scales." One factor was the convergence of multiple environmental pressures into distinctive problems at that level. Another was "political experience" that "increasingly demonstrates that it is at subnational scales that civil society is most energetically and effectively mobilizing to reassert democratic answers to how development and environment should be balanced for particular people and groups." If this is the shape of things to come, "our thinking about the nature of globalization and its implications for governance will have a long . . . way to go in keeping up with events."[174]

In 2003, Thomas Dietz, Elinor Ostrom, and Paul Stern reviewed the literature on "The Struggle to Govern the Commons." They observed that in his seminal article in 1968, Garrett Hardin had "missed the point that many social groups, including the herders on the commons that provided the metaphor for his analysis, have struggled successfully against threats of resource degradation by developing and maintaining self-governing institutions. Although these institutions have not always succeeded, neither have Hardin's preferred alternatives of private or state ownership."[175] Similarly, they observed that contemporary "[g]lobal and national environmental policy frequently ignores community-based governance and traditional tools, such as informal communication and sanctioning, but these tools can have significant impact."[176] Among many well-established "general principles for robust governance institutions for localized resources," the coauthors singled out three as particularly relevant to the governance of larger-scale resource problems including global climate change:

- ▪ "*Analytic deliberation.* Well-structured dialogue involving scientists, resource users, and interested publics, and informed by analysis of key information about environmental and human-environment systems, appears critical."
- ▪ "*Nesting.* Institutional arrangements must be complex, redundant, and nested in many layers. Simple strategies for governing the world's resources that rely exclusively on imposed markets or one-level, centralized command and control and that eliminated apparent redundancies in the name of efficiencies have been tried and have failed."
- ▪ "*Institutional variety.* Governance should employ mixtures of institutional types (e.g., hierarchies, markets, and community self-governance) that employ a variety of decision rules to change incentives, increase information use, and induce compliance."[177]

Community self-governance was conceived as part of a global and hierarchical system of governance that presents alternatives for officials with larger jurisdictions: "Larger scale governance may authorize local control, help it, hinder it, or override it. Now, every local place is strongly influenced by global dynamics." The coauthors also recognized that "[t]oo many strategies for governance of local commons are designed in capital cities or by donor agencies in ignorance of the state of the science and local conditions."[178]

More recently, in 2007, Thomas Wilbanks considered cross-scale linkages in climate policy for sustainable development. His synthesis of points "known at a relatively high level of certainty" included the following:[179]

- ▪ Personal interaction: "We know that decision-making based on broad societal participation requires personal interaction. . . . Clearly, spatial proximity contributes to decision-making as a social process, connecting with such issues as empowerment, constituency-building, and public participation." Information technologies have enabled electronic networks, relaxing the physical constraints of spatial distance.
- ▪ Mismatched boundaries: "We know that in many cases existing spatial-administrative frameworks, emerging from other concerns, are not necessarily a good fit with sustainable development systems and processes." As an example Wilbanks cited boundaries defined by rivers mismatched

to basin-wide problems. We would add mosaics of federal agencies mismatched to adaptation problems in local communities.

- Complementary potentials: "Within this complex pattern of often incompatible mosaics, we know that different scales tend to have different potentials and different limitations for action. To oversimplify considerably, local scales offer potentials for participation, flexibility, and innovativeness, while larger scales offer potentials for resource mobilization and cost-sharing."

Wilbanks noted the opportunity to integrate local-scale and larger-scale decision making, but emphasized cognitive and constitutive barriers: "[I]n fact integration is profoundly impeded by differences in who decides, who pays, and who benefits; and perceptions of different scales by other scales often reflect striking ignorance and misunderstanding."[180] Among the key factors for overcoming these barriers, in addition to mutual trust and leadership, were approximate equivalents to boundary organizations and networks: "Roles of intermediary third parties, facilitating cross-scale interactions through personal relationships and associated structures" and "Infrastructures for identifying and disseminating information about success experiences, so that individual cases generate benefits beyond their own narrow boundaries."[181] Success experiences are the more important ones because they motivate and inform action. Failures discourage action, especially if no better alternatives are available in the attention frame.

Wilbanks envisioned the integration of information and influence from the bottom up and assistance from the top down, perhaps to capitalize on complementary potentials: "Impacts (or concerns about impacts) at local and regional scales join together to push for actions at national and global scales; in fact, without such bottom-up encouragement, effective actions by larger scales tend to be limited in democratic systems of government. Large-scale actions are then shaped and fine-tuned in association with smaller-scale stakeholders and, in fact, in large part implemented through smaller-scale actions. . . ."[182] This would supply the simple democratic feedback suggested by the question posed earlier: "But how does it play in Peoria?" On the significance of answers, consider Herman Karl and colleagues who acknowledged as we do the need for a framework of national laws and regulations. "However," they continued, "the top-down approach that calls upon government to tell people what to do without meaningfully consulting them can cause resentment and create obstacles to creative solutions and durable policy because it exacerbates rather than reduces conflict."[183] They recommended collaborative approaches, requiring meaningful participation of citizens and government agencies, based on experience in natural resource and environmental policy in the United States. Consider also Madeleine Heyward's analysis of international climate change negotiations, which highlighted "the interconnectedness of countries' perceptions of equity and their particular national circumstances, and indicates the extreme difficulty of reaching international agreement on an effective coordinated approach to climate change." She suggested that "'bottom-up' approaches are currently more likely than the formal international framework to marshal the support necessary to achieve effective environmental outcomes."[184]

Some contributions to the literature on scale in climate policy claim that bottom-up initiatives are just as problematic as top-down initiatives, but for different reasons. Regardless of the validity of this claim, the sources cited above underscore the dependence of each kind of initiative on the other. Because top-down frameworks enabling mitigation initiatives are already in place, there is more to be gained in reducing net losses from climate change by developing the potential of community-based initiatives for both mitigation and adaptation from the bottom up. However, adaptation is problematic for scientific management because there is no single metric for gauging gains and losses across multiple value outcomes in each community or aggregating somewhat different values and value priorities across communities. Ian Burton and his colleagues understood metrics as a handicap for adaptation compared to mitigation. "The mitigation agenda has so readily infiltrated public policy thus far because it can be simplified to a common currency—the emission of GHGs—around which targets can be based."[185]

The Constitutive Process

At a deeper level, theories of the constitutive process—for decisions about decision making—corroborate decentralization in the climate-related literature reviewed above and provide additional insights. Herbert Simon, among others, claimed that complex systems tend to self-organize into hierarchies built up from elementary units. For Simon, complex systems were those "made up of a large number of parts that have many interactions. . . . [I]n such systems the whole is more than the sum of the parts in the weak but important pragmatic sense that, given the properties of the parts and the laws of their interaction, it is not a trivial matter to infer the properties of the whole."[186] Agent-based numerical models are complex systems in this sense; so are social systems built up from elementary units: "Almost all societies have elementary units called families, which may be grouped into villages or tribes, and these into larger groupings, and so on. If we make a chart of social interactions, of who talks to whom, the clusters of dense interaction in the chart will identify a rather well-defined hierarchic structure."[187] Dense clusters of interaction arise in part because no one has the cognitive capacity to attend to everyone else in a large social system, or even to come close. Consequently, the structure of a large social system tends to be nearly decomposable into subsystem components. In the short run the behaviors of the components are nearly independent of each other; "in the long run the behavior of any one of the components depends in only an aggregate way on the behavior of the other components."[188] Local communities formally subsumed under the climate change regime have this nearly decomposable structure, which supports factoring the global climate change problem into many smaller and more tractable local problems. Decentralization reduces local dependence on the center but still allows for aggregate flows of information, funds, and other resources between local communities and up and down across levels.

These aggregate flows of resources support scaling up successful community-based initiatives, to make their contributions significant at national or international levels. Rationales for scaling up were included in the climate-related literature reviewed in the previous section and in previous chapters.[189] President Clinton added another rationale in this observation: "Nearly every problem has been solved by someone, somewhere. The frustration is that we can't seem to replicate [those solutions] anywhere else."[190] However, many successful cases of scaling up have been documented; harvesting experience from them can improve success rates in quasi-evolutionary processes. Consider, for example, a review of scaling up public and nonprofit social programs that also considered their for-profit analog, commercial franchises. It found that "[i]n practice . . . replication is anything but a cookie-cutter process. The objective is to reproduce a successful program's *results*, not to slavishly recreate every one of its features."[191] Moreover, staged learning by doing cuts the challenge down to size: "[R]eplication is basically a process of planned evolution. Many replication efforts begin with expansion to a handful of sites, which can then provide useful lessons for broader initiatives. Learning from the planned—and unplanned—experiments that occur along the way is an important part of the implementation process."[192] Similarly, another review in international development policy concluded,

"The practical challenge is to identify a promising innovation or intervention for scaling up; to identify those elements that are context specific and those that are universal; to assure the universal elements are applied, but leaving room for local adaptation; and to evaluate, learn and change the approach as scaling up proceeds."[193]

Innovations diffuse through both centralized and decentralized systems, although the latter were discovered much later in this field of study. According to one leader in the field, Everett Rogers, the classical or centralized model dominated thinking and doing for decades, in large part because of the success of agricultural extension services. "In this model, an innovation originates from some expert source (often an R&D organization). This source then diffuses the innovation as a uniform package to potential adopters who accept or reject the innovation. The individual adopter of the innovation is a passive accepter."[194] However, Rogers reported, he "gradually recognized diffusion systems that did not operate at all like centralized diffusion systems. Instead of coming out of formal R&D systems, innovations often bubbled up from the operational levels of a system, with inventing done by certain users. Then the new ideas spread horizontally via peer networks, with a high degree of re-invention occurring as the innovations are modified by users to fit their particular conditions. Such decentralized diffusion systems usually are not run by technical experts. Instead, decision making in the diffusion system is widely shared, with adopters making many decisions. In many cases, adopters served as their own change agents."[195] Once again, we find complementarities between top-down and bottom-up flows of resources and opportunities for the climate change regime. For example, successful local models for the implementation of existing low-carbon technologies might be diffused to other local communities and to central authorities; in turn, the latter might use such models to develop more user-friendly and advanced low-carbon technologies to be diffused from the top down.

The emergence of networks for decentralized diffusion and adaptation of innovations has been recognized in recent studies of practitioners. Etienne Wenger and William Snyder called them "communities of practice," which are "groups of people informally bound together by shared expertise and passion for a joint enterprise." In contrast to the specific tasks of teams, the purposes of such groups are to "develop members' capabilities" and "to build and exchange knowledge."[196] Communities of practice are "as diverse as the situations that give rise to them."[197] Interactions within them are intermittent or regularly scheduled, face-to-face or by electronic network, but are somewhat more organized than networks of friends and acquaintances. From the experience of diverse cases, Wenger and Snyder concluded the following:

- "Communities of practice . . . are informal—they organize themselves, meaning they set their own agendas and establish their own leadership. And membership in a community of practice is self-selected. In other words, people . . . tend to know when and if they should join. They know if they have something to give and whether they are likely to take something away."[198]
- "The participants in these communities were learning together by focusing on problems that were directly related to their work. In the short term, this made their work easier or more effective; in the long term, it helped build both their communities and their shared practices—thus developing capabilities critical to the continuing success of the organizations" represented in the community of practice.[199]
- "In general, we have found that managers cannot mandate communities of practice. Instead, successful managers bring the right people together, provide an infrastructure in which communities can thrive, and measure the communities' value in nontraditional ways. These tasks of cultivation aren't easy, but the harvest they yield makes them well worth the effort."[200]

Evaluation of community activities in the scientific management tradition was not recommended because the results are often delayed and typically show up in the organizations represented by members rather than in the community of practice itself. Also, "it's often hard to determine whether a great idea that surfaced during a community meeting would have bubbled up anyway in a different setting." Wenger and Snyder recommend that "[t]he best way for an executive to assess the value of a community of practice is by listening to members stories, which can clarify the complex relationships among activities, knowledge, and performance." [201] The constructive role for higher executives is encouraging, guiding, and supporting community-based initiatives, but not controlling them.

Similar conclusions have been drawn from the emergence of international networks. According to Anne-Marie Slaughter, "We live in a networked world. . . . In this world, the measure of power is connectedness." Power based on connectedness "is not the power to impose outcomes. Networks are not directed and controlled as much as they are managed and orchestrated.

Multiple players are integrated into a whole that is greater than the sum of its parts—an orchestra that plays differently according to the vision of its conductor and the talent of individual musicians." In this networked world, "the focus of leadership should be on making connections to solve shared problems. . . . [G]overnment officials must . . . learn to orchestrate networks of these actors and guide them toward collaborative solutions." From Slaughter's perspective, governments are "gradually moving toward a more networked structure," although they have been slower than businesses and non-governmental organizations to understand twenty-first century challenges and to reform themselves accordingly. . . ."[202]

Networks enhance redundancy, which is vital for any kind of organization in a changing world. In his classic critique of scientific management, Martin Landau concluded that redundancy "provides safety factors, permits flexible responses to anomalous situations and provides a creative potential for those who are able to see it."[203] Redundancy depends on duplication and overlap. Duplication refers to independent, parallel components of a system performing the same function. For example, Barrow's distributed policies are parallel means to reduce local vulnerability to future storms; each is more or less independent of the other and of the outcome of the joint feasibility study. Similarly, the climate-related policies of industrial or Annex I nations are parallel means to reduce global greenhouse gas emissions; based on the diversity of emissions outcomes reported in Chapter 2, each nation's policies are indeed more or less independent of the other. Overlap refers to multiple functions that a component has the potential to perform. For example, heavy earth moving equipment served a transportation function when flooding during the great storm in Barrow made evacuation by other means impossible. Green building ordinances, among other urban policies, serve multiple functions including climate-change mitigation and adaptation.[204]Landau insisted, "If there is no duplication, no overlap, no ambiguity, an organization will neither be able to suppress error nor generate alternate routes of action. In short, it will be most unreliable and least flexible, sluggish, as we now say."[205] Yet for many decades, Landau complained, "such revitalization movements as Taylorism and scientific management . . . demanded the wholesale removal of duplication and overlap as they pressed for 'streamlined organizations' that would operate with the absolute minimal number of units that could possibly be employed in the performance of a task."[206]

Landau also recognized "more than one kind of 'rationality,' including the rationality of redundancy." He considered the makers of the U.S. Constitution "eminently 'rational'" in this sense. "They knew they were 'organizing' a system in the face of great uncertainty. . . . and sought a system which could perform in the face of error—which could manage to provide a stable set of decision rules for an exceedingly unstable circumstance."[207] The outcome, including formal amendments and judicial review, has served for at least 220 years. In Landau's characterization, the Constitution is "a patent illustration of redundancy. Look at it all: separation of powers, federalism, checks and balances, double legislatures, overlapping terms of office, the Bill of Rights, the veto, the override, judicial review, and a host of similar arrangements. Here is a system that cannot be described except in terms of duplication and overlap—of a redundancy of channel, code, calculation, and command."[208] In Landau's view, "the 'rationality' of politics derives from the fact that a system can be more reliable (more responsive, more predictable) than any of its parts" by arranging for the parts to keep each other in place through conflicts inherent in checks and balances and similar constitutive arrangements.[209] But no constitutive arrangements are perfect. In these American arrangements, "What is missing, because the framers did not provide for it, is a constitutional process for readily resolving these conflicts."[210] According to Robert Dahl, a leading democratic theorist, the result is a "new American political (dis)order" characterized by a weakening of institutions for encouraging conflicting interest groups to negotiate mutually beneficial policies and a lack of institutions for promoting representative and well-considered public opinions. Local community-based initiatives could provide the missing pieces, through decentralization that invites their fuller participation in those decisions most directly affecting them.

Landau insisted, "It is not possible . . . to determine whether a choice is rational except in terms of systemic context and goal."[211] The climate change regime as formalized in the UNFCCC authorizes both multiple goals including adaptation and redundant strategies including the precautionary principle, "no regrets," and development policies. The Conference of the Parties consists of a large number of nation-states with informal regional and other linkages among them. If the regime were reconceived as a nearly decomposable system, existing duplications and overlaps could be cultivated to improve the regime's overall performance and reduce its dependence on the success of any of its parts.[212] However, despite two decades of disappointing outcomes, the regime continues to monopolize attention and concentrate other scarce resources on one of its parts, as if it were the "one best way," if not the only way: the quest for mandatory international targets and timetables for the reduction of greenhouse gas emissions. The quest so far amounts to a *single point failure*, and will continue to fail so long as major nations withhold the resources necessary to control greenhouse gas emissions. Meanwhile, the regime's potential for evolutionary growth, recognized by Bodansky and others long ago, remains largely unexploited, and valuable suggestions have been largely ignored. For example, recall Granger Morgan's suggestion to rely more on independent initiatives by Norway, the Netherlands, and other leading nations to inform and motivate action elsewhere. Consider in this context Thomas Friedman's column on Denmark, which responded to the 1973 oil embargo with a portfolio of energy policies "in such a sustained, focused and systematic way that today it is energy independent." As part of its success, Denmark now uses wind turbines to supply about 20% of its electricity and manufactures about one-third of the world's wind turbines. The president of the Danish company that is the world's leading manufacturer of wind turbines suggested the potential, both tapped and untapped, for diffusing this path to reduce greenhouse gas emissions worldwide: "We've had 35 new competitors coming out of China in the last 18 months . . . and not one out of the U.S."[213]

To make the most of decentralized networks to reduce net losses from climate change, consider the dynamics of diffusion and adaptation of innovations. We have used these dynamics throughout this book as a heuristic springboard for insights into opening the climate change regime to other approaches, including adaptive governance.[214] Although consistent with practical prudence from earlier eras and other cultures, they were formulated as specifications of the maximization postulate by Lasswell and colleagues in 1952. After reviewing theories of political myth starting with Plato, the authors asked, "Under what circumstances is one myth rejected and another accepted? Under what circumstances is an established myth successfully transmitted?"[215] Their answers were propositions (technically, hypothesis schemas) on the probability, direction, and nature of change that apply to perspectives less basic than political myths. Thus, for "political myth" in the following propositions, one can substitute the basic climate change narrative shared within the regime's epistemic community, constructs of scientific management or adaptive governance, or various specific components of them including technologies, policies, and cases. In the propositions, "deprivation" refers to a decrease in value position and potential as assessed by the person or group in question—for example, losses of any kind experienced by victims of a big storm, including any missed opportunities they might have recognized. Conversely, "indulgence" refers to an increase in value position and potential, including gains of any kind and losses avoided.[216]

Consider the first proposition on the probability of change:

■ *"The probability of the rejection of a political myth is increased if the adherents experience deprivations rather than indulgence; if attention is directed toward a new myth whose adherents are indulged; and if early adherence to the new myth is followed by relative indulgence."*[217]

Each part of this proposition has practical implications for reform. The first part suggests that opening the climate change regime depends on those who feel deprived or dissatisfied. The dissatisfied include participants in the regime who assess the disappointing outcomes documented in Chapter 1 as significant deprivations for themselves and those with whom they identify, including future generations; people on the ground who assess the actual or impending impacts of extreme weather events on themselves or their communities as significant net losses; and, potentially, people relatively unconcerned about climate change but predisposed to accept mitigation or adaptation alternatives on a "no regrets" basis to address their priority concerns about economic, security, or moral problems, for example. (Shellenberger and Nordhaus considered the latter essential allies for action on climate change.) Conversely, it would be futile to rely on those outside or inside the regime who are satisfied with business as usual in response to climate change. Inside the regime are people for whom climate change issues are primarily means to other ends, such as publications and grants for scientific research, increases in budgets or authorities for climate-related programs, participation in meetings and negotiations on climate change, election to public office, or simply occupying the moral high ground. In any case, the first part of the proposition answers to common sense: Humans act in the expectation of reducing deprivations with respect to their goal values. For practical purposes, these values and supporting expectations must be matters of empirical inquiry into particular contexts.[218]

The second part of the proposition, on directing attention, also answers to common sense; the equivalent practical maxim for reformers is that "you can't beat something with nothing." A procedural implication is that cultivation of horizontal or peer networks can facilitate the diffusion and adaptation of innovations among those dissatisfied, and empower them at higher levels through organization on behalf of shared needs. Conversely, a strategy for centralized control of dissidents that can often work in the short term is to restrict horizontal networks and monopolize vertical channels of communication. A substantive implication of the proposition is that the quest for mandatory, legally binding targets and timetables since at least the Toronto Conference Statement in 1988 will tend to dominate, despite disappointing outcomes to date, unless and until other policy alternatives come to the attention of those dissatisfied with business as usual. One alternative, carbon taxes in various forms, has begun to emerge in the attention frame from economists and others skeptical of emission trading systems on empirical, administrative, or other grounds. A big push on advanced low-carbon technology R&D and geoengineering are additional alternatives that have begun to reemerge from scientists and engineers alarmed that the quest for targets and timetables will not generate sufficient greenhouse gas reductions in time to avert catastrophic climate change. Securing human rights at risk in climate change is a different approach, one that has only begun to emerge from an international reform movement that invokes the UN Universal Declaration of Human Rights.[219] And of course steps toward adaptive governance have begun to emerge at the periphery of the attention frame from those interested in the local opportunities or impacts of global climate change. This book attempts to integrate such steps as mutually reinforcing aspects of a promising approach, adaptive governance, and bring them to the attention of those dissatisfied with business as usual.

The third part of the first proposition underscores the importance of reinforcing the promise of any approach with actual payoffs. The equivalent practical maxim is that "nothing succeeds like success." This is part of the reason why, within the limited evidence available, we have focused on the payoffs of action for the people involved in various cases, from the Pacific islanders advised by PEAC to the villagers in Ashton Hayes, and included people who reduced Barrow's vulnerability to storm damage through distributed policies. Some initiatives in adaptive governance have failed in the past, and no doubt others will fail in the future. But sustaining progress does not depend on the success of any one initiative, or on all initiatives, for reasons Landau explained. Sustaining progress depends on enough successes to motivate and inform distributed processes of innovation, diffusion, and adaptation. The more comprehensive and detailed the documentation on such cases, the less likely it will be perceived as just more propaganda with little or no effect on predispositions supporting business as usual. Direct personal experience of an extreme weather event that damages a community is often more influential than comprehensive and detailed documentation in changing those predispositions.[220]

This leads to the second proposition on the direction of change:

■ *"The line of diffusion is from* communities *which are most indulged to the ones less indulged.* In terms of power, this implies that the strong will be copied by the weak."[221]

The equivalent practical maxim is that "no one copies a loser" as assessed by the participants in question. (From the standpoint of reformers, the less constructive maxim is "to get along, go along" with established practices.) In terms of skill, this implies that communities perceived as more proficient in mitigation or adaptation to climate change will be copied by communities that are less proficient in these tasks. In particular, this applies to policy innovations appraised as successful in any community and perceived by other communities as relevant to their own needs. Recall that communities considering themselves successful innovators are often motivated to publicize their successes, and those considering themselves worse off (or deprived) are often open to alternatives. Central information clearinghouses can help put them together and at the same time help inform higher-level officials of realities on the ground. However, the diffusion process malfunctions when innovators or consultants hype or otherwise falsify claims of success to promote themselves or their services, or when claims of success are uncritically accepted by those desperately seeking solutions to pressing problems. The practical implication is to include a third-party appraisal function in central information clearinghouses, to flag any undependable or misleading claims in the diffusion process.

This leads to the third proposition on the nature of change, defense through partial incorporation:

■ "The process of diffusion is complicated by the mechanism of 'partial incorporation,'" which "*occurs when an elite expects to reduce the external threat to its power position by accepting a comparatively small reduction of power at home by means of the incorporation."*

The corollary is that "[t]he symbols most eligible for partial incorporation are those which have *internal support, but are not the monopoly of counterelites connected with the external threat.*"[222] The symbols of adaptive governance most eligible for partial incorporation by defenders of business as usual are symbols of democracy—"the people," "public engagement," and the like. For example, according to Herman Karl and his colleagues, in the United States "many public agencies still advocate the traditional approach best character-ized by the phrase 'inform, invite, and ignore.' These traditional techniques prohibit meaningful discussion, discourage discourse, and fuel further con-flict."[223] This occurs despite the U.S. National Environmental Policy Act, which requires public participation in the preparation of an Environmental Impact Statement. (The EIS itself is a practice that has been widely diffused and adapted among nations.) The practical implication for reformers and other activists is to distinguish between symbols (words, rituals, and their equivalents) and deeds, and to emphasize the latter.[224] However effective it may be in the short term, defense through partial incorporation poses risks for an established elite in the long term: It legitimizes and intensifies demands like public participation advanced by outsiders with grievances; to the extent it succeeds in defending business as usual, the technique frustrates redress of those grievances. As frustrations continue to accumulate, people lacking better alternatives will attempt to empower themselves. The North Slope Iñupiat sought to do that nonviolently in the 1960s and early 1970s on the model of other minorities in the American Civil Rights movement. With some support from outside, including government officials, they succeeded in reforming the constitutive process to allow for their fuller participation. This was another step toward fuller realization of human dignity for all, the basic value proclaimed in the U.S. Declaration of Independence, the Pledge of Allegiance, and other authoritative sources.

To guide constitutive decisions in the evolution of the climate change regime, consider goals for the world constitutive process developed in the policy-oriented jurisprudence of Myres McDougal, Harold Lasswell, and Michael Reisman. For these scholars and practitioners, the "fundamental goal . . . is that of human dignity: the inherent and equal value of every human being."[225] Box 4.3 is our overview of their specifications of the funda-mental goal; as an overview, it provides a comprehensive framework of ques-tions and working answers but omits many valuable specifications. Readers might find the goals in the summary or the original useful for clarifying their own general preferences for constitutive process and for reconsidering in that light the constitutive process in the climate change regime. As an intro-duction to that introspective exercise, consider certain detailed specifications under "Arenas" that are directly relevant to decentralized decision making in the climate change regime. The authors urged "that constant attention be given to the identification of those matters which are most effectively decided in inclusive arenas." But they emphasized "the inclusive benefits accruing from encouraging the *decentralization* of initiative, expression, and decision...."[226] In other words, decentralization of some decisions in globally inclusive arenas such at the UNFCCC can provide global benefits. More specifically, "In some contexts decentralization may maximize the potential for democratic participation at lower levels of interaction, permit the most rapid decision, encourage the establishment of appropriate specialized arenas, be most sensitive to the special circumstances and conditions prevailing in sub-arenas, and allow for the integration of the widest range of diverse cultural forms in constitutive decision." The authors also recommend "maximum participation by individuals in those decisions having the most intense and intimate effect upon their lives...."[227] Those decisions include the allocation of resources from the top down to local communities that have been most damaged by climate change impacts or are most vulnerable to them.

Robert Dahl corroborated the emphasis on decentralization within a larger structure of decision making. Like other democratic theorists, Dahl accepted "rule by the people," but unlike most of them, he asked, "What do we mean by 'the people'?" He reported, "Having puzzled over the problem for years . . . I have become persuaded that there is no theoretical solution to the puzzle, but only pragmatic ones."[228] An interim conclusion in his reasoning was that "stages of government fitting together rather like the components of a Chinese box are necessary if 'the people' are to 'rule' on matters important to them, whether a neighborhood playground, water pollution, or the effective prohibition of nuclear war."[229] In the end, his reasoning suggested some pragmatic principles. The first of these was decentralization: "If a matter is best dealt with by a democratic association, seek always to have that matter dealt with by the smallest association that can deal with it satisfactorily."[230] We have argued that the smallest association that can deal with climate change adaptation satisfactorily, but not exclusively, is a local community like Barrow. Similarly, the principle of "subsidiarity," as promoted by the Assembly of European Regions, "holds that government should undertake only those initiatives which exceed the capacity of individuals or private groups acting independently. Functions of government, business, and other secular activities should be as local as possible." The principle has some moral force in *Rerum Novarum* by Pope Leo XIII (1891), legal authority in the European Union's Treaty of Maastrict (1992) and Treaty of Nice

Box 4.3. Goals for Constitutive Processes

Participants

"Who acts or participates . . . in the process which culminates in decision?"

Inclusivity means that "at every level of intergroup interaction and within each group, universal participation is to be sought." There are two complementary subgoals:

Representativeness "requires unimpeded access of all individuals to all decision arenas in accord with their capabilities for participation."

Responsibility "requires that all who in fact affect world constitutive process . . . should be made subject to that process and required to conform to community standards of responsibility."

Perspectives

What are the significant identifications, demands, and expectations of participants?

Common Interest. "The perspective most indispensable . . . is the demand for the continuing clarification and implementation of common interests in regard to all values."

Arenas

"Where and under what circumstances are the participants interacting?"

Balancing. "Goals in regard to arenas are best expressed in terms of an ongoing balancing process between centralized and decentralized, organized and unorganized, specialized and nonspecialized arenas, and temporally continuous and discontinuous arenas."

Availability and Compulsoriness. "In brief, arenas must be open and available to all participants who wish a role in decision commensurate with their legitimate interest in the content of that decision; and arenas must be sufficiently compulsory to constrain the participation of all those whose presence is necessary. . . ."

Base Values

What effective means do different participants have for achievement of their objectives?

Proportionality. "The policies relevant to the acquisition and management of base values in the constitutive process are principally policies of proportional-

---

(2003), and political influence in the assembly's efforts to highlight the role of regions, including the exchange of regional experience, within the EU's larger effort on climate change.[231]

Finally, it is worth noting that the regime has little to lose through decentralization, and perhaps much to gain. As noted in Chapter 2, the regime does not control concentrations of greenhouse gas emissions in the atmosphere. Nevertheless, it did put climate change on policy agendas worldwide. It helped to catalyze effective action to reduce emissions by activists predisposed to act on a voluntary, often "no regrets," basis in local communities such as Ashton Hayes, Samsø, Toronto, and Portland, in states such as California, in countries such as Norway and the Netherlands, and in firms

ity to the domain, range and scope of decision." For example, more inclusive decisions require more base values to enforce.

Authority. "The maintenance of an effective constitutive process . . . requires an allocation of sufficient authority to general community representatives both to protect common interests and to reject claims of special interest."

Control. "In addition to authority, allocate base values of sufficient magnitude to enable authority to become controlling." As a guiding principle, "the necessity of control is roughly in proportion to the intensity and scope of individual commitment to world public order."

Strategies

"In what manner are means or base values managed?"

Minimize Coercion. Coercion is to be applied only "as a last resort," "only in the common interest," and by inclusive decision-making procedures; "where immediate inclusive application is not feasible, its use [must] be subject to eventual inclusive review as to its lawfulness."

Economy. "Economy in application of available resources involves a high degree of flexibility in the fashioning of strategic programs . . . to realize community goals within specific contexts"; "all strategic instruments must be considered in the formation of any particular strategy."

Outcomes and Effects

What is the immediate result—decision affecting value outcome—of the process of interaction? What are the effects, with differing duration, of the process and outcome? See Box 4.2 on outcomes and criteria for evaluating intelligence and appraisal outcomes.

Source: Selected and adapted from Myres S. McDougal, Harold D. Lasswell, and W. Michael Reisman, "Policy Goals for World Constitutive Process," in McDougal and Reisman, Eds., *International Law Essays* (Mineola, NY: Foundation Press, 1981), 201–222; for the questions, see p. 200.

such as BP, Johnson & Johnson, and Frito-Lay. And it helped to catalyze effective action on climate change adaptation as well as mitigation. The upshot is that the regime can influence effective action to reduce net losses from climate change by persuasive means, even if it cannot control effective action through mandates.

The regime's influence is based in large part on its control of the focus of attention. Representatives of the regime control the IPCC's *Summaries for Policymakers*, including the prohibition of active or passive policy recommendations, through line-by-line review and revision prior to publication. The regime also controls climate change research to a considerable extent through national programs of research established and administered by representatives of the regime. Research proposals that challenge standards applied by existing programs, or fall in the gaps between existing programs, are not usually viable—leaving their advocates to choose between conforming in various ways or withdrawing. Similarly, policy proposals that challenge the quest for mandatory targets and timetables are not normally viable.

The regime could enhance its influence by drawing attention to distributed policies that have already succeeded on modest scales in reducing vulnerabilities to extreme weather events and climate change. This would provide field-tested models to inform people concerned about climate change and to catalyze local activists among them to organize for collective action where they live and work. This would also open the regime to the fuller participation of those affected by climate-related policy decisions—decentralization, in other words. To the extent that decentralization pays off from their perspectives, the regime will gain supporters and eventually more control.

REFRAMING THE CONTEXT

5

In previous chapters, we have presented the main empirical and theoretical reasons behind the proposals for adaptive governance introduced in Box 1.1. The proposals outline one approach to opening the established climate change regime, to help advance the common interests of the world's many diverse communities. It should be reemphasized that opening the regime does *not* mean replacing it. In this concluding chapter, we promote careful consideration of the proposals in the continuing evolution of the climate change regime, recognizing that they will be controversial in some quarters but not in others. Careful consideration begins with an introduction to issue expansion and contraction in climate change politics. Issue expansion promotes the aspirations of scientific management, while issue contraction serves to defend them against adaptive governance.[1]

Beginning more than two decades ago, expansion of the issue into a globally irreducible problem helped promote climate change as a claim on attention, funds, and other resources in competition with other issues—just as in an earlier era, "workers of the world unite!" expanded and promoted the relatively modest demands of labor unions. The outcome, a global framing, sustains demands for models of the total earth system and international cooperation as *prerequisites* for a global solution—a solution based on mandatory targets and timetables legally binding on all nations producing significant greenhouse gas emissions.[2] Within this global framing, it is rather easy to dismiss the potential of the Peorias of the world to reduce net losses from climate change, to belittle the individually modest accomplishments of local communities that have demonstrated enough progress to serve as models, and to ignore them. Models for action on mitigation include the village of Ashton Hayes, the island of Samsø, Toronto and other cities affiliated with ICLEI, the state of California, among others, and various corporations. Models for action on adaptation to climate change include Sisimiut in west Greenland, the Pacific islands assisted by PEAC, Melbourne in Australia, and distributed policies in Barrow in Alaska.[3] As an example of dismissal, consider one review that set aside the proposals, reduced the reasons for them in the Barrow case, and dismissed the case as irrelevant to a global solution: "I am not persuaded that as Barrow goes, so can go other communities around the globe. The case is too unique to be persuasive; unique in location, isolation, culture, economics, and most of all size and scale of complexity."[4] We expect more issue contraction—as well as uncertainty absorption, and partial incorporation—in defense of business as usual against adaptive governance and other approaches to opening the regime.

To promote more careful consideration, this chapter reframes the proposals for adaptive governance as matters of collective *action* in the larger context of a transition from the relevant past to possible futures. The first section outlines next steps that might be taken by people interested in advancing the common interest through adaptive governance in the communities in which they live or work. The next steps are not necessarily obvious in a culture saturated with scientific management. The second section reviews the relevant past of scientific management beyond climate change. The emergence of critiques and exceptions to scientific management in the evolution of the climate change regime, as reviewed in Chapter 2, is part of a much larger pattern of community development in which adaptive governance cannot be easily dismissed. The third section considers possible futures that will shape, and be shaped by, choices and decisions yet to be made in response to climate change. To help us all choose and act with our eyes open, we recommend the comparative evaluation of specific action alternatives in light of explicit criteria and evidence that is detailed and comprehensive within practical constraints, of course. The overriding question in climate change science, policy, and decision making remains open: Which interests will we individually choose and collectively decide to serve?

Next Steps

As the overriding goal of next steps toward adaptive governance, we recommend advancing the common interest of each community—which for us means reducing net losses of the many things valued by community members to extreme weather events and climate change. This requires flexibility to adapt policy alternatives to the unique context of the community and to the experience of what works and what does not in that community, and others like it, as events unfold. In this section, we consider next steps to take from the bottom up and the top down, and the need for appraisals of community-based initiatives regardless of their original leadership. We have no master plan, only adaptive governance as one approach to perceiving and capitalizing on existing opportunities. Where any aspect of adaptive governance has found a niche, the basic strategy is to move toward the other aspects as the need or opportunity arises. If Box 1.1 is an abstract map of the "space" to be covered, it suggests three important questions to be answered in each context: Where are we now? Where should we go from here? And how can we get there? The basic strategy capitalizes on latent synergies among the various aspects of adaptive governance.

For answers to these questions, we would be pleased to defer to people who know their own contexts better than we do. Meanwhile, to illustrate the basic strategy, we draw upon cases of adaptive governance from previous chapters. Of course these cases represent only a small fraction of relevant cases. They are part of what environmentalist Paul Hawken called "the largest social movement in all of human history" focused on ecological sustainability and social justice, including the rights of indigenous peoples. As he characterized it, this diverse movement "engages citizens' localized needs. [Its] key contribution is the rejection of one big idea in order to offer in its place thousands of practical and useful ones." It is also "eminently pragmatic. And it is impossible to pin down. Generalities that seek to define it are largely inaccurate."[5]

From the Bottom Up

For west Greenland, Lawrence Hamilton and colleagues developed a model of long-term adaptation centered on the region and two local communities, Sisimiut and Paamiut. Perhaps that model's potential to contribute to procedurally rational policy and decentralized decision making in the region has yet to be developed. If so, as a next step, outside researchers, entrepreneurs, and environmental or development activists might cooperate with people in local communities to compare the model with local knowledge and revise it as warranted. The revised model might generate additional insights into present policies for local people, based on fuller appreciation of past problems with economic overspecialization and past payoffs from entrepreneurial activities and local self-reliance. Discussions could raise opportunities for networking communities to share local experience and for organizing to influence national and international sources of support from the bottom up.

With the assistance of PEAC, some Pacific islands developed and implemented local policies in response to the 1997–1998 El Niño. Perhaps their subsequent experience in reducing losses from periodic droughts has yet to be harvested for self-empowerment of island communities in decentralized decision making. If so, as a next step, local policymakers and researchers could cooperate in intensive inquiry to clarify what has worked on particular islands, what has been diffused and adapted across islands, and what has effectively informed higher-level decisions in support of their drought mitigation policies. The results could inform action to improve policies, networks, and organization among the islands.

In Melbourne, perhaps the ongoing intensive inquiries in support of procedurally rational policies have been unsatisfactory in maintaining the reconciliation between water demand and supplies across the whole state of Victoria, particularly in rural areas and small communities.[6] If so, the next step could be scaling up to networks that expand the range of informed choices and coordinate action across the whole state. The Melbourne model is particularly valuable as a demonstration of how the public can be persuaded to identify with the community effort and to accept adaptations that entail personal costs, partly by connecting the personal experience of droughts with the need for action in response to climate change.

In Barrow, perhaps declining property tax revenues from Prudhoe Bay and Senator Stevens's defeat in the November 2008 elections warrant closer consideration of policy alternatives less expensive than the Army Corps of Engineers's evolving comprehensive plan for storm damage reduction. If so, the utility corridor and the landfill might be reconsidered separately as priorities for distributed decisions, along with passive alternatives that have been relatively neglected in the past, especially planning and zoning and selective relocation. Such efforts could be informed by relevant experience elsewhere—especially if new leadership builds on initial steps taken by Senator Stevens to organize Alaska Native villages vulnerable to erosion and flooding to organize as networks.

Cases of mitigation share one relatively specific goal, the reduction of greenhouse gas emissions, and are more numerous than *documented* cases of adaptation. This supports comparative inquiry, from which several insights stand out. First, while extensive science and R&D on advanced low-carbon technology can and should be justified for the long-term goal of decarbonization, they are *not* the key factors limiting immediate mitigation of climate change on the ground. Available science and technology have been sufficient for significant progress on a selective community-by-community basis—indeed, for achieving carbon neutrality in less than a decade on the island of Samsø, for example. The key limitations on effective deployment of existing science and technology have been human factors, successfully addressed without coercion by some local community leaders. Søren Hermansen on Samsø, Garry Charnock in Ashton Hayes, and Tony O'Donohue in Toronto, for example, each brought the issue home. Working with others in their communities, they showed what can be done within local interests and capabilities to mitigate climate change. Because human factors have been largely absorbed and neglected through extensive research, the next step is more intensive inquiry. In particular, what additional ways of overcoming limiting factors have been successfully field tested in more and different communities? Second, the policies field tested in local communities are sometimes highly complementary, as well as innovative and diverse. Samsø achieved carbon neutrality and more through renewable energy production, while Ashton Hayes made impressive progress in a few years through reductions in energy demand primarily. The next step may be intensive inquiry centered on identifying and correcting malfunctions in networking. In particular, what stands in the way of making connections between such communities with complementary interests and experiences? Third, it is not obvious that the aspirations of ICLEI and others to influence higher-level policies from the bottom up have succeeded. Perhaps the next step is to harvest the experience available. For example, what has Toronto done about limiting factors beyond its control that reportedly accounted for shortfalls with respect to its initial emissions reduction goal? Why did the U.K. Department for Environment, Food, and Rural Affairs award a grant to the Carbon Neutral project to in effect help diffuse and adapt the experience in Ashton Hayes?

Opening climate science to intensive inquiry on questions like these is one response to the growing concern that climate science is "over," in the sense that only the details are expected to change; the basic story told in current data and models of the earth system and their policy implications will remain the same. Opening climate science to intensive inquiry is also the main opportunity for bridging the "valley of indecision and delay" between climate science and climate policymaking. In particular, local activists could use the results to engage their neighbors, adapt field-tested policies to their distinctive interests and capabilities, and cultivate identifications with complementary efforts at the national and international levels. The mass marketing of a planetary emergency based on scientific assessments has been insufficient to cultivate those identifications on a broad scale.

From the Top Down

Starting at the international level, the Tsho Rolpa case as reported by Working Group II in IPCC's Fourth Assessment Report is a precedent for opening climate science to intensive inquiry. The next step could be collections of case studies on one or more priority adaptation problems, defined by high winds, floods, droughts, wildfires, and heat waves, for example, or by geographic region. For anyone interested, these case studies would bring into the picture uncertainties otherwise absorbed through extensive research, provided that the peer-reviewed scientific literature on selected cases is sufficiently comprehensive and detailed. (It can be supplemented by official documents and other sources in any case.) For local policymakers, these case studies would invite participation in the climate change regime, providing models to motivate and inform action on their own adaptation problems, perhaps with encouragement and support from higher levels. (The government of Nepal apparently initiated action to alleviate the problem at Tsho Rolpa, with participation by local communities at risk and funding from the Netherlands.) For the Conference of the Parties to the UNFCCC, the IPCC might encourage comparative inquiry on case studies and local community networks to clarify policy issues and means for decentralizing decision making where appropriate. For example, what might the parties do to help the Peorias of the world and their networks contribute more to the overriding goal in Article 2 of the UNFCCC? A precedent for expansion of IPCC's present mandate is the Response Strategies Working Group in the First Assessment Report in 1990. As reported in Chapter 2, it clarified policy issues and alternatives for negotiations on centralized decision making at the international level, leading to constitutive decisions formalized in the UNFCCC.

At the national level, one precedent to build upon for mitigation is the U.S. Climate Change Action Plan initiated by President Clinton and Vice President Gore in 1993 but partially terminated by the Bush Administration. Recall that CCAP's voluntary programs provided advice to businesses, cities, and federal facilities on technologies available for reducing greenhouse gas emissions and energy costs. Could it be revived and restructured in light of feedback from the bottom up? For adaptation the additional precedents to build upon include the NOAA initiatives, first PEAC and then the RISA program. However, it is difficult to know where RISA is with respect to adaptive governance, or where it might go from here, without systematic appraisals. Nevertheless, we can imagine parallel initiatives by other federal agencies involved in climate change, including those participating in the U.S. Climate Change Science Program. Consider the following steps toward adaptive governance that might be taken by a federal agency that aspires to take the lead on climate change adaptation. The sequence is rather generic; it could be adapted without much difficulty to a second U.S. National Assessment, to climate change mitigation initiatives from the top down, and to agencies in governments other than the United States.

1. *Define the Basic Program.* Suppose that the basic objective of a lead agency's new program is based on the common interest in reducing net losses of things valued by communities to climate-related events—high winds, floods, droughts, wildfires, heat waves—in a region of the United States. (This objective differs from, but is not necessarily incompatible with, marketing the lead agency's existing products or services or the agency itself.) The basic strategy then would be to guide, coordinate, and support local communities and other agencies sharing that objective; to maximize cooperation and minimize competition between them; and to integrate policy research and action on a continuing basis. The critical resource is the experience of local communities in climate change adaptation. They serve, individually and collectively, as laboratories providing intelligence on what works on the ground for both federal and local policymakers. The latter become participants, not merely stakeholders in the lead agency program.

2. *Select Local Communities* for intensive inquiry. Start with a small number of communities that share a similar kind of vulnerability and already understand the need for action to reduce their vulnerability. (Avoid those communities not yet willing to take on their own climate-related problems.) Among additional criteria, consider communities that have already made considerable progress and those that are especially vulnerable for any reason. The former have valuable experience to harvest for the latter, where it might

do the most good; and both kinds of communities have valuable experience for federal officials seeking to understand problems and working solutions on the ground, to inform federal policies from the bottom up. Candidate communities might be identified through recommendations from associations such as the National League of Cities, the U.S. Conference of Mayors, and the National Association of Counties; more specialized associations of local governments, businesses, and nonprofit organizations such as the Association of State Floodplain Managers or the American Indian and Alaska Native Climate Change Working Group; and agencies in contact with local communities such as NOAA's Coastal Services Center in South Carolina. Collaborate with those communities that accept an invitation to participate in the program, initially a pilot program, in return for the agency's help in reducing their vulnerabilities. Imagine, for example, an initial collection of a half-dozen or so local communities on U.S. coasts that have been damaged by recent hurricanes and are motivated to reduce their vulnerability to future ones. Large national samples of communities would be unnecessary and inconsistent with program objectives and resource constraints.

3. *Construct Case Studies* in collaboration with people on the ground. For each community, focus the case study narrative on the primary outcomes relevant to the basic objective—past losses since a historical baseline and present vulnerabilities—and on policy responses to those outcomes. In the narrative, and within practical constraints, develop the context for understanding the outcomes in detail and comprehensively, including primary and related outcomes and the factors responsible for them as assessed by differently informed people on the ground. Avoid a lead-agency assessment; it is neither necessary nor constructive for the local community. As examples, consider separately the nine brief case studies in the GAO report *Alaska Native Villages: Most are Affected by Erosion and Flooding, but Few Qualify for Federal Assistance*, which are too brief for this purpose, and the case study of Barrow with the great storm of October 1963 as the historical baseline in Chapter 3, which is more detailed and comprehensive than necessary for some purposes. Looking across communities, expect to find many outcomes and many factors accounting for them—both natural and human factors, some overlapping and some idiosyncratic, some quantified or quantifiable but many qualitative. A collection of detailed and comprehensive case studies should not be limited to or dominated by a standard format that would suppress innovations or absorb uncertainties, idiosyncrasies, and surprises.

4. *Disseminate the Experience* harvested in individual case studies as intelligence for the adaptation of policies at multiple levels of decision making. For each selected community, simply pulling together and publicizing its own relevant history may provide interim payoffs, informing and motivating further engagement and action—provided that the historical information is dependable and the lead agency defers to local policymakers on local decisions. Expect some of that information to be picked up and used in distributed policies by different parts of the community, as happened in Barrow. Assist representatives of the selected communities in coming together for face-to-face meetings, followed where appropriate by teleconferences, newsletters, online interactions, and the like, to encourage networking that would expand the range of informed policies for each community, clarify shared needs for assistance from outside state and federal governments, and update their local experience. Supplement such networks with a central information clearinghouse to match requests from communities with the supply of intensive case studies, emphasizing models of procedurally rational policy and decentralized decision making. In addition, use the clearinghouse to identify gaps in the supply of case studies and problems of dependability in information distributed. Third parties might be commissioned to correct problems of dependability or to augment the supply, to expand the range of working models relevant to each community.

For the lead agency, the case studies and network discussions could be used to identify state and federal agencies, laws and regulations, and forms of assistance that have been productive and counterproductive in reducing past losses and present vulnerabilities on the ground, and to identify gaps in external support for what works there. This amounts to an agenda of reforms for the lead agency to pursue. For example, if some communities are unable to identify and integrate resources available from existing state and federal programs to meet their needs on the ground, perhaps the indicated reform is to assign a coordinator to each community or to set up a direction center to give each community one-stop access to available resources—and to update the agency's intelligence on community needs at the same time. There are precedents.[7] Such reforms are likely to be supported by networked community representatives who have an interest in them and by other agencies that share the basic objective of the program. Of course, harvesting experience and disseminating intelligence from the bottom up still leaves the political task of reconciling different interests within communities at each level and across levels. But the previous steps could ground those politics in local experience that otherwise would be discounted or ignored and could open participation to more representative and responsible people who have a stake in adaptation.

5. *Iterate and Expand* the program as warranted by policy outcomes relevant to the basic objective. Go back to step 2, working with the same selected communities, and/or moving on to select other communities or kinds of problems according to better estimates, informed by the previous iteration, of the most promising investments. Avoid the quest for "the one best way" or "best practices" as if they were context independent. Avoid indecision and delay as if that would somehow clear up outstanding uncertainties. The priority for communities is to act incrementally and modestly but promptly in response to pressing problems—in part to clear up uncertainties—while leaving open opportunities to revise or terminate failed policies in light of experience. The context typically allows for distributed decisions that address nearly decomposable parts of the overall problem, even in a small community like Barrow. Because there are many pathways for reducing net losses, both within and between communities, a comprehensive national policy is best conceived as guiding, coordinating, and supporting action on modest policies as the need or opportunity arises.

Overall, the task is a matter of action: to build on what works in practice on the ground, to reduce past losses and present vulnerabilities compared to local historical baselines or comparative norms, and to modify or set aside what does not work. This requires appraisals of policy outcomes and processes that are sensitive to problems likely to arise in a culture of scientific management. (For more on appraisals, see below.) One such problem is a tendency toward centralized planning, as if climate change experts at the top monopolized the relevant expertise and mass marketing of that expertise would stimulate enough individual and collective action. This ignores the multiple and often distinctive interests of people on the ground and bypasses the local knowledge and experience often essential for program planning, support, and management. Another problem is a tendency toward premature program expansion, as if the requisite understandings were already in hand and program expansion was equivalent to success in the competition for more resources among agencies. This squanders scarce resources, constrains learning by doing, and eventually traps program officials in a defensive political mode. Premature program expansion is rarely justified; if the problem is urgent and widespread, consider launching multiple programs to proceed in parallel. Still another problem is a tendency to subordinate collective problem solving to maintaining or asserting control from the top down, by discouraging direct contacts among local communities or by other means. This suppresses innovation, error detection, and program adjustments, as Landau understood, and invites active or passive political resistance from the bottom up.

Appraisals

In adaptive governance, appraisal is *the* critical function, whether the initiative for a program comes from the top down or the bottom up. Box 4.2 provides guidance for thinking through appraisals in any policy area. In climate change policy, the agreed-upon policies are adaptation and mitigation to reduce losses of the things we value. The respective criteria for these policies are reductions in net losses and present vulnerability to climate-related events, and reductions in greenhouse gases to prevent further and more dangerous changes in the climate system. Observations for appraisal purposes should be focused on these primary outcomes but not limited to them. Judgments of success and failure also depend on observations of related outcomes, including costs. Explanations of formal and effective responsibility for multiple relevant outcomes also depend on additional observations. In other words, the main outcomes must be understood in context.

Adaptive governance cannot rely exclusively on quantitative measures for several reasons. First, metrics for the main outcomes in climate change are not necessarily dependable or constructive, as documented in a half-century of research on the dysfunctional consequences of quantitative performance measures.[8] On the mitigation side, we have been unable to find much detailed or current information on the dependability or consequences of emissions measures used for policy purposes, but there are a few exceptions. In 1997, the U.S. General Accounting Office reported, "The incomplete, unreliable, and inconsistent data on emissions prevent a complete assessment of Annex I countries' efforts to limit greenhouse gas emissions. . . ." For example, to facilitate meeting its target, "Denmark adjusted its 1990 inventory level upward to show what emissions would have been if imported hydroelectric power had been generated domestically with fossil fuels."[9] In 2008, a GAO report provided some technical details on carbon offsets traded in the U.S. voluntary market: "To be credible, an offset must be additional—it must reduce emissions below the quantity emitted in a business-as-usual scenario—among other criteria." This "additionality," together with limited information on quality assurance mechanisms, is a problem for buyers in the market. For some, "concerns about the credibility of offsets could compromise the environmental integrity of a compliance system."[10] In February 2007, PricewaterhouseCoopers reported that EU-wide standards for verification of emissions and accreditation of third-party verifiers had not been implemented, as noted in Chapter 2. From its survey of leading emissions trading schemes around the world, PricewaterhouseCoopers concluded, "Despite good intentions across the board, the general picture is one of new and immature markets, inconsistent and complex compliance frameworks, and, consequently, risk."[11] A recent report noted a "vexing challenge . . . that surface inventory assessments—based on measuring forests, agricultural fields and smokestack emissions, for instance—generally do not agree with atmospheric measurements."[12] Under such circumstances there are ample opportunities for getting the numbers right as a substitute for reducing emissions. This is goal substitution, one of the recurring dysfunctional consequences of performance measures.

On the adaptation side, metrics for the main outcomes also are problematic. In Barrow, for example, dollar estimates of damage from the series of big storms are not comparable, and in-kind reports of property damage are incomplete. Despite a household survey of property exposed to storm damage along the coast in Barrow, the Army Corps of Engineers's storm damage reduction project overestimated the benefits of its proposed alternatives, according to the Independent Technical Review at higher levels in the organization. Among other things, different assumptions were involved in putting a dollar value on property observed and reported. Any measure of reduced vulnerability from the location or design of the new Barrow Arctic Science Center, hospital, sewage treatment plant, or the evacuation road would depend on counterfactual assumptions—estimates of losses that would have occurred without these distributed policy decisions. Any measure of the present vulnerability of the utilidor and the landfill, arguably the highest priorities for storm damage reduction in Barrow, would be contingent upon a host of questionable assumptions about the significant details of these structures, future storms, and emergency management responses to them. In view of such complications, including goal substitution, nonlocal officials might heed Sir Josiah Stamp's warning in 1929: "The government is very keen on amassing statistics. They collect them, add them, refer to the nth power, take the cube root, and prepare wonderful diagrams. But you must never forget that every one of those figures comes in the first instance from the [village watchman], who just puts down what he damn pleases."[13]

Second, metrics for the main climate change outcomes are incomplete. Related outcomes and explanatory factors also matter in each context, even if they are difficult or impossible to measure satisfactorily. (Our concern is with reducing *net* losses, or "total impact" in terms of Box 4.2.) On the mitigation side, the flow of additional long-term investments into new low-carbon infrastructure may be a more important outcome than short-term compliance with an emissions reduction target in a cap-and-trade system. This is relevant to the European Union's cap-and-trade system, especially in light of the new coal-fired power plants planned without carbon capture and sequestration technology. Cleaner air, improvements in public health, and reduced energy costs are important preferred outcomes often related to emissions reductions. From the standpoint of people in Ashton Hayes, Samsø, and Toronto, the important preferred outcomes include working together and taking pride in the community's accomplishments and leadership, even if officials at higher levels narrowly focused on emissions reductions discount or ignore such intangibles. Creative local leadership is among the most critical factors for understanding preferred outcomes in the cases we have examined. It should not be discounted or ignored simply because it is difficult, impractical, or impossible to measure satisfactorily.

On the adaptation side, the lack of appropriate metrics sometimes defers systematic and comprehensive appraisal. The RISA program is a case in point, but not an isolated case. In a comprehensive review of local, state, federal, and international experience relevant to the proposed Government Performance and Results Act of 1993, the Congressional Budget Office concluded that it is typically difficult to find satisfactory performance measures for government agencies and that in any case officials typically prefer to be unfettered by the measures available.[14] In Barrow, the important factors involved in storm damage reduction, as motivating factors or valued outcomes in themselves, included sustaining the cultural heritage, local jobs, and self-determination. In the Pacific islands assisted by PEAC and in Melbourne, virtually every valued outcome in the community depends on a sustainable water supply. The official notes from a workshop conducted in 2006 on developing a community resilience index identified hundreds of factors contingently involved, as well as an anecdote illustrating the challenge, if not futility of measuring them all: Three workshop participants from the Washington, D.C., area reported that their neighbors, relying on local knowledge, opened clogged culverts with their own shovels on their own initiative, and thereby reduced neighborhood damage during recent floods.[15] Similarly, a person with a flashlight could have detected and opened a clogged culvert during a flood in Toronto one night, and thereby prevented erosion of a major road that cost $40 million (CDN) to repair.[16]

Third, measurement can divert resources from the main goal, reducing losses from the impacts of climate change. In many contexts, finding and disseminating field-tested alternatives to reduce vulnerability could be the better investment. In the Washington and Toronto cases, for example, the resources available for measuring resilience or a reduction of vulnerability might better be invested in harvesting local knowledge, neighborhood by neighborhood, to find mundane but cost-effective measures like clearing blocked culverts, and then spreading the word about them. Metrics evidently make more sense on mitigation at the local level. One reason Samsø was selected for Denmark's renewable energy pilot in 1997 was the relative ease of measuring energy inflows and outflows from an island. Judging from the available accounts, much more of the effort was invested in making Samsø the renewable energy island than in measuring the results. The measurement and estimation of emissions reductions through an annual household survey appear to have been a larger part of the effort in the Carbon Neutral project in Ashton Hayes. The documentation of its measurement methods for other communities suggests that a considerable investment is involved. A conference in Brussels in September 2008 on "EU ETS Compliance: The Way Forward" may indicate the difficulty of harmonizing emissions reporting for compliance. After several years of effort, one of the objectives still was to "create among Member States and regional [competent authorities] an awareness of, and commitment to, the degree of harmonization of implementation needed to safeguard the environmental integrity." The main deliverable from the conference was a plan to "build capacity" in the compliance strategy.[17]

Alternatives to exclusive reliance on quantitative performance measures have emerged from community-based initiatives in other policy areas. These initiatives engage representatives of diverse, often conflicting interest groups from the outset in *multiparty monitoring* to evaluate and redirect their ongoing efforts and to supplement more tangible, measurable results as they eventually become available. As understood by practitioners of community-based forestry in the Pacific Northwest, "Multiparty monitoring and evaluation are not equivalent to scientific research projects, which are designed to test hypotheses that can lead to generalized results and often take much longer to complete. Multiparty monitoring is designed to provide more timely feedback on whether ecological, economic, and administrative goals are being achieved in a specific place. It is a practice-based approach to improving restoration decisions. For practitioners, multiparty monitoring is the primary oversight mechanism for these projects. It is designed to capture lessons learned and to build relationships based on credibility and trust by engaging diverse groups to work together."[18] For example, a district forest ranger in northern New Mexico "engaged in practice-based management by taking people out in the field to witness concrete problems and conditions. If downed fences on a grazing allotment were the issue, then everyone was invited to visit the site together to agree on the factual basis of the issue." The lesson of experience was that "[u]ntil we learned what was really true on the ground it didn't make any sense to proceed" with meetings to resolve the issue.[19]

Multiparty monitoring and evaluation recognize that the same facts mean different things to people with different perspectives. (As reported in Chapter 2, IPCC discovered this the hard way, through protests, after it evaluated a life in the Third World at 1/15 the value of a life in the industrialized world.) If groups involved in a community-based initiative can resolve their differences in a program intended to advance their common interest, they gain a political advantage in seeking external resources for the program. For example, the director of a program to recover endangered fish in the Colorado River observed, "When we go back to Congress and ask for funding for the Recovery Program, the water users, the enviros, the Park Service, Fish and Wildlife Service, and others stand toe to toe and fight together for funding the program. I never thought that we would see it happen . . . we are going to join arms and fight for this common goal."[20] Thus, participants in the Recovery Program identified with it, and their solidarity helped secure an appropriation of $100 million over 10 years. On a tour of a community-based forestry project in the Pacific Northwest, a U.S. Senate staffer proclaimed that "this is such a breath of fresh air—a bigger idea than traditional either/or politics. We can have healthy communities and healthy forests. . . . If the people say that's what we want, a bigger piece of the pie will go to natural resources."[21] On the House floor, Rep. David Obey went further to demand consensus alternatives from the bottom up: "I have seen intractable differences on forestry matters in my own area resolve themselves in 6 weeks when people are legitimately willing to sit down, deal with each other in an honorable fashion, and recognize that each side has legitimate interests. And I think we have a right as legislators to go to groups on both sides of this issue and say, we have had it, fellows. Get together. Work it out."[22] Working it out may be difficult, but the potential payoff is more power and better access to other resources.

Relevant Past

To clarify the possible significance of these next steps toward adaptive governance in climate change policy, it is worthwhile to trace the origins of scientific management back as far as the rise of early modern states in Europe. In the following constructions of the relevant past, various forms of scientific management have improved the efficiency of means for a bewildering variety of ends, from improving standards of living to colonialism to killing. The outcomes have not been an unmixed blessing for people on the ground in industrialized nation-states of the West or for people elsewhere who have been colonized or left behind by modernization. By one account, the 1.1 billion people who now survive on less than a dollar a day are just as poor as 80% percent of the world's population 250 years ago.[23] The deprivations of modernization often motivate resistance and a search for alternative paths of development—alternatives that have included community-based initiatives consistent with adaptive governance, especially in recent decades. This is part of the *relevant* past for at least three reasons:

- Responding to climate change is just another aspect of community development from the standpoint of people on the ground *and* those scientists and policymakers who emphasize mainstreaming climate change adaptation or mitigation policies.
- The evolution of the climate change regime as documented in Chapter 2 and of Barrow's responses to climate change in Chapter 3 are parts of a larger pattern of community development at multiple levels that shed additional light on climate change issues.
- More generally, ignorance or denial of past experience in connection with climate change and community development is a major factor sustaining preoccupations with scientific management and inhibiting alternative approaches better adapted to our time.

In this larger historical context, scientific management is contingent, not inevitable. And adaptive governance cannot be easily dismissed through issue contraction. One does not have to accept every claim in the following constructions to appreciate the main point: Exclusive reliance on scientific management in the climate change regime is questionable at best.

Early Modern Roots

In *Seeing Like a State* published in 1998, James Scott constructed the relevant past to account for "the failure of some of the great utopian social engineering schemes of the twentieth century." Scott found that "much of early modern European statecraft seemed similarly devoted to rationalizing and standardizing what was a social hieroglyph into a legible and administratively more convenient format."[24] In short, "Where the premodern state was content with a level of intelligence sufficient to allow it to keep order, extract taxes, and raise armies, the modern state increasingly aspired to 'take in charge' the physical and human resources of the nation and make them more productive. These more positive ends of statecraft required a much greater knowledge of the society."[25]

For example, in early modern Europe, "the crown's interest in forests was resolved through its fiscal lens into a single number: the revenue yield of the timber that might be extracted annually." This interest ignored nonrevenue uses of the forest by people on the ground, such as "foliage [for] fodder and thatch; fruits, as food for people and domestic animals; twigs and branches, as bedding, hop poles, and kindling; bark and roots, for making medicine and tanning; sap, for making resins; and so forth."[26] The Crown's interest led to precise measurements of trees in representative plots in a grid and eventually to elaborate tables organized by the size and age of trees under specified conditions. "By radically narrowing his vision to commercial wood, the state forester had, with his tables, paradoxically achieved a synoptic view of the entire forest. This restriction of focus reflected in tables was in fact the only way in which the whole forest could be taken in by a single optic." The tables enabled stripping the forest down for experimental and routine commercial purposes: clearing underbrush, planting uniform trees in rows, then cutting and extracting them. "The fact is that forest science and geometry, backed by state power, had the capacity to transform the real, diverse, and chaotic old-growth forest into a new, more uniform forest that closely resembled the administrative grid of its techniques."[27] But the aspirations of scientific forestry were thwarted by variations in topography and natural contingencies—"fires, storms, blights, climatic changes, insect populations and diseases"—and by humans living nearby who depended on the forest.

"Although, like all utopian schemes, it fell well short of attaining its goal, the critical fact was that it did partly succeed in stamping the actual forest with the imprint of its designs."[28]

Similarly, according to Scott, the state in absolutist France in the seventeenth century sought "to extract from its subjects a reliable revenue that was more closely tied to their actual capacity to pay."[29] Taxes beyond that capacity risked flight, evasion, and even revolt by subjects. However, for tax purposes, "local practices of measurement and landholding were 'illegible' to the state in their raw form. They exhibited a diversity and intricacy that reflected a great variety of purely local, not state, interests. That is to say, they could not be assimilated into an administrative grid without being either transformed or reduced to a convenient, if partly fictional, shorthand."[30] A field, for example, might be "a significant source of bedding straw, gleanings, rabbits, birds, frogs, and mushrooms" for local users, but it needed to be reduced to an administrative shorthand, such as commercial wheat or hay, for tax purposes.[31] Not surprisingly, "The struggle to establish uniform weights and measures and to carry out cadastral mapping of landholdings . . . required a large, costly, long-term campaign against determined resistance. Resistance came not only from the general population but also from local power-holders. . . ."; "The logic behind the required shorthand was provided, as in scientific forestry, by the material interest of rulers: fiscal receipts, military manpower, and state security. . . . Backed by state power through records, courts, and ultimately coercion, these state fictions transformed the reality they presumed to observe, although never so thoroughly as to precisely fit the grid."[32] In conclusion, Scott quoted *The Quantifying Spirit in the Eighteenth Century*: "The need for the increasingly bureaucratic state to organize itself and control its resources gave an impulse to the collection of vital and other statistics; to forestry and rational agriculture; to surveying and exact cartography; and to public hygiene and climatology."[33] But "the map is not the territory," as Alfred Korzybski famously remarked.

In the twentieth century, these transformative state simplifications combined with other factors to account for such great human tragedies as the "Great Leap Forward in China, collectivization in Russia, and compulsory villagization in Tanzania, Mozambique, and Ethiopia. . . ."[34] As Scott summarized it, "the legibility of a society provides the capacity for large-scale social engineering, high-modernist ideology provides the desire, the authoritarian state provides the determination to act on that desire, and an incapacitated civil society provides the leveled social terrain on which to build."[35] Among these factors, the quest for legibility persists in the development of data and models to make the total earth system legible for mandatory climate change policies at the national and international levels. However, the power to impose such policies from the top down has been lacking so far. The high modernist ideology nevertheless persists. "It is best conceived as a strong, one might even say muscle-bound, version of the self-confidence about scientific and technical progress, the expansion of production, the growing satisfaction of human needs, the mastery of nature (including human nature), and, above all, the rational design of social order commensurate with the scientific understanding of natural laws."[36] In Scott's view, it is a matter of ideological faith: "uncritical, unskeptical, and thus unscientifically optimistic about the possibilities for the comprehensive planning of human settlement and production."[37] Like Rayner and Malone and others in the field of climate change, he constructed a case for human institutions, "a case that contrasts the fragility of rigid, single-purpose, centralized institutions to the adaptability of more flexible, multipurpose, decentralized social forms."[38] In the end, Scott found the progenitors of twentieth-century totalitarian plans primarily responsible for their failures. They "regarded themselves as far smarter and farseeing than they really were and, at the same time, regarded their subjects as far more stupid and incompetent than *they* really were."[39]

Scott's progenitors can be included among *Voltaire's Bastards*, the title of a book by John Ralston Saul published in 1993. Saul also constructed the relevant past, "clearing away the underbrush to lay bare the obscured overall pattern" that is "merely the evolution of the Age of Reason."[40] In Saul's assessment, the differences between the Enlightenment, Romanticism, Neo-Classicism, Nihilism, and Modernism, to name just a few, "all blend into one another when we stand back far enough to get a good look." Indeed, we "disguise from ourselves the fact that we have taken in that long period but one clear step—away, that is, from the divine revelation and absolute power of church and state. That very real struggle against superstition and arbitrary power was won with the use of reason and scepticism."[41] As the leading figure in this struggle Saul celebrated Voltaire, who "had a genius for deflating the credibility and thus destroying the legitimacy of established power. His weapon was words so simple that anyone could understand and repeat them."[42] A complicated man, Voltaire aspired to influence ruling elites, first at Versailles and then at Sanssouci, but from an early age he courted censure with his satirical attacks on the church and the aristocracy. "Having failed to influence the monarchs and men of power to whom he had access, Voltaire turned towards the citizenry and became the defender of human rights and the most ingenious advocate of practical reforms."[43] When Voltaire returned to Paris in triumph in 1778, shortly before his death, all classes in the city seemed to use his return "as an excuse to demonstrate that they had been converted to the idea of government by a new philosophical coalition of *reason balanced with humanism*."[44] But little more than a decade later, the aftermath of the revolution in Paris corroborated Voltaire's earlier skepticism, expressed in his private letters, about republics.[45] In Saul's assessment, "Reason began, abruptly, to separate itself from and to outdistance the other more or less recognized human characteristics—spirit, appetite, faith and emotion, but also intuition, will and, most importantly, experience. This gradual encroachment on the foreground continues today."[46]

Saul traces the separation of reason and humanism back to Machiavelli and *The Prince* in 1513. Machiavelli was the first modern political thinker who saw politics as a profession not bound by religious faith. He "was indifferent to moral questions. He was a modern courtier in search of employment. Vitality, not Virtue, was the characteristic he sought in a political leader. At the centre of everything he wrote was the theme of political efficiency."[47] Machiavelli was followed by Ignatius Loyola, the principal founder of the Society of Jesus (commonly called the Jesuits) in 1540. Indeed, "reason was stronger in his hands than it had been in those of the [Protestant] reformers. What he had created was a flexible, unfettered weapon, free of all obligation either to morality or to specific ideas."[48] The pattern as Saul perceived it was that "[t]he Inquisitors, Machiavelli, and Loyola were all devoted to a priori truths and the service of established power. . . . Out on the cutting edge of social and political reform, methodology was becoming a mercenary for hire."[49] Though not a Jesuit himself, Cardinal Richelieu, the chief minister of the king of France from 1624 to 1642, was influenced by and promoted Jesuitical thinking. Richelieu firmly believed in the institution of monarchy as a rational system of government that restrained the irrational tendencies of the people, the judiciary, and Parliament.[50] Saul characterizes him as "the first modern statesman to apply an integrated rational method to that new concept—the nation-state" under the absolute formal authority of Louis XIII.[51]

The French Revolution replaced Voltaire and others like him with a cynical new elite, the technocrats, whose absolutist and inaccessible administrative methods divorced from common sense morality conflicted with the democratic reflexes of citizens. "The general frustration created by these obscure battles produced a new type of leader—the Hero." Saul posits, "He was a facile combination of the democratic and rational approaches—simultaneously popular and efficient. He was popular thanks to the combining of the majesty proper to kings with the worship proper to God in order to twist public opinion into adulation. He was efficient because his power left him free to administer without social constraint."[52] According to Saul, "Napoleon was the first and is still the definitive model" of modern absolute dictators, combining an emotive personality with efficient methods and rational argument.[53] "Trapped between these technocrats and Heroes were the reasonable men who thought of themselves as true men of reason—men who held firmly to their common sense morality." Thomas Jefferson is one such model, among others.[54] In Saul's judgment, "The host of those who served the good cause, and still serve . . . is legion. But they are not the ones who have defined the main line of rule, fighting a rearguard action in defence of humanism."[55] He concluded that the legacy of this evolution in our time is "a system determined to apply a kind of clean, unemotional logic to every decision, and this to the point that the dictatorship of the absolute monarchs has been replaced by that of reason."[56]

In recent decades, "Robert McNamara is the individual who most dramatically fits the role of the man of reason in flamboyant decline." McNamara brought modern methods of rational management from the Ford Motor Company to the public sector as Secretary of Defense beginning in 1961 and as president of the World Bank from 1969 to 1981. Saul's sketch of McNamara is tragic: "He is a man who believes in the forces of light and darkness. A man of honour. He resigned as Johnson's secretary of defense because he felt the Vietnam War was spinning out of control. As head of the World Bank, he attempted to save a desperate Third World by sending a flood of money in its direction. He believes that the application of reason, logic and efficiency will necessarily produce good. And yet his actions have resulted in uncontrollable disasters from which the West has still not recovered."[57] For example, his rationalization of officer training "changed the motivation of officers from self-sacrifice to self-interests. . . . Getting killed, after all, is not logical, rational, or efficient or what a businessman would perceive as being in his rational self-interest." His decision "to limit arms costs by producing larger runs of each weapon and selling the surplus abroad" contributed to "the largest arms market in the history of the world which," Saul claims, "is now the largest market of any kind anywhere in the world economy."[58] To translate American nuclear superiority into usable military power, McNamara helped impose the Flexible Response strategy on NATO allies. Among other unintended consequences, this increased the risk of all-out nuclear war initiated by the use of small nuclear weapons in Europe, and it undermined European confidence in American leadership.[59]

Saul considered McNamara a product of his time; he rose to elite positions because he conformed to the dictatorship of reason.[60] His characterization of the modern man of reason provides some insight into persistence of the pattern.

> The movement of history is the great enemy of someone like McNamara. History is linear memory and, as such, beyond organization and indifferent to reason. The characteristic common to the modern man of reason is this loss of memory; lost or rather, denied as an uncontrollable element. And if it must be remembered, then that evocation of real events is always presented as either quaint or dangerous. The past, when it involves a failed system, disappears from the mind. The past is always ad hoc. The future is always optimistic, because it is available for unencumbered sloganeering. And the present lies helpless beneath his feet, just begging to be managed.[61]

When the past disappears from the mind, the feedback and appraisal necessary for adaptation to differences or changes in circumstances is missing or denied. The problem is compounded by denial of internal contradictions and the suppression of outside dissent. In a technocracy, "Opposition becomes a refusal to participate in the process. It is irrational. And this trivialization of those who criticize or say no from outside the power structure applies not only to politics but to all organizations."[62] Thus, not surprisingly, "Successive absolute solutions are provided for major public problems and then slip away without our consciously registering their failure."[63] This makes it difficult to hold authorities and experts accountable for their actions. In short, "Our society contains no method of serious self-criticism for the simple reason that it is now a self-justifying system which generates its own logic."[64]

Saul provided his own summary: "At the heart of our problem lies the belief in the idea of single, all-purpose elites using a single all-purpose methodology. We have developed this in search of a social cohesion based on reason. Certainly, there is [also] an essential need to find common ground on which an integrated moral view can be built. Without that, society can't function."[65] As an example of "a relatively integrated moral outlook" for a healthy Western civilization, Saul suggested "agreement on democratic principles" to bring "a myriad of ideas and methods . . . face to face. Through civilized conflict the society's assumed moral correctness is constantly tested. This tension—emotional, intellectual, moral—is what advances the society. These contradictions are what make democracy work, but they also create technological advance."[66] Without a working democracy, the appearance of contradictions may be abolished by a single elite with a single method. In addition, Saul suggested we might learn from history and historical figures. "Somehow we must do today what Voltaire once did—scratch away the veneer in order to get at the basic foundations. We must rediscover how to ask simple questions about ourselves."[67] For those of us dissatisfied with the outcomes of climate policies to date, one simple question is "What can be done differently and better?" For Thomas Jefferson and philosophers of the eighteenth century, "Analysis was a means for hunting out falsehood and superstition. But a clear, practical line back into past experience was the foundation upon which the rational man was to base his abstract examination. . . . Jefferson . . . came back again and again to an analytical and scientific approach based upon a full and conscious assumption of the past."[68] The education and encouragement of elites competent to do this "will differ from country to country. The same is true for basic, general education. There is no need to search for global solutions, apart from an absolute necessity to destroy the idea that such things exist."[69]

High Modern Era

In parallel with developments in Europe, Frederick Winslow Taylor extended the viewpoint of *Seeing Like a State* to the industrial workplace in the United States, employed the dehumanizing logic of *Voltaire's Bastards* to maximize prosperity and efficiency, and justified the intended outcomes as good for workmen whether they liked it or not. Taylor formally introduced American engineers to scientific management in 1895 and published *The Principles of Scientific Management* in 1911. Consider his abbreviated synopsis of "the four great underlying principles of scientific management: *First*. The development of a true science. *Second*. The scientific selection of the workman. *Third*. His scientific education and development. *Fourth*. Intimate friendly cooperation between the management and the men."[70] The development of a true science assumed that "the one best way" exists: "[I]n each element of each trade there is always one method and one implement which is quicker and better than any of the rest."[71] Finding it was the task of managers who assumed "the burden of gathering together all of the traditional knowledge which in the past has been possessed by the workmen and then of classifying, tabulating, and reducing this knowledge to rules, laws, and formulae. . . ."[72] Workmen were incompetent in this regard: "[T]he science which underlies each act of each workman is so great and amounts to so much that the workman who is best suited to actually doing the work is incapable of fully understanding this science, without the guidance and help of those who are working with or over him, either through lack of education or through insufficient mental capacity." Consequently, "to work according to scientific laws, the management must take over and perform much of the work which is now left to the men. . . ."[73] The fourth principle glossed over differences in the interests of management and the men. So did the "very foundation" of scientific management, "the firm conviction that the *true* interests of the two are one and the same. . . ."[74] But Taylor indirectly acknowledged different interests, and possible resistance based on them, in his emphasis on enforcement: "It is only through *enforced* standardization of methods, *enforced* adoption of the best implements and working conditions, and enforced cooperation that this faster work can be assured. And the duty of enforcing the adoption of standards and of enforcing this cooperation rests with the *management* alone."[75]

Taylorism and related forms of scientific management worked well enough by their own standards to diffuse throughout the industrial world in the twentieth century. Scott noted that a Taylorized factory "was often remarkably efficient, as in the early Ford plants; it was always, however, a great boon to control and profit."[76] It was nevertheless incomplete: "No Taylorist factory can sustain production without the unplanned improvisations of an experienced workforce."[77] For Scott it was remarkable that "educated elites who were otherwise poles apart politically" accepted scientific management. Evidently, "The vision of society in which social conflict was eliminated in favor of technological and scientific imperatives could embrace liberal, socialist, authoritarian, and even communist and fascist solutions."[78] Similarly, Saul noted that "it would be foolish to deny that Taylorism played a major role in early-twentieth-century industrial advances."[79] Moreover, Taylorism diffused across conventional boundaries as it was refined through the century. "Directly or indirectly Taylorism has dominated business school methodology and changed business structures around the world. But it was also adopted in varying forms by both the Soviet and the Nazi regimes. . . . However, the astonishing thing about scientific management is that it has never gone wrong by its own standards. It has simply been more or less controlled by different civilizations."[80] Judith Merkle summed up the politics: "Scientific Management, translated into politics, advocated the development of the state as an organ of national planning and allocation according to a rationally derived system of priorities; it glorified a monolithic rational-technical order in place of the weak democratic forum that compromised among the interests of power groups."[81]

But the deprivations imposed by scientific management call its continued dominance of the modern era into question. Consider, for example, Vaclav Havel, the Czech statesman and playwright writing in 1992, shortly after the collapse of the Soviet bloc. "The modern era has been dominated by the culminating belief, expressed in different forms, that the world—and Being as such—is a wholly knowable system governed by a finite number of universal laws that man can grasp and rationally direct for his own benefit. . . . Communism was the perverse extreme of this trend. It was an attempt, on the basis of a few propositions masquerading as the only scientific truth, to organize all of life according to a single model, and to subject it to central planning and control regardless of whether or not that was what life wanted." Communism was defeated "by life, by the human spirit, by conscience, by the resistance of Being and man to manipulation. It was defeated by a revolt of color, authenticity, history in all its variety and human individuality against imprisonment within a uniform ideology." This defeat undermined the ideological foundations of the modern era according to Havel: "The fall of Communism can be regarded as a sign that modern thought . . . has come to a final crisis. . . . It is a signal that the era of arrogant, absolutist reason is drawing to a close. . . ." In this opening, Havel endorsed among other alternatives a search for "new scientific recipes, new ideologies, new control systems, new institutions, new instruments to eliminate the dreadful consequences" caused by previous ones. He also called for rehabilitation of "faith in the importance of particular measures that do not aspire to be a universal key to salvation. . . . We must see the pluralism of the world, and not bind it by seeking common denominators or reducing everything to a single common equation."[82]

Whether its "final crisis" is under way, scientific management continues to evolve amid increasing numbers of emerging alternatives that bear a family resemblance to adaptive governance. As noted in Chapter 1, this pattern has been documented in connection with natural resource policy in the American West.[83] We documented the pattern in connection with the climate change regime in Chapter 2 and on a local scale in Barrow in Chapter 3. William Easterly, former senior research economist at the World Bank for 16 years, documented a similar pattern in connection with the international development policies of the West since World War II. In The White Man's Burden (2006), Easterly described the Planners' approach prevalent in the tradition of scientific management: "Setting a beautiful goal such as making poverty history, the Planners' approach then tries to design the ideal aid agencies, administrative plans, and financial resources to do the job." His characterization of the results, while oversimplified, merits consideration as a critique of conventional wisdom: "Sixty years of countless reform schemes to aid agencies and dozens of different plans, and $2.3 trillion later, the aid industry is still failing to reach the beautiful goal. The evidence points to an unpopular conclusion: Big Plans will always fail to reach the beautiful goal."[84] He contrasted the Planners with the Searchers, who are "agents for change in the alternative approach." In this approach, "A Searcher hopes to find answers to individual problems only by trial and error experimentation." The Planners and Searchers are best understood as ideal types, oversimplified by design.[85]

Easterly put differences between them in historical context: "The debate between Planners and Searchers in Western assistance is the latest in a long-standing philosophical divide in Western intellectual history about social change. . . . Karl Popper described it eloquently as 'utopian social engineering' versus piecemeal democratic reform. This is pretty much the same divide as the one Edmund Burke described in the late eighteenth century as 'revolution' versus 'reform' (the French Revolution was a bloody experiment in utopian social engineering)." Like Scott, he found utopian social engineering schemes "in such diverse contexts as compulsory resettlement of Tanzanians into state villages and Communist five-year plans to industrialize in the Soviet Union and Eastern Europe." He also found them in "shock therapy" instead of gradualism in the transition from communism to capitalism in the former Soviet bloc; in comprehensive "structural adjustment" reforms in Africa and Latin America required by the International Monetary Fund and the World Bank beginning in the 1980s; and in the UN's well-intentioned Millennium Project, the current Big Push to end world poverty, which "shows all the pretensions of utopian social engineering."[86] Meanwhile, "Some success stories show that aid agencies can make progress on [local] problems. . . . Aid agencies could do much more on these problems if they were not diverting their energies to utopian Plans and were accountable for tasks such as getting food, roads, water, sanitation, and medicines to the poor."[87]

Community Development

In the postwar period many Searchers' initiatives consistent with adaptive governance have emerged and realized enough progress to serve as models. One of the earliest was the Cornell–Peru Project in Vicos, led by Allan Holmberg and his team of anthropologists from Cornell University, which coincided with the initial Big Push in the 1950s. The story can be told in some detail because it has been well documented. It is worth telling to surface the complexities that lie beneath general suggestions or principles, including Easterly's, that share a family resemblance with adaptive governance. The Vicos experience and generalizations based on experiences like it can be used to inform next steps in adaptive governance for climate change. Others have also reframed climate change as a matter of community development.[88]

In the early 1950s, Vicos was a typical hacienda in the remote highlands of Peru and home to about 1,700 people. Variously called serfs, peons, peasants, and Indians, the Vicosinos (or "people of Vicos") were impoverished, malnourished, and powerless, obligated to work collectively on hacienda lands three days per household per week in addition to separate cultivation of small subsistence plots. Since 1594 the hacienda's absentee landlord was The Public Benefit Society of Huaraz, which leased Vicos to a patron "who held absolute control over the peons, giving or denying them permission to travel to market or elsewhere in search of paid employment, seek education or even to get married. . . . If asked, the people of Vicos described themselves as slaves, owned by the landlords."[89] Then the patron's payments on a 10-year lease lapsed. The Cornell team took over the remaining five years of the lease on the first day of 1952, with prior approval of the Peruvian government's Indian Institute. Mario Vázquez, a Peruvian anthropologist, had prepared the way through earlier field research in Vicos as part of Cornell's comparative study of modernization of peasant societies. In several months of preparation for the takeover, the project made plans to advance the human dignity of the Vicosinos.[90] Holmberg later wrote, "In designing our program and a method of strategic intervention, we were very much aware of two, among many, guiding principles stemming from anthropological research: first, innovations are most likely to be accepted in those aspects of culture in which people themselves feel the greatest deprivations; and second, an integrated or contextual approach to value-institutional development is usually more lasting and less conflict-producing than a piecemeal one."[91] The contextual approach took into account interconnections among power and seven other values—and thereby avoided reduction of the project's ends to any single target, or the project's means to any single intervention.

The initial goals specifying what was meant by human dignity in each of the eight value categories guided the project throughout its history.[92] However, as team member Paul Doughty explained, the "initial strategy . . . was only a starting point in terms of the actual steps and actions which evolved as all parties to the project gained experience."[93] Using Vázquez's research, the project focused rather quickly on addressing major deprivations felt by the Vicosinos. One was the "obligation of the Indian households to supply the extra, free services (such as swineherds and cooks) to the manor."[94] The project abolished this obligation, and then paid volunteers to perform the services as needed. The project also reimbursed each serf for back pay owed by the previous patron; the cost, roughly 3 cents per week over the previous three years, was deducted from the project's payment for transfer of the lease. "Through such small but immediately reinforcing interventions, a solid base for positive relations with members of the community was first established."[95] It helped that Vázquez already was known and trusted by nearly every Vicosino. No doubt other project personnel also helped: "From the very beginning, for example, an equality of salutation was introduced in all dealings with the Vicosinos; they were invited to sit down at the tables with us; no segregation was allowed at public affairs; Project personnel lived in Indian houses."[96]

Meanwhile, the onset of the project "coincided with a failure of the potato harvest of both the Patron and the serf community due to a blight which had attacked the crop. The poor of the community were literally starving, and even the rich were feeling the pinch. Complaints about the theft of animals and food were rife."[97] Having funds for research only, not capital investments, the project relied on available resources. "The principal resources . . . were the labor of the Indian community and the lands which had been formerly farmed by the overlord. By employing this labor to farm these lands by modern methods (the introduction of fertilizer, good seed, pesticides, proper row spacing, etc.), and by growing marketable food crops, capital was accumulated for enlarging the wealth base. Returns from these lands, instead of being removed from the community, as was the case under the traditional system, were plowed back into the experiment to foment further progress toward our goals."[98] Further progress included the development of educational facilities and new skills in the population in addition to increases in agricultural productivity. "At the same time, new techniques of potato production and other food crops, first demonstrated on Project lands, were introduced to the Indian households which, within a couple of years, gave a sharp boost to the Indian community. In short," according to Holmberg, "by 1957, when Cornell's lease on the land expired, a fairly solid economic underpinning for the whole operation had been established, and the goal of enlarging the wealth base had been accomplished."[99] The Vicos community took over payments on the lease in 1957.

To initiate the devolution of power, the project from the outset reversed past policies by meeting with leaders of the traditional governing body, encouraging their participation, and listening to them. At the same time, the weekly gathering of serfs to issue work orders also became a forum where they were advised of project activities and encouraged to offer their views. "In the first five years of the Project . . . decision-making and other skills had developed to the point where responsibility for running the affairs of the community was largely in indigenous hands."[100] In mid-1956, project staff consulted Vicosinos, individually and in groups, "about what they wished for the future of Vicos upon the conclusion of the university's rental contract. The opinion of the majority was to purchase the manor, and, to this end, they suggested collective work on the manor lands with the proceeds of this labor going to pay the cost of the purchase."[101] To organize the effort, a community council was established with elected representatives from 10 geographic zones corresponding to the population centers. After the first elections in October 1956, council officers with outside help obtained a resolution from the government authorizing purchase or expropriation of the manor. This initiated "a long and sterile struggle . . . between the Vicosinos and the Public Benefit Society of Huaraz, owner of the manor. . . . In this period of uncertainty and struggle, the people of Vicos became more united and cohesive behind their elected representatives." Also in this period, "the Vicosinos were aided and protected by a 'power umbrella' consisting of the Cornell Peru Project and the Peruvian Indian Institute."[102] Assisted by the U.S. ambassador to Peru and Ted Kennedy, they opened serious negotiations in the summer of 1961 on a purchase price and persuaded the government to approve the sale. On July 13, 1962, the Vicosinos purchased the hacienda and their freedom with the fruits of their own labor.[103] Thus, a decade after the Cornell–Peru Project began, the Vicosinos were governing and sustaining their own process of development as part of modern Peru. No longer needed, all personnel of the Cornell–Peru Project had left Vicos by the end of 1963 and returned from time to time only to see how the Vicosinos were doing and to renew acquaintances.

Success in Vicos activated interests in diffusion and scaling up but also incited opposition. "As Vicosinos mastered modern techniques, for example, they were approached by their Mestizo compatriots in the surrounding area, seeking advice on how to improve their own crops."[104] As word spread to every officially registered indigenous community, "the project was besieged by community delegations to its Lima office and multi-page letters written in elementary Spanish arrived recounting abuses and requesting project assistance, to 'do what you did in Vicos.'" At the same time, publicity about the progress of Vicosinos and their efforts to purchase the manor activated opposition from local landowners, who found it "inconceivable . . . that such a property might be sold back to the indigenous inhabitants." They employed many legal and political means to block the sale, and repressed— sometimes violently—other indigenous inhabitants who wanted to be like Vicos.[105] The influence of Vicos from the bottom up depended on shifts in national politics that opened windows of opportunities for supporters at higher levels. Inspired by Vicos, two ministries of the Peruvian government established a program in 1957 to adapt the approach elsewhere.[106] In August 1969, Vázquez was recruited by the new military junta. "With extensive powers he was able to implement, on a national scale, many of the kinds of measures tested and proved during the Vicos experiment toward developing a prototype for improving peasant life by increasing the peasant share of power."[107] But such reforms were not necessarily lasting, and opponents somehow restricted the influence of the Vicos experience on international development policies. Nevertheless, the Vicosinos have sustained their own process of empowerment and development. They are much better off than they were in all value categories, as documented by Doughty in a return visit to Vicos in July 1997.[108]

For Holmberg in 1970, "The major lesson of Vicos, for Peru as a whole, is that its serf and suppressed peasant populations, once freed and given encouragement, technical assistance, and learning, can pull themselves up by their own bootstraps and become productive citizens of the nation. The Vicos experience also proved that this development did not require, in essence, vast investments of capital from outside sources."[109] Doughty supplied the details: "With the continuous involvement of the people, the 'hands-on' research-and-development strategy that coordinated the efforts of various agencies and experts already assigned to such tasks, the final costs of the project (an estimated $711,000 or about $35.00 per capita per year) were modest when compared to the large-scale 'trickle-down' development schemes so much in favor, with per capita expenditures many times that amount."[110] Holmberg concluded that the Vicos experience "showed that with the appropriate opening of opportunity and strategic intervention from the outside (particularly technical assistance), such an impoverished and exploited community as the serfs of Vicos could actually finance, to a large degree, the changes necessary to provide for a greater sharing of human values. The element of power proved to be the key that permitted the Cornell Peru Project to open the door to change; the devolution of power to the people of Vicos proved to be the mechanism which made the new system viable."[111] Harold Lasswell affirmed power as the key to success. He noted that the supporting coalition did not include every peasant, scientist, or politician, but only "those with a self-commitment to human dignity. . . . [They were] self-selected on an informal basis."[112] Self-selection is an important insight into the transferability of the Vicos strategy and adaptive governance.

Similarly, the Inter-American Foundation has emphasized the devolution of power. With appropriations from the U.S. Congress, it has funded a few dozen proposals for grassroots development each year for more than 30 years. According to IAF Vice President Patrick Breslin, these projects "are conceived and managed by local people trying to solve their own community's problems, not by outsiders who decide that what is needed is family planning or education or hydroelectric dams or health care or farm-to-market roads or any of the other magic-bullet ideas that have surfaced and sunk in the last half century."[113] It is not that any of these ideas is inherently wrong. "What is wrong is assuming that you can change one factor in a complex system and then predict the outcome. . . . Over and over, ambitious development schemes are planned with impressive internal coherence, and then imposed from the top down on a complex human system."[114] But the results are seldom predicted or controlled for reasons that Breslin found articulated in scientific studies of chaos and complexity: Because "it's effectively impossible to cover every conceivable situation, top-down systems are forever running into combinations of events they don't know how to handle. . . ."[115] In contrast, IAF's grassroots development approach "relies more on the capacity of poor people to understand their own problems and craft their own solutions—often in dialogue with local technicians—than it relies on projects designed from the outside."[116]

In Breslin's judgment, the failures of top-down development projects stemmed from linear and mechanistic thinking about human society powerfully shaped by the metaphor of Newtonian physics. He asked, "What would a nonlinear development model look like? Metaphors from chaos and complexity studies suggest that it would look very much like what we call grassroots, participatory, bottom-up development."[117] Breslin nevertheless insisted, "This contrary approach to development was shaped not by an intellectual paradigm, but by experience."[118] He outlined what top-down development policy would look like if we stopped thinking in linear and mechanical terms and took post-Newtonian scientific metaphors more seriously:

- "[O]ne of the first things to disappear would be the illusion of control inherent in projects designed from the top down."
- "[P]ower must be ceded and dispersed downwards to permit adaptive behavior and the emergence of new patterns."
- "Projects would be smaller . . . and more numerous. . . . The individuality and uniqueness of projects would be recognized."
- "Goals would be clear, but the focus would be much broader than the scorecard or checklist marking their attainment."
- "Evaluation would become less a measurement of progress toward externally set goals and more of a feedback mechanism into a human group's evolution."
- "Development workers would necessarily spend more time out of their offices and move closer to the grassroots." They would stick around long enough to see their mistakes.
- "Development workers . . . would become facilitators, enabling representatives of other communities to visit and see first hand what in the successful project they would wish to replicate."[119]

All of this is consistent with adaptive governance as we conceive it. Breslin reassured those involved in conventional approaches to development that "moving resources to a nonlinear model would not mean the end of government-to-government, or international bank-to-government, foreign assistance."[120] Assistance based on the linear model, such as a vaccination program, is often appropriate. Grassroots development, like adaptive governance, is a matter of opening the established frame, not throwing it away. Breslin also projected that the evolution and diffusion of IAF's grassroots development approach along these lines would counter "the tendency to blame failed development efforts on the 'target population's supposedly inferior culture.'" That can happen when the population's preferred roles differ from roles imposed by outsiders. "Grassroots experience, on the other hand, teaches that often it is precisely the values of the local culture that drive successful projects."[121]

Poor people have been key figures in other development strategies adapted to their needs by outside aid workers through direct and sustained contacts on the ground. William Easterly, the former World Bank economist, has concluded, "Only the self-reliant efforts of poor people and poor societies themselves can end poverty, borrowing ideas and institutions from the West when it suits them to do so. But aid that concentrates on feasible tasks will alleviate the sufferings of many desperate people in the meantime."[122] As a consultant to the World Bank, Frank Penna organized poor musicians in Senegal and helped them secure royalties from commercial performances of their recordings, reduce piracy of their recordings, and participate in the government's application for a World Bank loan. He has diffused and adapted this model to other forms of intellectual property and to other countries within and outside Africa.[123] Working for UN Habitat's Urban Rehabilitation Project, Samantha Reynolds organized neighborhood women in Mazar-i-Sharif in northern Afghanistan to parlay their traditional skills of weaving, tailoring, embroidery, and hairstyling into a series of small businesses that helped support a community meeting place, literacy courses, a playground and kindergarten, and eventually led to an elected council for neighborhood governance. This model was diffused, adapted to other neighborhoods in Mazar at their request, and scaled up to include other communities in Afghanistan. Afghanis sustained it by going underground when the Taliban took over and aid workers were forced out in September 1997. Afghanis advocated this model in negotiation of the Bonn Agreement after the end of Taliban rule late in 2001.[124] Similarly, Paul Polak, a psychiatrist and entrepreneur who founded International Development Enterprises (IDE) in 1981, insisted that "making it possible for very poor people to invest their own time and money in attractive, affordable opportunities to increase their income is the only realistic path out of poverty for most of them."[125] Polak claimed that "IDE has already sold more than 2 million treadle pumps to dollar-a-day families, increasing their net annual income by more than 200 million dollars a year and creating a multiplier effect on poor villages of at least $500 million a year."[126]

Thus, multiple strategies for community-based development have adapted to differences and changes in context, but nevertheless share a family resemblance.[127] For those interested in mainstreaming responses to climate change as a matter of community development, three additional lessons are worth emphasizing. First, outside assistance is most likely to be accepted, and to leverage needed local resources, when it addresses specific deprivations experienced by people on the ground. Second, outside assistance that improves one value outcome is more likely to have lasting effects if integrated with improvements in other value outcomes in the same community. Each improvement can increase the resources available to invest in the next step in a self-sustaining process. Third, progress in the development of one community tends to activate interests in scaling up the model by diffusing and adapting it in other communities. These interests, however, may be blocked by special interests defending the status quo. The first two lessons were known well enough to plan the Cornell–Peru Project more than a half-century ago. All three lessons were corroborated by subsequent experience in Vicos. Since then, such lessons have been rediscovered time and again to remedy problems arising from scientific management in various forms. The problem is not lack of general knowledge about what works and what does not. Allowing for exceptions, the problem is an apparent inability to apply in specific contexts existing knowledge that lies outside the established frame of reference.

Possible Futures

In conclusion, we turn briefly to the possible futures of community-based initiatives consistent with adaptive governance and of business as usual in the tradition of scientific management. Of course these do not exhaust the possible futures of responses to climate change, but they do help clarify important individual choices and collective decisions that will be made in the years ahead. We can be brief because there are no observations on the future, only projections based on the relevant past the underlying dynamics. Outcomes will be subject to many interacting contingencies; the most important of these for minimizing net losses from climate change are human choices and decisions.

Community-Based Initiatives

The basic trend toward adaptive governance consists of the rise of community-based initiatives in favorable circumstances here and there. Some have made considerable progress, and some have not, according to outcomes assessments that are interim, not final: Documentation is limited in most cases, and development in response to climate change (or any other challenge) is an open-ended process. In this book relatively successful cases of adaptation are represented by Tsho Rolpa in Nepal, Sisimiut in west Greenland, the Pacific islands assisted by PEAC, Melbourne in Australia, and distributed policies in Barrow, Alaska. Relatively successful cases of mitigation are represented by Ashton Hayes in the United Kingdom, Samsø in Denmark, Toronto, Portland

and other cities affiliated with ICLEI, as well as California and other states. These cases represent only a small fraction of the trend. Recall that 546 local governments were involved in ICLEI's Cities for Climate Protection project in 2006. And each year thousands of local communities must respond, adaptively or not, to losses from extreme weather. Much of the available experience lies below thresholds of national and international attention and remains undocumented. The exceptions tend to be presented as stand-alone cases, not part of a larger pattern, even if general lessons are drawn.[128]

The trend also includes the organization of overlapping networks of community-based initiatives. As noted in Chapter 2, ICLEI is the oldest and largest network focused on climate change mitigation, and it includes several affiliated networks. The nine RISA programs and the Alaska Native villages vulnerable to coastal erosion and flooding are at least potential networks focused on climate change adaptation. As an illustration of the many networks being formed worldwide, Box 5.1 lists 10 additional networks focused mostly on adaptation. The first, the Urban Leaders Adaptation Initiative, recently reported premises, processes, and lessons quite consistent with adaptive governance as we understand it.[129] Its experience corroborates our point that context matters: "Each of the 10 Urban Leaders partners started with different resources and histories of climate change action, and are in various stages of development and implementation of adaptation strategies. . . . Despite the similarities, each partner city or county has exemplified unique planning and implementation strategies and approaches."[130] Both within and between networks of community-based initiatives, there is no single formula, no "one best way," that can accommodate all significant differences. Outcomes are contingent on the interaction of many factors in each context.

The underlying dynamics in the cases we have examined help clarify possible futures for adaptive governance. Within the constraints of available information, it appears that leaders in each community-based initiative have been motivated by deprivations (including missed opportunities) with respect to existing interests. For climate change activists, those interests typically include averting adverse impacts expected from climate change in the near term or the long term, taking the moral high ground on behalf of the local if not global environment, and other interests more broadly shared within the community. The more broadly shared interests sometimes include preventing further loss of life and limb, property, and other things of value in the aftermath of damaging storms, floods, fires, and droughts, and "no regrets" interests relatively independent of climate change that are even more broadly shared—for example, providing more local jobs, profiting from local

Box 5.1. Organizing Networks: 10 Recent Initiatives

This is an illustrative rather than exhaustive list of recent initiatives in organizing networks in response to climate change.

The *Urban Leaders Adaptation Initiative* at the Center for Clean Air Policy is a partnership with Chicago, King County, Los Angeles, Miami–Dade County, Milwaukee, Nassau County, Phoenix, San Francisco, and Toronto formed in 2006 to assist in mainstreaming local climate change resilience strategies, disseminating information on what works locally, and developing recommendations for state and national governments to support local adaptation efforts.

The *Clinton Climate Initiative* is helping 40 of the world's largest cities develop and implement large-scale projects to modify existing buildings, transit systems, lighting, and waste management to reduce energy use and greenhouse gas emissions. For example, CCI has negotiated discounted pricing agreements with manufacturers of energy-efficient products.

The *Urban Climate Change Research Network*, formed in 2007 at the time of the C40 Large Cities Climate Summit, plans an assessment report in 2009 to cover both the state-of-knowledge and the state-of-action mitigation and adaptation in cities and to highlight the role of communities in cities with vulnerable populations as participants in formulating responses.

The *National Indigenous Climate Change Working Group* was formed by indigenous Australians to clarify their aspirations for inclusion in national climate change policies. The group seeks intelligence and support from corporations, government, and research institutions and is working toward commissioning pilot projects in selected communities.

Many Strong Voices is a community-based collaborative promoting the sustainability of coastal communities in the Arctic and small island developing states in response to climate change. The

investments, reducing energy costs, increasing energy self-sufficiency, working together as a community, and gaining respect and recognition.

At the outset, leaders tend to rely primarily on such intangible resources as trust, respect, skill in interpersonal relations, and knowledge of the particular community; the political support, funds, and other resources needed to implement specific projects typically come later. Their task is to organize community members by discovering through personal interactions the shared interests that might support projects to reduce losses from climate change. Shared interests in turn provide the basis for identifying with the community effort and accepting some near-term sacrifices on its behalf to realize larger rewards later on. That can lead to consensus and action on specific projects, but, of course, there is no guarantee of success. Much depends on whether a project is designed and implemented well enough to succeed, five-year action plan includes comparative case studies, outreach for training, networking and fundraising, and the development of strategies to influence decision makers.

The *Climate Witness Program* of the World Wildlife Fund collects observations of climate change based on indigenous knowledge to support adaptations. In response to such observations, the WWF in partnership with the Fiji government and a Japanese charity sent water tanks to Kabara Island, Fiji, to mitigate freshwater shortages in March 2006.

The *Community Based Adaptation Exchange* supports the exchange of information on community-based climate adaptation in the United Kingdom, produced by the Institute of Development Studies and the International Institute for Environment and Development.

Transition Towns, begun in Ireland and the United Kingdom, is a loose, mostly online affiliation of community organizations in 14 countries addressing climate change and peak oil issues. The nonprofit Transition Network Ltd. supports them.

The *Global Climate Change Collaborative* (G3C), initiated by the MIT–U.S. Geological Survey Science Impact Collaborative (MUSIC) "is a network of institutions around the world that conduct action research projects to help communities, planners, and policy makers develop adaptive management and adaptive governance processes to prepare for the impacts of climate change." (http://web.mit.edu/dusp/epp/music/G3C/index.html)

World Wide Views on Global Warming, initiated by The Danish Board of Technology and The Danish Cultural Institute, is a project to organize citizens around the world to contribute their views on issues to be addressed by negotiations of the Conference of the Parties to the UNFCCC meeting in Copenhagen in December 2009. (http://www.wwviews.org/)

whether failed projects are terminated to free up resources for creative policy alternatives, and whether successful projects capitalize on opportunities to augment the resources available for new initiatives. Many factors can be the one critical factor limiting or killing an initiative. Doing all of them well enough to sustain progress cannot be a *prerequisite* for action; no one ever knows enough at the outset. Realistically, sustained progress can only be the *outcome* of a series of interactions in which leaders often become followers, and vice versa, on various parts of the collective effort. All of this has been done at the local level, where it is much easier to do than at regional, national, and global levels. Harvesting experience from more cases, particularly successes, is a major research priority for improving success rates.

When they become aware of their shared interests, leaders of community-based initiatives or their partners tend to organize networks. One shared interest is swapping experience on what has worked and what has not, both on the ground in their own communities and in their interactions with officials at higher levels. Each community could use a handful of field-tested models relevant to its own circumstances to consider for possible adaptation. But each also should be aware of what has *not* worked, to avoid replication without modification. Another shared interest is seeking external resources to support what works on the ground, by informing and lobbying higher-level officials willing and able to help. Leaders can expect to meet more of their common needs for external resources by working together than separately. Finally, harvesting experience from existing climate change networks to identify and correct malfunctions is a major research priority. Aside from improving success rates among community-based initiatives, such research can help scale up successful initiatives to make significant reductions in net losses from climate change on larger scales, at regional, national, and international levels.

If this understanding of the dynamics is approximately correct, then recent developments seem more likely to change the mix of motivations and opportunities that have sustained the trend toward adaptive governance than to reverse it. The economic recession that took hold in the United States in late 2007 and rippled around the world in 2008 tends to further subordinate climate change interests to economic interests for most people. But for climate change activists at all levels, it provides additional opportunities to reduce greenhouse gas emissions by harnessing some of those economic interests on a "no regrets" basis. For example, energy costs can be reduced in the short term through improvements in conservation and efficiency, as in Ashton Hayes. Development of renewable energy resources for a low-carbon future and energy independence, as on Samsø, can be financed by increases in public investments intended primarily to boost aggregate demand and employment. If there is less private capital available for mitigation of climate change, there will be more public funds available to assist projects that reduce greenhouse emissions, at least in the United States.[131] The effective use of those funds will depend on multiple limiting factors, including the leadership of local activists to mobilize the consent and participation of their relatively indifferent neighbors. The election of Barack Obama as president of the United States renewed hopes in the United States and worldwide for action on climate change mitigation from the top down through U.S. participation in a successor to the Kyoto Protocol and a U.S. cap-and-trade system. But the president, once a community organizer in Chicago, also believes that economic and political change comes from the bottom up, and he campaigned on that basis using the Internet to mobilize grassroots financial and political support.[132] The extent to which that approach to campaigning carries over to governing, including climate change mitigation and adaptation, remains to be seen.[133]

Meanwhile, local communities and national governments will have to respond one way or another to the "normal" load of climate-related disasters, even if projections of increased losses turn out to be mistaken. If not, expect more such disasters to intensify and disseminate demands to reduce losses through climate change adaptation and mitigation, and to give these policy alternatives an advantage at the margin over other problems in the competition for attention and other resources at each level. Moreover, there are growing concerns about the progress and prospects for climate change mitigation among leaders in the regime. For example, in June 2008, at a briefing on the twentieth anniversary of James Hansen's famous testimony, members of Congress called Hansen a "hero" and "a latter-day climate-change Paul Revere." Hansen replied that "our actions to deal with climate change over the past 20 years have really been minuscule and we're really running out of time." Earlier the same day, in response to a standing ovation at the National Press Club, Hansen said, "Actually, it's not a time to celebrate. Although the issue has become popular, the fact of the matter is that the emissions are continuing, basically unfettered."[134] In general, such heightened concerns can motivate the concentration and intensification of efforts on business as usual, on the expectation that the level of effort alone has been the limiting factor; therefore, repeating the message more frequently at higher volume will succeed. Or such heightened concerns can motivate a search for policy alternatives within business as usual and in other approaches beyond, on the expectation that some major efforts have been misinformed or misdirected. Among the other approaches available is adaptive governance.

Attention is a scarce resource for all of us who are less than omniscient, and a critical resource for all reform movements. Reform alternatives cannot gain broader support unless they are brought to the attention of people who might be favorably predisposed toward them. But attention for the most part is controlled by established institutions. Realizing the full potential of adaptive governance depends on reallocating attention at the margin from the regime's agenda to community-based initiatives that have succeeded in reducing net losses from climate change and to their supporting networks. This would help people already engaged in some aspect of adaptive governance understand their individual efforts as part of a larger movement, organize accordingly, and hold their own in the competition for constituencies and other resources. Reallocating attention also could motivate and inform larger constituencies, including people who are more committed to reducing net losses from climate change than to defending the established approach and policies, and people who are predisposed to become engaged on a "no regrets" basis.

For at least two decades, however, the attention frame in climate change policy has been more or less monopolized by the regime's scientific management agenda—primarily scientific assessments by the IPCC and others, the quest for mandatory, legally binding targets and timetables to reduce greenhouse gas emissions under the UNFCCC, and defenses of the quest against critiques by climate change skeptics. But not all the critiques come from skeptics. Consider, for example, the relative neglect of three major critiques, introduced in Chapter 1, of the established way of thinking and doing in the regime. Each was published in a prominent scientific journal but has been only rarely cited in the scientific literature. Rayner and Malone's commentary on "Zen and the Art of Climate Maintenance" was published in *Nature*, one of the most prominent scientific journals, in 1997, but was cited only 12 times through 2008, according to the ISI Web of Knowledge. Some articles in *Nature* from that year have been cited more than 1,000 times. Similarly, David Cash's viewpoint in "Distributed Assessment Systems" as an emerging paradigm was published in *Global Environmental Change* in 2000, but was cited only eight times through 2008.[135] Hendrik Tennekes's complaint about the hubris implied by "Managing Planet Earth" and similar aspirations was published in *Weather* in 1990, but was cited only six times through 2008; three of those citations were by one of us. A subsequent major critique, "A Madisonian Approach to Climate Policy," by David Victor and colleagues was published in *Science* in September 2005. It was cited only 12 times through 2008.[136] The degree of attention given to these major critiques is not enough to support the self-organization of a new specialization in climate change science and policy that is self-identified as "adaptive governance" or an equivalent symbol of convenience such as "Madisonian." However, the possibility of a new specialization is suggested by the many scientists cited in these pages for their contributions toward adaptive governance.

Similarly, major exceptions to scientific management in the evolution of the regime have been relatively neglected. We found no references to the Pacific ENSO Applications Center (PEAC) in "major world publications" as that category is defined in LexisNexis Academic,[137] and only five references in the scientific literature through 2008. PEAC's work in connection with the 1996–1997 El Niño nevertheless provides a valuable model for action on climate change adaptation initiated from the top down. Samsø is a valuable model for mitigation from the bottom up, but was mentioned only 10 times in major world publications since the beginning of 1990. The Samsø Energy Academy's Web site notes additional coverage of the renewable energy island and reports that Hermansen was named a "Hero of the Environment 2008" by *Time* magazine.[138] These are welcome steps toward fuller recognition of what has and can be achieved locally. ICLEI, the most prominent network, has been mentioned in only 76 stories in the same sources since the beginning of 1990, but it is encouraging that 28 of them occurred in 2007 or 2008. In comparison, in one month alone—December 2008—major world publications mentioned the Kyoto Protocol in 347 stories and emissions trading in 646 stories.[139] Evidently, these publications expect climate change policies to come from the top down. It would be extremely difficult for climate change activists interested in proceeding from the bottom up to find each other through these publications.[140]

Attention is one barrier to realizing the full potential of adaptive governance to reduce net losses from climate change. Careful consideration is another. It cannot be taken for granted, given examples of premature dismissal like the one introduced at the outset of this chapter and given the likelihood of political backlash. Backlash will occur as more attention begins to be allocated to examples of intensive science, procedurally rational policies, and decentralized decision making; the examples are recognized as integral parts of an emerging approach; and the approach gains support. However, like all proposals for reform, the proposals for adaptive governance are debatable, at least until there is experience sufficient to resolve whether the means employed will advance the ends sought at a reasonable cost. Until then, and so long as the outcomes of the established approach continue to be disappointing, if not frustrating as assessed according to the ultimate objective of the UNFCCC, adaptive governance and other approaches to opening climate change regime are worth debating. And that requires careful consideration, to distinguish constructive criticisms from others. By "careful consideration," we mean comparative evaluation of established policies in climate change science, policy, and decision making with reform policies in light of explicit criteria and evidence that is detailed and comprehensive. This is nothing more than critical thinking to deter the substitution of scientific management criteria for common interest criteria and to encourage the evaluation of both established and reform policies by the same criteria. Critical thinking is necessary to avoid the circular and self-justifying logic of *Voltaire's Bastards*. For those of us genuinely concerned about climate change, the issue is not how to defend this or that policy alternative but how to improve upon the disappointing outcomes to date. The appropriate standpoint takes "reforms as experiments" in D. T. Campbell's sense.

From this standpoint, a comparative evaluation is important because established policies, especially the mainstay quest for mandatory international targets and timetables, provide the baseline for considering how to improve outcomes. They need to be compared with reform alternatives that might work better. Exempting the established policies amounts to a defense of business as usual. The more specific the alternatives considered, the easier it will be to evaluate them dependably and to find consensus on what to do next. The alternatives field tested in cases are more specific than the next steps suggested earlier in this chapter; the next steps are more specific than the proposals in Box 1.1; and the proposals are more specific than adaptive governance as a whole. Explicit criteria are necessary for assessing what constitutes an "improvement" in outcomes. Our criteria are made explicit in the section The Common Interest in Chapter 1. The ultimate objective of the UNFCCC is approximately equivalent, if emphasis is placed on preventing what is "dangerous" to the multiple interests authorized in the treaty, and the means go beyond stabilization of greenhouse gas concentrations in the atmosphere to include adaptation, which is partially recognized in the second part of Article 2. Leaving criteria implicit helps special interests displace the common interest, however conceived. Detailed and comprehensive evidence is necessary for dependable evaluations of established and reform alternatives from a common interest standpoint. Otherwise, as noted in Chapter 4, debate deteriorates into competing lines of censorship and propaganda that does less to inform thinking and doing than to control them. Censorship restricts attention to reform policies; propaganda attacks or defends through issue expansion and contraction, the absorption of uncertainty, and partial incorporation, among other tactics. To "inform, invite, and ignore" community-based initiatives is an example of partial incorporation; it accepts community-based initiatives in principle but subordinates them in practice.

In conclusion, we expect more community-based initiatives to arise here and there in response to actual and expected losses from climate change, and some will organize as networks. Whether they can realize their potential to contribute significantly in reducing net losses on regional, national, and international scales is contingent primarily on whether they gain attention and careful consideration—enough to allow scientists and practitioners already involved in some aspect of adaptive government to communicate and collaborate effectively, and to attract the interest and support of others who are frustrated by the disappointing past or projected outcomes of business as usual. We can envision the possibility of a tipping point of the kind that occurred around 1900 in the United States. At the time, federal bureaucracies were a collection of separate reform efforts motivated by failures of established institutions to cope satisfactorily with problems of industrialization and related social changes. "Between 1877 and 1900, America's state-building vanguard waged a series of losing struggles to restructure American government around a national bureaucratic regimen."[141] After the turn of the century, these efforts came together: "Administrative reformers catapulted from their position as institutional outcasts struggling for recognition at the periphery of power into coalitions with major institutional actors willing to alter the basic determinants of governmental operations."[142] Similarly, activists deprived of civil rights in the United States struggled as institutional outcasts after World War II but gradually organized as a movement and gained recognition under the leadership of Martin Luther King, Jr., among others: President Truman ordered desegregation of the military, the Supreme Court ordered desegregation of public schools, President Eisenhower enforced the order, and so on. Perhaps the tipping point came with the Civil Rights Act of 1964 and the Voting Rights Act of 1965 under the leadership of President Johnson. Such reforms occur through an incremental political process in which human choices and decisions are the critical contingencies.

Business as Usual

As documented in this book, the basic trend in business as usual for the last two decades has been the quest for mandatory international targets and timetables to reduce greenhouse gas emissions. The quest has been stymied for the most part by a lack of political will in one part of the climate change regime, governments in the Conference of the Parties acting collectively, despite the persistent efforts of another part of the regime, the epistemic community. The quest nevertheless has been sustained by the aspirations of scientific management, which, as we have seen, have a long history in Westernized parts of the world and were institutionalized in the climate change regime in the few years around 1990. Under these circumstances, most scientists, reporters, public officials, and other contributors to public discourse on climate change find it easier and more rewarding to elaborate the established agenda, especially if they are unaware of other possibilities.[143]

Those who direct attention to other possibilities run the risk of inattention or rejection if their critiques or challenges are not rationalized as consistent with business as usual. Others in the regime have actively defended the quest and enforced the agenda against critiques and challenges, on the expectation that deviations threaten the existing distribution of resources if not the eventual success of the quest. For example, until about a decade ago, adaptation to climate change was generally considered a deviation from the quest and a threat to it. Adaptation is more widely accepted today but still relatively neglected; it is less amenable to scientific management.

Meanwhile, with increasing concerns that progress has been miniscule and time is running out, the still-dominant consensus on the quest has been coming apart at the margin in the United States. This is manifest in the rise of additional mitigation policy alternatives in the attention frame in recent years. Apart from the election of Barack Obama, those concerns have not been alleviated by the most recent developments. In June 2008, the Senate defeated a motion to close debate on the Lieberman–Warner bill, the leading attempt to mandate a national system to cap-and-trade emissions. The outcome, 48 for and 36 against, fell short of the required 60 votes and thereby deferred further action. Ten Democrats who voted to close debate said they would have opposed the bill if it had come to a final vote.[144] In December 2008, at the meeting of the Conference of the Parties in Poznan, Poland, Sen. John Kerry said that China and other fast-growing developing nations exempt from the Kyoto Protocol will have to accept some kind of emissions reduction target for the United States to ratify an emissions reduction treaty.[145] A few days later, an overview of the energy and climate situation confronting the new administration affirmed that "while Mr. Obama will enjoy a larger Democratic majority than Mr. Clinton did in his two terms, the Senate has long made such steps a prerequisite for its required consent to any climate treaty." Moreover, "The intense ideological and regional rivalries that have stalled climate change legislation in Congress for years have not suddenly melted away. And even though Mr. Obama promises to give energy legislation a high priority, he first must stabilize an economy that is shedding jobs by the hundreds of thousands each month."[146]

Meanwhile, the meeting of the Conference of the Parties in Poznan reportedly "ended . . . with few signs of progress on the main goal, limiting emissions of heat-trapping gases without hampering economic development. It is widely felt that if the United States does not demonstrate concrete domestic steps to curb its emissions from burning fossil fuels, fast-growing developing countries will continue to balk at taking on obligations to cut their emissions."[147] The United States and nearly all other nations have made commitments to complete a new global climate treaty in Copenhagen in December 2009, but fulfilling those commitments will be further complicated by the global economic crisis. According to Executive Secretary of the UNFCCC Yvo de Boer, "You can't pick an empty pocket. . . . I think that the financial crisis is going to make it more difficult for industrialized countries to make public resources available for cooperation with developing countries."[148] Instead, the implication was that industrialized countries will turn inward to invest in economic recovery at home. That is significant from de Boer's perspective because "a pledge of northern investment in developing countries, for 'green' economic growth and for adapting to droughts, floods and other impacts of warming, would be essential to get poorer nations to sign onto a new global climate agreement."[149]

The other policy alternatives that have been gaining attention in the United States in recent years are variations on business as usual, within the constraints of scientific management. One is a revenue-neutral gasoline or carbon tax as an alternative to a cap-and-trade system. For example, in June 2008 it was reported that James Hansen "disagrees with supporters of 'cap and trade' bills to cut greenhouse emissions, like the one that foundered in the Senate this month. He supported a 'tax and dividend' approach that would raise the cost of fuels contributing to greenhouse emissions but return the revenue directly to consumers to shield them from higher energy prices."[150] At the end of 2008, Thomas Friedman contended, "Without a higher gas tax or carbon tax, Obama will lack the leverage to drive critical pieces of his foreign and domestic agenda. . . . And he could make it painless: offset the gas tax by lowering payroll taxes, or phase it in over two years at 10 cents a month." In this view, a higher gas or carbon tax is a "win, win, win . . ." or common interest alternative: It would boost demand for fuel-efficient cars, attract more investment in storage battery technology to help expand wind and solar industries, reduce demand for oil imports, dry up petrodollar financing of terrorists, increase leverage against petro-dictatorships, reduce the current account deficit, and strengthen the dollar. In addition, such a tax "reduces U.S. carbon emissions driving climate change, which means more global respect for America."[151] Meanwhile, according to two conservatives, "Conservatives don't support tax increases that are veiled as 'cap and trade' schemes for pollution permits. But offer us a tax swap, and we could become the new administration's best allies on climate change. A climate-change bill withered in Congress this summer because families don't need an enormous, and hidden, tax increase. If the bill's authors had instead proposed a simple carbon tax coupled with an equal, offsetting reduction in income taxes or payroll taxes, a dynamic new energy security policy could have taken root."[152]

Two of President Obama's economic advisors "have both argued that a tax on carbon emissions from burning gasoline, coal and other fuels might be a more economically efficient means of regulating pollutants than a cap-and-trade system. . . ." However, President Obama and his environmental advisor, Carol Browner, reportedly have "ruled out a straight carbon tax, perhaps mindful of the stinging political defeat the Clinton administration suffered in 1993, when, prodded by Mr. Gore, it proposed one."[153] In his campaign, candidate Obama promised "to use a cap-and-trade bill for curbing heat-trapping gases as both the means of shifting investments away from energy sources that cause emissions of such gases and also as the source of the $15 billion a year he promised to invest in advanced energy technology. That figure may be dwarfed by spending on stimulus programs, including so-called green projects like building wind farms and making buildings more energy efficient."[154]

A Big Push on advanced low-carbon technologies has been another policy gaining attention. For example, according to Bjorn Lomborg in 2007, "We should commit ourselves to spending 0.05 percent of GDP in such R&D of non-carbon-emitting energy technologies. This approach would cost about $25 billion per year. It would increase funding to R&D about ten times." Evidently, "we" refers to the people or the nations of the world, as if they already were organized and committed to cooperate for this purpose. In any case, Lomborg's approach would support "research of all sorts, exploratory and applied" including "pilot programs to test and demonstrate promising new technologies. . . ."[155] For this skeptical environmentalist, a Big Push is the alternative to the main quest: "The fundamental problem with today's climate approach is that ever stricter emissions controls—as in Kyoto and a possible, tighter Kyoto II—are likely to be unworkable. . . . Kyoto is at the same time impossibly ambitious and yet environmentally inconsequential."[156] For others, a Big Push is an essential complement to cap-and-trade systems in the main quest. In April 2008 Andrew Revkin reported a shift in the global warming debate: "[N]ow, with recent data showing an unexpected rise in global emissions and a decline in energy efficiency, a growing chorus of economists, scientists and students of energy policy are saying that whatever benefits the cap approach yields, it will be too little and come too late."[157] Economist Jeffrey Sachs spelled out the rationale: "Even with a cutback in wasteful energy spending, our current technologies cannot support both a decline in carbon dioxide emissions and an expanding global economy. If we try to restrain emissions without a fundamentally new set of technologies, we will end up stifling economic growth, including the development prospects for billions of people."[158] The general consensus was that the development of CO_2 capture and sequestration technology, solar electric plants, and plug-in hybrid cars, for example, "will only come about with greatly increased spending by determined governments on what has so far been an anemic commitment to research and development. A Manhattan-like Project, so to speak."[159]

To put this in historical perspective, as reported in Chapter 2, investments in technological research have been a cornerstone of the federal climate change strategy since the 1960s. In FY 2006 alone, appropriations for the Climate Change Technology Program ($2.773 billion) exceeded appropriations for the Climate Change Science Program ($1.709 billion) and for any other part of the federal climate change strategy. Meanwhile, others emphasize the deployment of available technology. For example, Adil Najam, coauthor of an IPCC report on policy options, cautioned, "You can do a tremendous lot with available technology. . . . It is true that this will not be enough to lick the problem, but it will be a very significant and probably necessary difference."[160] People in Darmstadt, Germany, developed and diffused the passive house, which licked the problem of heating homes without furnaces.[161] People on Samsø licked the problem of achieving carbon neutrality (and more) using available technology; it was not a limiting factor in that context. Their leader, Søren Hermansen, drew attention to local consent as the limiting factor: "The crucial point is that we have shown that if you want to change how we generate energy, you have to start at the community level and not impose technology on people."[162] Lomborg's expansive and massive R&D program overlooks local consent. It does, however, include top-down "policy incentives to encourage adoption of existing technologies, which in turn foster incremental learning and innovation that often lead to rapidly improving performance and declining costs."[163] In short, learning by doing is an important part of R&D.

More ambitious geoengineering policy alternatives also have been gaining attention. These include any "large-scale, intentional engineering of our environment for the primary purpose of controlling or counteracting changes in the chemistry of the atmosphere."[164] For example, mirrors launched into space or sulfate aerosols dispersed into the atmosphere could reflect solar radiation to counteract global warming; the growth of plankton that absorb CO_2 in the oceans could be artificially stimulated to control concentrations of CO_2 in the atmosphere. The history of geoengineering alternatives in U.S. climate policy has been traced back at least to 1965.[165] More recently, it includes a special section of a report by the National Academy of Science in 1992, a symposium at the annual meeting of the American Association for the Advancement of Science (AAAS) in 1994, and a special issue based on the symposium in *Climatic Change* in 1996. In that special issue, Stephen Schneider reasoned that "should the middle to high end of the current range of estimates of possible inadvertent climatic alterations actually begin to occur, we would indeed face a very serious need for drastic, unpopular action to change consumption patterns in a short time—a politically dubious prospect." Thus, "somewhat reluctantly" Schneider concluded that "engineered countermeasures need to be evaluated."[166] Geoengineering remained marginal in scientific research until 2006, when another special issue of *Climatic Change* with an editorial by Nobel Laureate Paul Crutzen sparked further intense scrutiny.[167] Like Schneider and Hansen, Crutzen expressed frustration about miniscule progress in reducing greenhouse gas emissions and suggested that geoengineering alternatives "might again be explored and debated."[168] Among other influential scientists, John Holdren, in his presidential address to the AAAS in 2007, called for "a serious program of research to determine whether there are geo-engineering options . . . that make practical sense."[169] The consensus in the climate science community is that "these schemes are last-resort solutions at best and contain many large and unknown consequences for human society."[170] Imagine the Environmental Impact Statement.

Researchers connected with the Departments of Defense and Energy have promoted geoengineering in more concrete terms. In 1997, Edward Teller, Lowell Wood, and Roderick Hyde explored different approaches to scattering 1% of incoming solar radiation to cancel global warming. They concluded that "research along several lines to study the deployment and operation in sub-scale . . . is *justified immediately* by considerations of basic technical feasibility and possible cost-to-benefit. . . . Today, our scientific knowledge and our technological capability already are likely sufficient to provide solutions to these problems. . . . Whether exercising of present capability can be done in an internationally acceptable way is an undeniably difficult issue, but one seemingly far simpler than securing international consensus on near-term, large-scale reductions in fossil fuel–based energy production."[171] Lowell Wood, a veteran of four decades at the Lawrence Livermore National Laboratory, was a key speaker at the NASA-sponsored conference "Managing Solar Radiation" in November 2006. As quoted by historian James Fleming, who participated in the conference, Wood said, "'Mitigation is not happening and is not going to happen.'... The time has come, he said, for 'an intelligent elimination of undesired heat from the biosphere by technical ways and means,' which, he asserted, could be achieved for a tiny fraction of the cost of 'the bureaucratic suppression of CO_2.' ... Wood calls his brainstorm a plan for 'global climate stabilization,' and hopes to create a sort of 'planetary thermostat' to regulate the global climate."[172] Fleming observed that Wood was "very sure of himself, he was very clear to the meteorologists who were in the group, that their expertise wasn't really relevant to this topic, it was, in his terms, all physics, he says, 'I understand the radiation budget, and I know how to attenuate these sunbeams.'"[173] In gambling terms, this is doubling down the bet on scientific management. This is also the epitome of hubris in climate policy.

That hubris has been balanced to some extent by skepticism like Fleming's and by opposition to geoengineering among climate research specialists. In 2006 Ralph Cicerone acknowledged opposition to Crutzen's paper: "I am aware that various individuals have opposed the publication of Crutzen's paper, even after peer review and revisions, for various and sincere reasons that are not wholly scientific. Here, I write in support of his call for research on geo-engineering...."[174] In the 1996 special issue, Daniel Bodansky raised basic questions of international governance including this one: "Does one country have the right to make decisions about projects that are specifically intended to affect the entire world? And if not, how should decisions be made?"[175] In 2008 Alan Robock reviewed social and political problems along with scientific and engineering problems in an article titled "20 Reasons Why Geo-Engineering May Be a Bad Idea." He concluded, "Scientists may never have enough confidence that their theories will predict how well geo-engineering systems can work. With so much at stake, there is reason to worry about what we don't know."[176] The ambivalence about doing and publishing geoengineering research was perhaps best expressed by Canadian researcher David Keith: "[M]y view on it is not that I want to do it, I do not, but that we should move this out of the shadows and talk about it seriously, because sooner or later we will be confronted with decisions about this, and it's better if we think hard about it, even if we want to think hard about reasons why we should never do it."[177]

In conclusion, we expect that action on mandatory international targets and timetables to reduce greenhouse gas emissions, gasoline or carbon taxes, or a Big Push in energy technology R&D will be motivated primarily by economic recovery and energy security interests. (Research on geoengineering is a different matter.) At the same time, business as usual and variations on it—in proportion to their scale and opportunity costs—will continue to be stymied within and outside the climate change regime by a lack of political will. This will continue to be manifest in familiar ways: in the substitution of scientific assessments and technology R & D for actions that deploy existing knowledge and technology; in the official proclamation of long-term goals that defer the costs of realizing them to future office holders and their constituents; in the negotiation of emissions-reduction targets and timetables lacking sanctions severe enough to be enforceable; in limited capabilities to measure compliance and otherwise enforce those regulations that do have some teeth; and in the neglect of appraisals to terminate policies that have not worked and to improve those that have. However, the political will necessary to act collectively and effectively will be strengthened if and when the projected adverse impacts of global climate change materialize and become too large to discount or ignore. By then it will be too late to minimize net losses from climate change through mitigation.

Choices and Decisions

For those of us concerned about climate change, individually and collectively, one course of action is to defend business as usual, including variations on it, as the only way to proceed. Another is to open business as usual to policy alternatives that lie outside the constraints of scientific management, including adaptive governance. Steps toward adaptive governance already have been satisfactorily field tested on a small scale here and there using existing resources, including knowledge and technology, political will, and leadership. These resources have been augmented as consequences of modest actions that paid off, not as prerequisites. With enough attention and careful consideration, these resources can be augmented still further, to significantly reduce net losses from climate change on regional, national, and international scales. This would require a change in "our way of thinking," the third major factor in humanity's relationship to the earth, according to Al Gore in *An Inconvenient Truth*—a change that recognizes the progress on climate-related problems made by local, community-based initiatives, and the potential of similar strides among the Peorias of the world. We can invest in community-based initiatives now, as well as in new technologies or other variations on business as usual.

As an opportunity to be more specific about the basic choices and decisions, consider a lengthy op-ed by Al Gore, "The Climate for Change," published shortly after the November 2008 elections in the United States.[178] Gore began with a new development: "The inspiring and transformative choice to elect Barack Obama as our 44th president lays the foundation for another fateful choice that he—and we—must make this January to begin an emergency rescue of human civilization from the imminent and rapidly growing threat posed by the climate crisis." At one point Gore quoted Abraham Lincoln during another crisis in 1862: "The occasion is piled high with difficulty, and we must rise with the occasion. As our case is new, so we must think anew, and act anew." Gore proposed a "five-part plan to repower America with a commitment to producing 100 percent of our electricity from carbon-free sources within 10 years." However, the first four parts were familiar from the history of U.S. energy policy since the energy crises of the 1970s:

- "First, the new president and the new Congress should offer large-scale investment in incentives for the construction of concentrated solar thermal plants in the Southwestern deserts, wind farms in the corridor stretching from Texas to the Dakotas and advanced plants in geothermal hot spots that could produce large amounts of electricity."
- "Second, we should begin the planning and construction of a unified national smart grid for the transport of renewable electricity from the rural places where it is mostly generated to the cities where it is mostly used. . . ."
- "Third, we should help America's automobile industry (not only the Big Three but the innovative new startup companies as well) to convert quickly to plug-in hybrids that can run on the renewable electricity that will be available as the rest of this plan matures. . . ."
- "Fourth, we should embark on a nationwide effort to retrofit buildings with better insulation and energy-efficient windows and lighting. . . ."

Similar proposals have been on the U.S. energy policy agenda at least since October 1976, when Amory Lovins, in a famous article on energy strategy, distinguished the hard and soft paths. Gore's plan is a hybrid with elements of each. The hard path "relies on rapid expansion of centralized high technologies to increase supplies of energy, especially in the form of electricity." The soft path "combines a prompt and serious commitment to efficient use of energy, rapid development of renewable energy sources matched in scale and in energy quality to end-use needs, and special transitional fossil fuel technologies."[179]

The fifth proposal was familiar from the history of U.S. climate policy from the beginning of the Clinton–Gore Administration in 1993, if not before:

■ "Fifth, the United States should lead the way by putting a price on carbon here at home, and by leading the world's efforts to replace the Kyoto treaty next year in Copenhagen with a more effective treaty that caps global carbon dioxide emissions and encourages nations to invest together in efficient ways to reduce global warming pollution quickly, including by sharply reducing deforestation."

The justifications of the five-part plan were up-to-date, however: "It is a plan that would . . . create millions of new jobs that cannot be outsourced." Moreover, "the bold steps that are needed to solve the climate crisis are exactly the same steps that ought to be taken in order to solve the economic crisis and the energy security crisis." This would mainstream climate policy as a matter of community development at the national and international levels.

What is left out of the picture in "The Climate for Change" that might be important for advancing the common interest? First, much of the relevant past explaining why another plan is needed has been overlooked, perhaps because it lies outside the constraints of scientific management. In connection with the climate crisis, Gore implied that we need another plan because of failed presidential leadership. But there was no lack of commitment, at least at the top of the Clinton–Gore Administration, which faced formidable political opposition that has not disappeared.[180] In connection with energy security, Gore noted that President Nixon and all subsequent presidents have proclaimed some version of energy independence as a national goal. Nevertheless, U.S. dependence on imported oil has doubled from one-third to two-thirds of domestic consumption since Nixon's proclamation because, Gore suggested, some saw realizing the national goal as a problem of increasing domestic production of fossil fuels, based on the "cynical and self-interested illusion" that fossil fuels could be made inexpensive and "clean" through advanced technology.

A fuller explanation might take into account the conclusion of Lovins's technical analysis in 1976: "The barriers to far more efficient use of energy are not technical, nor in any fundamental sense economic. So why do we stand here confronted, as Pogo said, by insurmountable opportunities?

The answer—apart from poor information and ideological antipathy and rigidity—is a wide array of institutional barriers, including more than 3,000 conflicting and often obsolete building codes" at the beginning of Lovins' long list of limiting factors.[181] Similar institutional barriers to the diffusion of existing technologies persist as significant factors in IPCC assessments and other sources.[182] The diversity and complexity of the barriers alone are enough to frustrate centralized management from the top down. So are the diversity and complexity of soft energy technologies, if they are to be matched in scale and energy quality to end-use needs. After specifying criteria for soft technologies, Lovins explained, "The schemes that dominate ERDA's solar research budget—such as making electricity from huge collectors in the desert, or from temperature differences in oceans, or from Brooklyn Bridge–like satellites in outer space—do not satisfy our criteria, for they are ingenious high technology ways to supply energy in a form and at a scale inappropriate to most end-use needs. Not all solar technologies are soft."[183] In important ways, the scale and quality of energy in Gore's plan were not matched to end-use needs.

Second, people on the ground have been all but overlooked in "The Climate for Change," which takes the perspective of *Seeing Like a State* in the long tradition of scientific management. The policy proposed in the first part of Gore's plan is to be executed by "the new president and the new Congress"; the policy proposed in the fifth part is to be executed by them on behalf of "the United States." If the "we" in the other parts includes "all Americans," the assumption must be that "we" are prepared to act as one despite our differences. The need for the five-part plan is justified by appeal to the experts but not to the diverse perspectives or diverse needs of the American public. The experts include "[e]conomists across the spectrum [who] agree that large and rapid investments in a jobs-intensive infrastructure initiative is the best way to revive our economy in a quick and sustainable way." The experts also include the "world authority on the climate crisis, the Intergovernmental Panel on Climate Change [which] after 20 years of detailed study and four unanimous reports, now says the evidence is 'unequivocal.'" There is a concluding reference to "the young Americans, whose enthusiasm electrified Barack Obama's campaign." For Gore, "There is little doubt that this same group of energized youth will play an essential role in this project to secure our national future. . . ." But that essential role is unclear, except possibly to serve as foot soldiers in the implementation of a five-part plan on which they had no obvious influence. We return to the question raised earlier, "But how does it play in Peoria?" We find no answer in the op-ed, but the relevant past suggests that most of the American public is indifferent to action on climate change even if they are concerned about it. To those both indifferent and antagonistic, Gore's plea was simply "please wake up."

Third, adaptation to the adverse impacts of climate change has been overlooked in "The Climate for Change," which is focused on the mitigation of climate change through U.S. energy policy and U.S. leadership in the international quest for mandatory targets and timetables to reduce greenhouse gas emissions. Nevertheless, adaptation will be unavoidable in view of increasing losses from climate-related disasters and the continuing accumulation of greenhouse gases in the atmosphere that portend more losses to come. Which of the losses can be avoided by whom, when, where, and how? Specific answers to these questions are needed because policy problems and opportunities on the ground are diverse and complex, and because much of the action in adapting to climate change (and mitigating it) will take place on the ground. Specific answers will be part of a procedurally rational process, at best, and beyond the capabilities of national and international authorities and experts working alone. No one is omniscient.

Some of the former vice president's plan might work, if not for an emergency rescue of human civilization, then at least to reduce projected losses from climate-related disasters. However, by looking into what has been left out of the picture, those of us concerned about climate change might find possibilities for improving the odds. For example:

- The objective of Gore's plan is to produce 100% of the nation's electricity from carbon-free sources in 10 years. Perhaps something can be learned from Søren Hermansen and his neighbors, who achieved that on Samsø in about five years.
- Part of Gore's plan is to retrofit buildings. Perhaps something can be learned from passive house pioneers in Darmstadt and the people of Ashton Hayes who reduced their carbon footprint quickly and significantly through energy efficiency and conservation.
- These local activists and others like them might be recognized and engaged to identify institutional and other barriers on the ground, disseminate information on how they have been overcome here and there, and advise officials at the top on how they can help.
- People whose local communities have or will be devastated by climate-related disasters also might be engaged as leaders in adaptation to climate change impacts that we cannot avoid, and as allies in mitigating further climate change.

- Instead of assuming that such people will accept experts' risk analyses at face value, or that sound science can resolve our political differences, local activists might be engaged in a two-way "dialogue, not a diatribe," as Herman Karl and colleagues put it.

These possibilities illustrate the specific kinds of human choices that can be made by those concerned about climate change as a policy problem. As these possibilities suggest, the basic choice is not between scientific management *or* adaptive governance; as the saying goes, we can walk and chew gum at the same time.

Opening the climate change regime to adaptive governance or other approaches would be a significant change in "our way of thinking." It is not a threat to the climate system, or to the people who depend on it, or to the things they value that are vulnerable to climate change—even if opening the regime is perceived as a threat to the interests that have sustained business as usual for two decades. In clarifying our interests, it would be reasonable for us to distinguish between the values sought, for whom, and the expected consequences of the alternatives available. Have we overlooked some of our values, or some of those with whom we identify? Are our expectations consistent with the evidence available? Such introspection may reveal that some of our interests have been valid and appropriate, and some have not. In any case, our future choices and decisions will express our interests. And the choices and decisions acted upon, not merely proclaimed, will reveal our true interests.

It remains to be seen which interests we will individually choose and collectively decide to serve in response to climate change, in the continuing development of communities at all levels.

NOTES

Chapter 1

1. Emphasis added, quoted from *An Inconvenient Truth: A Global Warning*, directed by David Guggenheim and released by Paramount Classics in 2006. This film was an official selection of the 2006 Sundance Film Festival and the Cannes Film Festival, and won the Academy Award in 2007 for Best Documentary. The printed companion is Al Gore, *An Inconvenient Truth: The Planetary Emergency of Global Warming and What We Can Do About It* (New York: Rodale, 2006). Al Gore shared the 2007 Nobel Peace Prize with the Intergovernmental Panel on Climate Change (IPCC) for his leadership on global warming.

2. Bjorn Lomborg, e.g., emphasized the differences between Gore and the IPCC in "Ignore Gore—But Not His Nobel Friends," *The Sunday Telegraph (London)* (11 November 2007), 24. On the scientific consensus in the United States, see Jane A. Leggett, *Climate Change: Science and Policy Implications,* CRS Report for Congress RL33849 (Washington, D.C.: Congressional Research Service, Updated 2 May 2007). On the scientific consensus in the international community, see the widely publicized assessments of the IPCC that are introduced below and summarized in later chapters.

3. The Framework Convention and related sources can be accessed at the Gateway to the UN System's Work on Climate Change, http://www.un.org/climatechange/projects.shtml.

4. See the UNFCCC's background information on the Kyoto Protocol, accessed 5 September 2007, at http://unfccc.int/kyoto_protocol/background/items/3145.php. See also Susan R. Fletcher and Larry Parker, *Climate Change: The Kyoto Protocol and International Actions,* CRS Report for Congress RL 33836 (Washington, D.C.: Congressional Research Service, Updated 8 June 2007).

5. David A. King, "Climate Change Science: Adapt, Mitigate, or Ignore?" *Science* 303 (9 January 2004), 176–177. At the time King was science advisor to her Majesty's government in the United Kingdom. For details on the Kyoto Protocol, see http://unfccc.int/kyoto_protocol/background.php.

6. The figures in the text were calculated from 2009 UNFCCC data on total greenhouse gas emissions including LULUCF, in Gg CO_2 equivalent for the 1990 baseline year and 2006. (LULUCF refers to adjustments for Land Use, Land-Use Change, and Forestry.) The data were accessed in 2009 at http://unfccc.int/ghg_data/ghg_data_unfccc/time_series_annex_i/items/3842.php.

7. For example, if the 1990 baselines reported by the UNFCCC in 2006 are used, total GHG emissions for 1990 to 2006 including LULUCF go up to 18.7% for the United States, to 11.9% for Annex I industrialized countries excluding those in transition to a market economy, and to +1.6% for all Annex I countries. For the figures reported in 2006, see the *National Greenhouse Gas Inventory Data for the Period 1990–2004 and Status of Reporting,* FCCC/SBI/2006/26 (19 October 2006), 13, Table 5.

8. However, some data are starting to be analyzed. Michael R. Raupach, Gregg Marland, Philippe Ciais, Corinne Le Quéré, Josep G. Canadell, Gernot Klepper, and Christopher B. Field, "Global and Regional Drivers of Accelerating CO_2 Emissions," *Proc. National Academy of Sciences* 104(24) (12 June 2007), 10,288–10,293 (DOI 10.1073/pnas.0700609104), "Together, the developing and least-developed economies (forming 80% of the world's population) accounted for 73% of global emissions growth in 2004 but only 41% of global emissions and only 23% of global cumulative emissions since the mid-18th century."

9. Joseph Kahn and Jim Yardley, "As China Roars, Pollution Reaches Deadly Extremes," *New York Times* (26 August 2007), 1.

10. However, CO_2 emissions per person were 19.4 tons in America and only 5.1 tons in China. Elisabeth Rosenthal, "Booming China Leads the World in Emissions of Carbon Dioxide, a Study Finds," *New York Times* (14 June 2008), A5.

11. Recent analyses suggest that the peak and decline should happen earlier and at much lower concentrations than those estimated by the Intergovernmental Panel on Climate Change in its Fourth Assessment Report (2007). See the work by Malte Meinhausen and colleagues from *Nature* 458 (30 April 2009), 1158–1162, and *Global Environmental Change* 17 (May 2007), 260–280, the basis for the stabilization level presented here. Some researchers suggest that there is no level above current levels that can guarantee temperature increases of 2°C or less in this century. For example, Andrew Weaver and colleagues find that a 90% global emissions reductions, com-

bined with direct capture of already emitted greenhouse gases in the air, is required to meet this criterion. See *Geophys. Res. Lett.* 34 (October 2007), L19703.

12. IPCC, Working Group III, *Summary for Policymakers in Climate Change 2007: Mitigation* (Cambridge, UK: Cambridge University Press), 3: "Between 1970 and 2004, global emissions of CO_2, CH_4, N_2O, HFCs, PFCs and SF_6, weighted by their global warming potential, have increased by 70% (24% between 1990 and 2004), from 28.7 to 49 Gigatonnes of carbon dioxide equivalents. The emissions of these gases have increased at different rates. CO_2 emissions have grown between 1970 and 2004 by about 80% (28% between 1990 and 2004) and represented 77% of total anthropogenic GHG emissions in 2004."

13. Carbon cycle researcher Rachel Law has noted that there has been "some speculation that [the higher- than-average growth rates] could indicate a declining biospheric sink—but most likely it's dominated by increasing CO_2 emissions." Michael Raupach and colleagues have documented this trend in "Global and Regional Drivers of Accelerating CO_2 Emissions," noting that "[f]or the recent period 2000–2005, the fraction of total anthropogenic CO_2 emissions remaining in the atmosphere . . . was 0.48. This fraction has increased slowly with time . . . implying a slight weakening of sinks relative to emissions. However, the dominant factor accounting for the recent rapid growth in atmospheric CO_2 (>2 ppm y^{-1}) is high and rising emissions, mostly from fossil fuels."

14. Stefan Rahmstorf, Anny Cazenave, John A. Church, James E. Hansen, Ralph F. Keeling, David E. Parker, and Richard C. J. Somerville, "Recent Climate Observations Compared to Projections," *Science* 316 (May 2007), 709, DOI: 10.1126/science.1136843.

15. Martin Parry, Nigel Arnell, Mike Hulme, Robert Nicholls, and Matthew Livermore, "Commentary: Adapting to the Inevitable," *Nature* 393 (22 October 1998), 741; and Roger A. Pielke, Jr., "Rethinking the Role of Adaptation in Climate Policy," *Global Environmental Change* 8 (1998), 159–170.

16. Munich Re Group, *Topics 2000: Natural Catastrophes—the Current Position* (Special Millenium Issue), 43. The category "great natural catastrophes" includes floods, windstorms, and earthquakes. See also Association of British Insurers, *Financial Risks of Climate Change: Summary Report* (June 2005), available at http://www.abi.org.uk/climatechange.

17. Earth Policy Institute, *Hurricane Damages Soar to New Levels* (Update 58, 2006), accessed at www.earth-policy.org/Updates/2006/Update58.

18. This calculation excludes human losses from waves and surges including tsunamis, volcanos, slides, epidemics, and earthquakes. The source is *2005 Disasters in Numbers*, available from EM-DAT: The OFDA/CRED International Disaster Database (www.em-dat.net) at the Universite catholiqué de Louvain in Brussels, Belgium.

19. IPCC, Working Group I, *Summary for Policymakers*, in *Climate Change 2007: The Physical Science Basis* (Cambridge, UK: Cambridge University Press), notes in

Table SPM.2, 8, that increases in heavy precipitation events and tropical cyclone intensity are "likely" to have been observed and "more likely than not" to be attributable to human activities in whole or part. But both the observations and the linkages to human activities remain an area of active investigation and debate. See, e.g., Judith A. Curry, Peter J. Webster, and Greg J. Holland, "Mixing Politics and Science in Testing the Hypothesis that Greenhouse Warming Is Causing a Global Increase in Hurricane Intensity," *Bull.Amer. Meteor. Soc.* 87(8) (2006), 1025; Peter J. Webster, Greg J. Holland, Judith A. Curry, and Hai-Ru Chang, "Changes in Tropical Cyclone Number, Duration, and Intensity in a Warming Environment," *Science* 309 (2005), 1844–1846; Kerry Emmanuel, "Increasing Destructiveness of Tropical Cyclones Over the Past 30 Years," *Nature* 436 (2005), 686–699; and James B. Elsner, James P. Kossin, and Thomas H. Jagger, "The Increasing Intensity of the Strongest Hurricanes," *Nature* 455 (4 September 2008), 92–95. Scientists engaged in active debate on this topic signed a statement on the U.S. Hurricane Problem (25 July 2006), urging that the debate "should in no event detract from the main hurricane problem facing the United States: the ever-growing concentration of population and wealth in vulnerable coastal areas. These demographic trends are setting us up for rapidly increasing human and economic losses from hurricane disasters, especially in this era of heightened activity." The statement is posted at http://wind.mit.edu/~emanuel/ Hurricane_threat.htm.

20. Judith A. Merkle, "Scientific Management," in Jay M. Shafritz, Ed., *International Encyclopedia of Public Policy and Administration* (Boulder, CO: Westview Press, 1998), 2036–2040 at 2040. Of course the term "scientific management" is subject to various interpretations as it continues to evolve. See also James C. Scott, *Seeing Like a State: How Certain Schemes to Improve the Human Condition Have Failed* (New Haven, CT: Yale University Press, 1998); and John Ralston Saul, *Voltaire's Bastards: The Dictatorship of Reason in the West* (New York, NY: Vintage, 1993).

21. Ronald D. Brunner, Toddi A. Steelman, Lindy Coe-Juell, Christina M. Cromley, Christine M. Edwards, and Donna W. Tucker, *Adaptive Governance: Integrating Science, Policy, and Decision Making* (New York, NY: Columbia University Press, 2005), p. 2. For more on scientific management, see also Ronald D. Brunner, Christine H. Colburn, Christina M. Cromley, Roberta A. Klein, and Elizabeth A. Olson, *Finding Common Ground: Governance and Natural Resources in the American West* (New Haven, CT: Yale University Press, 2002).

22. We do not limit the term "bottom up" to calculating the potential savings in greenhouse gas emissions from the deployment of various technologies.

23. See Brunner et al., *Adaptive Governance*, and the sources cited therein on natural resource policy. Variants of adaptive governance elsewhere in the American context can be found in Hedrick Smith, *Rethinking America: A New Game Plan from the American Innovators: Schools, Business, People, and Work* (New York, NY: Random House: 1995); Lisbeth B. Schorr, *Common Purpose: Strengthening Families and Neighborhoods to Rebuild America* (New York, NY: Doubleday, 1997); and

Thomas Petzinger, Jr., *The New Pioneers: The Men and Women Who Are Tranforming the Workplace and the Marketplace* (New York, NY: Simon & Schuster, 1999). George Packer, "Knowing the Enemy," *The New Yorker* (18 December 2006), 60–69, emphasizes disaggregating (or factoring) security problems. The term "adaptive governance" is not used in these variants of the basic concept as we understand it. Conversely, the term is used for overlapping concepts in the following: Thomas Dietz, Elinor Ostrom, and Paul Stern, "The Struggle to Govern the Commons," *Science* 302 (12 December 2003), 1907–1912, which has a section titled "Adaptive Governance"; Carl Folke, Thomas Hahn, Per Olsson, and Jon Norberg, "Adaptive Governance of Social-Ecological Systems," *Annu. Rev. Environ. Res.* 30 (2005), 441–473; Lance Gunderson and Stephen S. Light, "Adaptive Management and Adaptive Governance in the Everglades Ecosystem," *Policy Sciences* 39 (2006), 323–334; and Rohan Nelson, Mark Howden, and Mark Stafford Smith, "Using Adaptive Governance to Rethink the Way Science Supports Australian Drought Policy," *Environ. Science & Policy* 11 (2008), 588–601. In short, a grounded interpretation of the term "adaptive governance" depends on the particular context in which it is used.

24. Several frames at a smaller scale, and more specialized to climate change politics in the United States, are surveyed by Matthew C. Nisbet and Chris Mooney, "Framing Science," *Science* 316 (6 April 2007), 56, in an effort to improve the communication of science. The frames are "scientific uncertainty" and "unfair economic burden" on the one hand, and on the other, a "Pandora's box" of catastrophes and "public accountability" from official censoring of climate science.

25. Julia Uppenbrink, "Arrhenius and Global Warming," *Science* 272 (24 May 1996), 1122. For more on the history of such predictions, see the letter by Neville Nicholls, "Climate: Sawyer Predicted Rate of Warming in 1972," *Nature* 448 (30 August 2007), 992.

26. Quoted by John Firor in *The Changing Atmosphere: A Global Challenge* (New Haven, CT: Yale University Press, 1990), 48. Note that the statement did not warn of the peril of the experiment; instead, it highlighted an opportunity for science, one acted upon during the International Geophysical Year from July 1957 to December 1958. A young geochemist, Charles David Keeling, began to measure carbon dioxide in the atmosphere in order to understand how carbon dioxide was exchanged between atmosphere and ocean. See our Fig. 1.1.

27. Bert Bolin and Erik Eriksson, "Changes in the Carbon Dioxide Content of the Atmosphere and Sea Due to Fossil Fuel Combustion," in Bert Bolin, Ed., *The Atmosphere and the Sea in Motion* (New York, NY: Rockefeller Institute Press, 1959), 130–142; and Bert Bolin, "Atmospheric Chemistry and Broad Geophysical Relationships," *Proc. Natl. Acad. Sci.* 45 (1959), 1663–1672. In the latter article, Bolin suggested that the first noticeable increase in carbon dioxide in the atmosphere was reported by G. S. Callendar in 1938 in the *Quart. J. Roy. Meteor. Soc.* in an article titled "The Artificial Production of Carbon Dioxide and Its Influence on Temperature." But in fact Irish physicist and contemporary of Charles Darwin, John Tyndall, was the

first to discover the so-called "greenhouse" effect. In the mid-19th century Tyndall showed in the laboratory that gases such as water vapor and carbon dioxide are transparent to light energy but absorb heat energy. More importantly, he realized that this property explained the average temperature at the surface of the earth, which is warmer than expected given the energy input from the sun, and that increases in these gases in the atmosphere would cause an increase in temperature.

28. The following sketch of the establishment of the climate change regime is adapted from Ronald D. Brunner, "Science and the Climate Change Regime," *Policy Sciences* 34 (2001), 1–33, and based on a variety of sources primarily including: Daniel Bodansky, "Prologue to the Climate Change Convention," in I. M. Mintzer and J. A. Leonard, Eds., *Negotiating Climate Change: the Inside Story of the Rio Convention* (Cambridge, U.K.: Cambridge University Press, 1994), 45–74; Daniel Bodansky, "The Emerging Climate Change Regime," *Annu. Rev. Energy Environ.* 20 (1995), 425–461; Sonja A. Boehmer-Christiansen, "Global Climate Protection Policy: The Limits of Scientific Advice," *Global Environ. Change* 4 (1994), Part I (June), 140–159, and Part II (September), 185–200; Sonja A. Boehmer-Christiansen, "A Scientific Agenda for Climate Policy?" *Nature* 368 (10 March 1994), 400–402; Alan D. Hecht and Dennis Tirpak, "Framework Agreement on Climate Change: A Scientific and Political History," *Climatic Change* 29 (1995), 371–402; Sheldon Ungar, "Social Scares and Global Warming: Beyond the Rio Convention," *Soc. Nat. Res.* 8 (1995), 443–456; Jill Jäger and Tim O'Riordan, "The History of Climate Change Science and Politics," in Tim O'Riordan and Jill Jäger, Eds., *The Politics of Climate Change: A European Perspective* (London and New York: Routledge, 1996), 1–31; and Shardul Agrawala, "Early Science-Policy Interactions in Climate Change: Lessons from the Advisory Group on Greenhouse Gasses," *Global Environ. Change* 9 (1999), 157–169.

29. Bodansky, "Prologue to the Climate Change Convention," 48.

30. Quoted in Boehmer-Christiansen, "Global Climate Protection Policy," Part I, 157. It was officially called the International Conference on the Assessment of the Role of Carbon Dioxide and Other Greenhouse Gases in Climate Variations and Associated Impacts.

31. Boehmer-Christiansen, "Global Climate Protection Policy," Part I, 157. Compare Agrawala, "Early Science-Policy Interactions in Climate Change," 168, on the significance of the AGGG: "It was the establishment of a standing advisory panel, the AGGG, and the group of experts that coalesced under its auspices that 'kept the climate flame alive' and eventually jogged the policy community into action."

32. Boehmer-Christiansen, "A Scientific Agenda for Climate Policy?" 401. The ICSU later changed its name to the International Council for Science but kept the acronym. For an overview of scientific organizations involved in the World Climate Research Program and their links to the U.S. Global Change Research Program, see *Our Changing Planet: The FY 1990 Research Plan: The U.S. Global Change Research Program*, A Report by the Committee on Earth Sciences (July 1989), pp. 24–25. These

organizational charts do not show the overlapping memberships of leading scientists such as Bert Bolin of Sweden.

33. Bodansky, "Prologue to the Climate Change Convention," 50.

34. Boehmer-Christiansen, "Global Climate Protection Policy," Part I, 157.

35. "The Greenhouse Effect: Impact on Current Global Temperature and Regional Heat Wave," Statement of James E. Hansen, Hearings before the Committee on Energy and Natural Resources, 100th Congress, 1st Session, on *The Greenhouse Effect and Global Climate Change* (23 June 1998), Part 2 (Washington, D.C.: Government Printing Office).

36. Quoted in Ungar, "Social Scares and Global Warming," 445.

37. Senator Wirth told this story in an interview on "Hot Politics," a *Frontline* program rebroadcast on 22 April 2008 by the Public Broadcasting System.

38. On the lack of political will, compare Joseph Stiglitz, Nobel Laureate in Economic Science, in "A New Agenda for Global Warming," *Economists' Voice* (July 2006), 1–4, at 4: "The well-being of our entire planet is at stake. We know what needs to be done. We have the tools at hand. We only need the political resolve." Compare also Jared Diamond, "What's Your Consumption Factor?" *New York Times* (2 January 2008), A10, on necessary and desirable trends in per capita consumption rates that have a bearing on climate change: "[W]e already know how to encourage the trends; the main thing lacking has been political will."

39. Boehmer-Christiansen, "Global Climate Protection Policy," Part II, 187.

40. Ibid., 189. On the selection of the chair for the IPCC's Fourth Assessment Report and perhaps a veto, see the letter by Bert Bolin, "Politics and the IPCC," *Science* 296 (17 May 2002), 1235. See also the memorandum "Regarding: Bush Team for IPCC Negotiations" (6 February 2001), from Randy Randol, senior environment advisor at ExxonMobil, to the White House CEQ. It asked, "Can [IPCC Chair Robert] Watson be replaced now at the request of the U.S.?" The memo was released by the Natural Resources Defense Council, which obtained it under the Freedom of Information Act. See http://www.nrdc.org/media/pressreleases/020419a.asp.

41. Boehmer-Christiansen, "Global Climate Protection Policy," Part II, 189. Compare Agrawala, "Early Science-Policy Interactions in Climate Change," 166: "By late 1990, experts working nominally under the AGGG had lost, to the IPCC, a Darwinian selection struggle imposed by the climate science and policy environment."

42. Boehmer-Christiansen, "Global Climate Protection Policy," Part II, 188.

43. From the Preface to the IPCC, *Second Assessment Synthesis of Scientific-Technical Information Relevant to Interpreting Article 2 of the UN Framework Convention on Climate Change* (1995). Working groups on "[i]mpacts and responses were combined in 1993, while a new group was set up to look at 'cross-cutting' aspects, with contributions from a wider range of social sciences" according to Boehmer-Christiansen, "A Scientific Agenda for Climate Policy?" 402. For a defense of the IPCC, see Richard H. Moss, "The IPCC: Policy Relevant (Not Driven) Scientific

Assessment," *Global Environ. Change* 5 (1995), 171–174, in the form of a comment on Boehmer-Christiansen's two-part article in the same journal, vol. 4 (1994).

44. Bodansky, "Prologue to the Climate Change Convention," 51.

45. We are grateful to Dr. Elbert W. (Joe) Friday, a participant in the events described here, for corroborating the history in the first three paragraphs of this section: "That is very much as I remember it" (personal communication, 20 September 2008). Friday went on to explain that he and his scientific colleagues sought to create a larger and more effective AGGG within the IPCC. In our assessment of the record, the IPCC turned out to be larger but less effective in policy arenas. Friday served in the U.S. government as assistant administrator for Weather Services and director of the National Weather Service from 1988 to 1997.

46. The exception was the U.S. Climate Change Action Plan (CCAP), a collection of nearly 50 new or expanded programs mainly relying on energy conservation to implement the nation's voluntary commitments to reduce its greenhouse gas emissions under Article 4(2) of the UNFCCC. The CCAP was announced by President Bill Clinton and Vice President Al Gore in October 1993. For more on the CCAP, see Chapter 2.

47. Boehmer-Christiansen, "Global Climate Protection Policy," Part II, 190.

48. The budget of USGCRP increased from $134 million in FY 1989 to $659 million in FY 1990 and exceeded $1 billion for the first time in FY 1992. Jane A. Leggett, *Climate Change: Federal Expenditures,* CRS Report for Congress RL33817 (Washington, D.C.: Congressional Research Service, 22 January 2007), 8.

49. See Agrawala, "Early Science-Policy Interactions in Climate Change," 160, on overlapping affiliations of several major figures in the AGGG, and the short biography of Richard Moss published in Moss, "The IPCC: Policy Relevant (Not Driven) Scientific Assessment," 171.

50. Paul N. Edwards, "Global Comprehensive Models in Politics and Policymaking: Editorial Essay," *Climatic Change* 32 (1996), 149–161, at 152.

51. Boehmer-Christiansen, "Global Climate Protection Policy," Part II, 195. Agrawala, "Early Science-Policy Interactions in Climate Change," 167, recognizes "the delicate trade-off between the need for continuity and instincts for self-preservation" as the most important lesson to be drawn from the experience of AGGG and the early IPCC.

52. Compare Al Gore, in a discussion of *An Inconvenient Truth as* quoted by David Neff in "Al Gore: Preacher Man," ChristianityToday.com (31 May 2006): "Science thrives on uncertainty; politics is paralyzed by uncertainty." Accessed at http://www.christianitytoday.com/movies/commentaries/algore.html.

53. Consider Bert Bolin, chair of the IPCC at the time, in "Next Step for Climate-Change Analysis," *Nature* 368 (10 March 1994), 94: "Just what is meant by the word 'dangerous' needs to be sorted out. IPCC will bring together basic knowledge of relevance in this context, but reaching agreement is a political issue and must therefore be achieved through intergovernmental negotiations within the framework of the

convention." Consider also Syukuro Manabe, a pioneering climate modeler: "Once I start promoting my political views, my credibility as a scientist is compromised. I would rather not take a position—I may have one—but I don't want to take it publicly." Quoted in Agrawala, "Early Science-Policy Interactions in Climate Change," 164.

54. IPPC, *Second Assessment Synthesis of Scientific-Technical Information Relevant to Interpreting Article 2 of the UN Framework Convention on Climate Change* (1995), sec. 1.9. For a more recent example of the global framing, see Thomas R. Karl and Kevin E. Trenberth, "Modern Global Climate Change," *Science* 302 (5 December 2003, 1719–1723, at 1722: "Climate change is truly a global issue, one that may prove to be humanity's greatest challenge. It is very unlikely to be addressed without greatly improved international cooperation and action." See also King, "Climate Change Science: Adapt, Mitigate, or Ignore?" 177: "But any alternative would need to accept that immediate action is required and would need to involve all countries in tackling what is truly a global problem."

55. Compare Boehmer-Christiansen, "A Scientific Agenda for Climate Policy?" 402: "Global policy on global warming [in 1994] is emerging from untidy political processes—not through technocratic design." She adds that "political battles over the knowledge base have also grown."

56. Quoted in David Malakoff, "Thirty Kyotos Needed to Control Warming," *Science* 278 (19 December 1997), 2048.

57. Compare the assessments in Boehmer-Christiansen, "Global Climate Protection Policy," Part II, 192; Bodansky, "The Emerging Climate Change Regime," 438; and Ungar "Social Scares and Global Warming," 448. These are summarized in Chapter 2.

58. See Andrew C. Revkin, "Years Later, Climatologist Renews His Call For Action," *New York Times* (23 June 2008); and Dana Milbrook, "Burned Up about the Other Fossil Fuel," *Washington Post* (24 June 2008), A3.

59. William C. Clark, "Managing Planet Earth," *Scientific American* 261 (September 1989), 47–54, at 47. On the history of the global faming, see Dimitris Stevis, "The Globalizations of the Environment," *Globalizations* 2 (December 2005), 323–333.

60. Hendrik Tennekes, "A Sideways Look at Climate Research," *Weather* 45 (1990), 67–68, at 68. The explanation of technocratic totalitarianism came in a letter from Hendrik Tennekes (21 July 1997) in response to a query from the first author.

61. All quotations in this paragraph are from Steve Rayner and Elizabeth L. Malone, "Commentary: Zen and the Art of Climate Maintenance," *Nature* 390 (27 November 1997), 332–334, at 332, 333. Their conclusion is "Ten Suggestions for Policymakers," in Steve Rayner and Elizabeth L. Malone, Eds., *Human Choice & Climate Change, Volume 4: What Have We Learned?* (Columbus, OH: Battelle Press, 1998), 109–138.

62. David W. Cash, "Viewpoint: Distributed Assessment Systems: An Emerging Paradigm of Research, Assessment and Decision-Making for Environmental Change," *Global Environ. Change* 10 (2000), 241–244, at 241, 242. A major source for

Cash was Thomas J. Wilbanks and Robert W. Kates, "Global Change in Local Places: How Scale Matters," *Climatic Change* 43 (1999), 601–628. Cash cites as another example of a distributed information-decision support system the U.S. Cooperative State Research Education and Extension Service. We consider this Progressive Era innovation to be more like a centralized system in the tradition of scientific management. This follows Everett M. Rogers, *Diffusion of Innovations,* 4th ed. (New York, NY: Free Press, 1995), 364–369, which distinguishes centralized and decentralized models of diffusion.

63. See the Preface for more on the history of the project and this book. Amanda Lynch, project principal investigator and atmospheric scientist, had developed regional climate models for the Arctic and conducted research in Barrow as a member of the faculty at the University of Alaska at Fairbanks and later the University of Colorado at Boulder. She and others in her group recruited as another coprincipal investigator Ron Brunner, a policy scientist at the University of Colorado with prior interests in decentralized energy, natural resources, and climate change policies. For early research relevant to this project, see, e.g., Amanda H. Lynch, W. L. Chapman, John E. Walsh, and Gunter Weller, "Development of a Regional Climate Model of the Western Arctic," *J. Climate* 8 (1995), 1555–1570; and Ronald D. Brunner, "Decentralized Energy Policies," *Public Policy* 28 (Winter 1980), 71–91.

64. We use the term "Eskimos" because Natives in Alaska are proud to call themselves as such. For example, one of the subsidiaries of the Native-owned village corporation in Barrow, Ukpeaġvik Iñupiat Corporation (UIC), is called SKG Eskimos Inc. However, some indigenous peoples elsewhere in the Arctic do not like the term.

65. The Bowhead Transportation Company is a wholly owned subsidiary of UIC. The announcement explained, "Because of the gradual shift in weather patterns and ice movement, Bowhead will accelerate its sailing departure [to Arctic coastal villages] by two weeks to take advantage of these changes." Like other coastal villages, Barrow depends on bulk cargo shipped by seasonal barges when sea ice permits. Barrow is not accessible by roads, and air transportation is much more expensive.

66. Another climate-related problem, subsistence bowhead whale hunting, had become more dangerous with sea ice less stable and more easily detached from shore. But the North Slope Borough, the principal unit of local government, had compensated with modern technology, including GPS locator beacons and a helicopter for search and rescue. Moreover, the North Slope Borough's Department of Wildlife Management already had an excellent record of wildlife-related policy research going back at least to the 1970s when the International Whaling Commission attempted to ban subsistence whale hunting. As noted in Chapter 3, subsistence bowhead whale hunting is a central focus of Iñupiat Eskimo culture.

67. The center was originally called the Barrow Global Climate Change Research Facility.

68. For a summary of project research, see Amanda H. Lynch and Ronald D. Brunner, "Context and Climate Change: An Integrated Assessment for Barrow, Alaska," *Climatic Change* 82 (2007), 93–111. The most extensive summary is Amanda Lynch and Ronald Brunner et al., *Barrow Climatic and Environmental Conditions and Variations—A Technical Compendium* (Boulder, CO: Cooperative Institute for Research in Environmental Sciences, 2004), 132 pages. This was circulated for corrections and comments in Barrow in the fall of 2004. It includes the chapter "Policy Responses," 67–132. On policy, see also Ronald D. Brunner and Amanda H. Lynch et al., "An Arctic Disaster and Its Policy Implications," *Arctic* 57 (December 2004), 336–346. For further information on the project, see http://nome.colorado.edu/HARC/index. html. We based the integrative and policy aspects of this research on the policy sciences; see Harold D. Lasswell, *A Pre-View of Policy Sciences* (New York, NY: Elsevier, 1971). There was no need to create yet another conceptual framework.

69. From the Preface to the Arctic Climate Impact Assessment, *Impacts of a Warming Arctic* (Cambridge, U.K.: Cambridge University Press, 2004), which continues with specifics: The Arctic Council "is comprised of eight arctic nations (Canada, Denmark/Greenland/Faroe Islands, Finland, Iceland, Norway, Russia, Sweden, and the United States of America), six Indigenous Peoples Organizations (Permanent Participants: Aleut International Association, Arctic Athabaskan Council, Gwich'in Council International, Inuit Circumpolar Conference, Russian Association of Indigenous Peoples of the North, and Saami Council), and official observers (including France, Germany, the Netherlands, Poland, United Kingdom, non-governmental organizations, and scientific and other international bodies)." As noted in Chapter 3, the Iñupiat in Barrow took the lead in establishing the Inuit Circumpolar Conference.

70. Harold D. Lasswell and Abraham Kaplan, *Power and Society: A Framework for Political Inquiry* (New Haven: Yale University Press, 1950), xxii–xxiii. Emphasis in original. For the authors at least, these ideal types are not ends in themselves. They are means of advancing the common interest as clarified near the end of this chapter.

71. IPCC, *Second Assessment Synthesis*, secs. 5.2 and 5.3.

72. Rayner and Malone, "Ten Suggestions," 126.

73. William D. Ruckelshaus, "Toward a Sustainable World," *Scientific American* 261 (September 1989), 166–174, at 166. Ruckelshaus was then a former administrator of the U. S. Environmental Protection Agency and member of the World Commission on Environment and Development.

74. IPCC, *Second Assessment Synthesis of Scientific-Technical Information relevant to interpreting Article 2 of the UN Framework Convention on Climate Change* (1995), sec. 1.9. More recently King, "Climate Change Science: Adapt, Mitigate, or Ignore?" 177, affirmed that "issues of justice and equity lie at the heart of the climate change problem."

75. This total is in constant 2007 dollars beginning with FY 1989, and includes the estimate for FY 2008 and the request for FY 2009. It is calculated from annual data in *Our Changing Planet: The U. S. Climate Science Program for FY 2009*, 4, accessible at http://www.usgcrp.gov/usgcrp/Library/ocp2009/.

76. This closely paraphrases the definition in the policy sciences and policy-oriented jurisprudence. On the former, see Lasswell and Kaplan, *Power and Society*, 23. Emphasis in original: "An *interest* is a pattern of demands and its supporting expectations. . . . The point that the definition aims to bring out is simply that an interest is neither a blind desire nor a knowing untinged by valuation. In every interest analysis discloses competent demands and expectations both." In policy-oriented jurisprudence, see Myres S. McDougal, Harold D. Lasswell, and W. Michael Reisman, *International Law Essays* (Mineola, NY: Foundation Press, 1981), 205: "By an interest, we refer to a value demand formulated in the name of an identity and supported by expectations that the demand is advantageous."

77. The Universal Declaration was adopted by General Assembly Resolution 217 A (III) in December 1948. It may be accessed at http://www.un.org/Overview/rights. html. The declaration denies exclusive rights claimed by the select few. "Everyone is entitled to all the rights and freedoms set forth in this Declaration, without distinction of any kind, such as race, colour, sex, language, religion, political or other opinion, national or social origin, property, birth or other status" (Article 2). Moreover, "Everyone has the right to take part in the government of his country, directly or through freely chosen representatives" and "The will of the people shall be the basis of the authority of government" (Article 21). Near its 60th anniversary, the declaration was invoked by Mary Robinson, the former president of Ireland and former UN High Commission for Human Rights, in connection with the Poznan meeting of the Conference of the Parties to the UNFCCC. See her comment, "Climate Change Is an Issue of Human Rights," *The Independent (London)* (10 December 2008). It was republished at about the same time in *The Australian, The Irish Times*, and *The Belfast Telegraph*. See also International Council on Human Rights Policy, *Climate Change and Human Rights: A Rough Guide* (Geneva: 2008), accessed at http://www. ichrp.org/files/reports/36/136_report.pdf.

78. Brunner et al., *Finding Common Ground*, 8. Emphasis added. For fuller development of the concept and applications, see pp. 8–18 and the literature cited therein. Compare McDougal, Lasswell, and Reisman, *International Law Essays*, 205: "In the most fundamental sense, international law is a process by which the peoples of the world clarify and implement their common interests in the shaping and sharing of values." Values are preferred outcomes.

79. Brunner et al., *Finding Common Ground*, 8.

80. For an example of goal displacement and substitution, see Edwards, "Global Comprehensive Models in Politics and Policymaking," 150, which argues that "the emergence of [an 'epistemic community'] is one major reason why global change has reached the political agenda of governments, and thus that comprehensive model-

building serves an all-important political purpose even if it does not and perhaps cannot serve the immediate needs of policy makers." For a reply, see Radford Byerly, Jr., "Editorial Comment," *Climatic Change* 32 (1996), 163–164, at 163: "For costs of more than a billion dollars a year . . . more than an 'epistemic community' is rightly expected from the USGCRP."

81. For guidance on the responsibilities of scientists in public policy, we recommend Harold D. Lasswell, "The Political Science of Science: An Inquiry into the Possible Reconciliation of Mastery with Freedom," *Amer. Political Sci. Rev.* 50 (December 1956), 961–979; and Harold D. Lasswell, "Must Science Serve Political Power?" *Amer. Psychol.* 25 (1970), 117–123. More generally, for guidance on answering the question, "What ought I to prefer," we recommend Harold D. Lasswell and Myres S. McDougal, *Jurisprudence for a Free Society: Studies in Law, Science and Policy* (New Haven, CT, and Dordrecht, Netherlands: New Haven Press and Martinus Nijhoff), 725–758.

82. Compare Bjorn Lomborg, *Cool It: The Skeptical Environmentalist's Guide to Global Warming* (New York, NY: Alfred A. Knopf, 2007), 9: "We need to remind ourselves that our ultimate goal is not to reduce greenhouse gases or global warming per se but to improve the quality of life and environment." Compare also Robinson, "Climate Change Is an Issue of Human Rights": "Urgently cutting emissions must be done in order to respect and protect human rights from being violated by the future impacts of climate change, while supporting the poorest communities to adapt to already occurring climate impacts is the only remedy for those whose human rights have already been violated."

83. See Michael Shellenberger and Ted Nordhaus, *The Death of Environmentalism: Global Warming Politics in a Post-Environmental World* (2004), which characterized environmentalism as a special interest and proposed opening global warming politics in the United States to enlist the support of nonenvironmental interest groups on whom progress depends. Their paper was accessed at http://www.thebreakthrough. org/ images/Death_of_Environmentalism.pdf. See also Thomas L. Friedman, "The Power of Green," *New York Times Magazine* (15 April 2007), 40.

84. See Ehsan Masood, "Temperature Rises in Dispute Over Costing Climate Change," *Nature* 378 (30 November 1995), 429. See also M. Granger Morgan, Milind Kandlikar, James Risbey, and Hadi Dowlatabadi, "Why Conventional Tools for Policy Analysis Are Often Inadequate for Problems of Global Change: An Editorial Essay," *Climatic Change* 41 (1999), 271–281, at 271, 273.

85. This paraphrases Stanford economist Paul Romer, who is quoted in Thomas L. Friedman, "It's a Flat World After All," *New York Times Magazine* (3 April 2005). Compare Rahm Emanuel—"you don't ever want a crisis to go to waste"—as quoted in Paul Krugman, "Franklin Delano Obama," *New York Times* (10 November 2008), A25.

86. And to some extent it already has been field tested, with mixed but promising results, in natural resource policy in the American West. See the case study chapters in Brunner et al., *Adaptive Governance* especially, but also case studies in Brunner et al., *Finding Common Ground*.

87. Stevis, "The Globalization of the Environment," notes that the image "Spaceship Earth" appeared as early as the mid-1960s. The image is current in Peter A. Corning, "From My Perspective: Why We Need a Strategic Plan for 'Spaceship Earth,'" *Technological Forecasting and Social Change* 72 (2005), 749–752, which attributes it to the economist Kenneth Boulding.

88. From "Think Conservation Before Water Supplies Kick the Bucket," an editorial in the *Sunday Age* (28 January 2007), 16.

89. Rayner and Malone, "Ten Suggestions," 113.

90. On the significance of details in the historical sciences, including policy sciences, see Chapter 4. On uncertainty absorption, see James G. March, Herbert A. Simon, and Harold Guetzkow, *Organizations* (New York, NY: John Wiley, 1958), 164–165: "Uncertainty absorption takes place when inferences are drawn from a body of evidence and the inferences, instead of the evidence itself, are then communicated. . . . Through the process of uncertainty absorption, the recipient of a communication is severely limited in his ability to judge its correctness. . . . To the extent that he can interpret it, his interpretation must be based primarily on his confidence in the source and his knowledge of the biases to which the source is subject, rather than on a direct examination of the evidence."

Chapter 2

1. IGBP explicitly distinguished climate change and global change. The latter includes climate change and all changes in the earth system that arise from human modification of the atmosphere, such as changes in land use, urbanization, and the exploitation of fisheries.

2. The International Geosphere–Biosphere Programme: A Study of Global Change, *The Initial Core Projects,* Report No. 12 (Stockholm, Sweden: IGBP Secretariat, Box 50005, SE-104 05, June 1990), 1–3.

3. Ibid., 1–4.

4. Vannevar Bush's report was submitted to President Truman in 1945 and republished as Vannevar Bush, *Science, the Endless Frontier* (Washington, D.C.: National Science Foundation, 1960). Bush wrote that our national health, prosperity, and security all depend on "essential, new knowledge [that] can be obtained only through basic scientific research." (p. 5). "As long as [centers of basic research] are free to pursue the truth wherever it may lead, there will be a flow of new scientific knowledge to those who can apply it to practical problems in Government, in industry, or elsewhere" (p. 12). For a critique of the linear model, see George E. Brown, Jr., "The Objectivity Crisis," *Amer. J. Physics* 60 (September 1992): 779–781, at 780–781: "[T]he scientific community must seek to establish a new contract with policy makers, based not on demands for autonomy and ever-increasing budgets, but on the implementation of an explicit research agenda rooted in [social] goals" such as "zero population

growth, less waste, less consumption of nonrenewable resources, less armed conflict, less dependence on material goods as a metric of wealth or success." See also Donald E. Stokes, *Pasteur's Quadrant: Basic Science and Technological Innovation* (Washington, D.C.: Brookings Institution, 1997), which advocates use-inspired basic research but accepts linearity.

5. International Geosphere–Biosphere Programme, *A Study of Global Change: The Initial Core Projects*, Report No. 12, 1–5. The world's decision makers were not yet organized as the Conference of the Parties to the UN Framework Convention on Climate Change (UNFCCC).

6. A separate International Human Dimensions Programme on Global Environmental Change (IHDP) was established in 1996 to work "at the interface between science and practice." For more information, see http://www.ihdp.uni-bonn.de/.

7. Ibid. This justification of the time scale conflicts with known limitations of the models envisioned. Like numerical models for weather forecasting, these models encounter strict limits on predictability at seasonal time scales.

8. Ibid., 1–3.

9. According to Will Steffen in a personal communication via email, 12 May 2007, "The regional research centers were never developed, but instead the START [or Global Change Systems for Analysis, Research and Training] was instituted to spearhead the capacity-building aspect of IGBP."

10. International Geosphere–Biosphere Programme, *Science Plan and Implementation Strategy*, Report No. 55 (Stockholm, Sweden: IGBP Secretariat, Box 50005, SE-104 05, 2006), 1, edited by B. Young, K. Noone and W. Steffen.

11. Ibid., 4.

12. Ibid., 20.

13. See W. Steffen et al., *Global Change and the Earth System: A Planet Under Pressure*, 1st ed. (Heidelberg, Germany: Springer-Verlag, 2004), 265, on the GAIM 23 Questions.

14. W. Steffen, personal communication, e-mail, 12 May 2007.

15. Steffen et al., *Global Change and the Earth System*, 32.

16. W. Steffen, personal communication, e-mail, 12 May 2007.

17. Quoted from the ESSP Web site http://www.essp.org/, accessed 9 May 2007. Publications of the ESSP comprise reports and strategic plans from the participating programs and hence the Web site remains the primary source of overarching plans.

18. Ibid. "Teleconnected" refers to climate signals in different parts of the globe that are correlated statistically and significantly.

19. Declaration of the attendees at the Challenges of a Changing Earth: Global Change Open Science Conference, Amsterdam, the Netherlands, 13 July 2001. Edited by Berrien Moore III (chair, IGBP), Arild Underdal (chair, IHDP), Peter Lemke (chair, WCRP), and Michel Loreau (cochair, DIVERSITAS).

20. G. Sawhill, "The Growth of Climate Change Science: A Sociometric Study," *Climatic Change* 48 (2001), 515–524.

21. *Our Changing Planet: A U.S. Strategy for Global Change Research.*, A Report by the Committee on Earth Sciences, To accompany the U.S. President's Fiscal Year 1990 Budget (January 1989); and *Our Changing Planet: The FY 1990 Research Plan*, The U.S. Global Change Research Program, A Report by the Committee on Earth Sciences (July 1989). For background on the USGCRP, see Roger A. Pielke, Jr., "Policy History of the US Global Change Research Program: Part I. Administrative Development," *Global Environ. Change* 10 (2000), 9–25, and "Policy History of the US Global Change Research Program: Part II. Legislative Process," *Global Environ. Change* 10 (2000), 133–144.

22. See *Our Changing Planet: The FY 1990 Research Plan*, 25 for the organization chart, Appendix B for proposed budget allocations across the agencies, and ii for a listing of committee members and their agency affiliations.

23. *Our Changing Planet: The FY 1990 Research Plan*, 8. Emphasis in original.

24. *Our Changing Planet: The FY 1990 Research Plan*, 5.

25. Ibid.

26. Ibid.

27. Ibid., xiii.

28. George E. Brown, Jr., "Global Change and the New Definition of Progress," *Geotimes* (June 1992), 19–21. Rep. Brown's critique is considered near the end of this section.

29. *Our Changing Planet: The FY 1990 Research Plan*, 7.

30. Ibid., 20. Emphasis in original.

31. *Our Changing Planet: The U.S. Climate Change Science Program for Fiscal Years 2004 and 2005*, A Report by the Climate Change Science Program and the Subcommittee on Global Change Research, A Supplement to the President's Budgets for Fiscal Years 2004 and 2005 (July 2004). The quotations in this paragraph can be found on pp. 10 and 6, respectively.

32. See ibid., 15, for a list of the SAPs, and Ch. 7, 119, for a discussion that includes the legislative authority.

33. From the first author's notes made at the workshop.

34. *Our Changing Planet: FY 2006*, 135.

35. Ibid., 14.

36. Ibid., 116.

37. These quotations are from the Executive Summary of IPCC, Working Group III, chaired by Frederick M. Bernthal, *Climate Change: The IPCC Response Strategies* (Washington, D.C.: Island Press, 1991), xxv.

38. Ibid., xlviii.

39. Ibid., xxiii.

40. Emphasis in original. IPCC, Working Group I, *Summary for Policymakers*, in *Climate Change 2007: The Physical Science Basis* (Cambridge, UK: Cambridge University Press), 10; Solomon, S., D. Qin, M. Manning, Z. Chen, M. Marquis, K. B. Averyt, M.Tignor and H. L. Miller, Eds.

41. See section 8, "Generic Assessment of Response Strategies," for the paragraphs, and p. 12 for the quotation.

42. Dave Griggs, personal communication, 29 February 2008.

43. Page 11, Report of the Conference of the Parties on its Seventh Session, held at Marrakesh from 29 October to 10 November 2001, Addendum Part Two: Action Taken By The Conference Of The Parties, Volume I.

44. http://unfccc.int/national_reports/napa/items/2719.php, accessed 9 May 2007.

45. Ian Burton and Maarten van Aalst, *Look Before You Leap: A Risk Management Approach for Incorporating Climate Change Adaptation in World Bank Operations* (February 2004), 41.

46. IPCC, Working Group I, *Summary for Policymakers*, in *Climate Change 2007: The Physical Basis*, 18.

47. Ibid.

48. IPCC, Working Group II, *Summary for Policymakers*, in *Climate Change 2007: Impacts, Adaptation and Vulnerability* (Cambridge University Press, Cambridge, UK), 7–22, at 8; M. L. Parry, O. F. Canziani, J. P. Palutikof, P. J. van der Linden and C. E. Hanson, Eds.

49. Ibid., 12.

50. Ibid., 17.

51. Ibid., 19.

52. Ibid.

53. Ibid., 20, adding in parentheses that "a list of these recommendations is given in the Technical Summary Section TS-6."

54. IPCC, Working Group III, *Summary for Policymakers*, in *Climate Change 2007: Mitigation* (Cambridge, UK: Cambridge University Press), 19; B. Metz, O. R. Davidson, P. R. Bosch, R. Dave, L. A. Meyer, Eds.

55. Ibid., 21.

56. Ibid., 18.

57. Ibid., 22.

58. David W. Cash, "Viewpoint: Distributed Assessment Systems: An Emerging Paradigm of Research, Assessment and Decision-Making for Environmental Change," *Global Environ. Change* 10 (2000), 241–244, at 241.

59. M. Granger Morgan, Milind Kandlikar, James Risbey, and Hadi Dowlatabadi, "Why Conventional Tools for Policy Analysis Are Often Inadequate for Problems of Global Change: An Editorial Essay," *Climatic Change* 41 (1999), 271–281, at 273.

60. Thomas J. Wilbanks and Robert W. Kates, "Global Change in Local Places: How Scale Matters," *Climatic Change* 43 (1999), 601–628.

61. This is both an empirical finding to be documented later in the Barrow case and a logical implication of the conjunction rule of probability.

62. Quoted in Andrew Revkin and Patrick Healy, "Global Coalition to Make Buildings Energy-Efficient," *New York Times* (17 May 2007), A18.

63. Naomi Oreskes, Kristin Shrader-Frechette, and Kenneth Berlitz, "Verification, Validation, and Confirmation of Numerical Models in the Earth Sciences, *Science* 263 (1994), 641–646, at 641.

64. Steve Bankes, "Exploratory Modeling for Policy Analysis," *Operations Research* 41 (1993), 435–449, at 435, 437.

65. IPCC, Fourth Assessment Report, Working Group I, Ch. 9, 697.

66. Examples include components of the U.S. Climate Change Action Plan, which is reviewed in the next section.

67. Bankes, "Exploratory Modeling for Policy Analysis," 437. For more on this, including vivid examples, see Orrin H. Pilkey and Linda Pilkey-Jarvis, *Useless Arithmetic: Why Environmental Scientists Can't Predict the Future* (New York, NY: Columbia University Press, 2007).

68. Victor R. Baker, "Flood Hazard Science, Policy, and Values: A Pragmatist Stance," *Technol.Soc.* 29 (2007), 161–168, at 167.

69. Herman A. Karl, Lawrence E. Susskind, and Katherine H. Wallace, "A Dialogue, Not a Diatribe," *Environment* 49 (January/February 2007), 20–34, at 22.

70. Scientific uncertainty remains a priority concern of the IPCC: "Although the science to provide policymakers with information about climate change impacts and adaptation potential has improved since the Third Assessment, it still leaves many important questions to be answered. The chapters of the Working Group II Fourth Assessment include a number of judgments about priorities for further observation and research, and this advice should be considered seriously (a list of these recommendations is given in the Technical Summary Section TS-6)." IPCC, Working Group II, *Summary for Policymakers*, in *Climate Change 2007: Impacts, Adaptation and Vulnerability*, 20.

71. On the distinction between ordinary and constitutive decisions, see Harold D. Lasswell, *A Pre-View of Policy Decisions* (New York, NY: Elsevier, 1971), Chs. 5 and 6, which include goals and criteria for the appraisal of ordinary and constitutive decision processes from a common interest standpoint. For more on the constitutive decision process, see Myres S. McDougal, Harold D. Lasswell, and W. Michael Reisman, "The World Constitutive Process of Authoritative Decision," in *International Law Essays* (Mineola, NY: Foundation Press, 1981), 191–286.

72. *Climate Change: The IPCC Response Strategies*, lv.

73. Ibid., lvi.

74. Ibid., li.

75. Bert Bolin, "The Kyoto Negotiations on Climate Change: A Science Perspective," *Science* 279 (16 January 1998), 330–331, at 331. DOI: 10.1126/science.279.5349.330.

76. From the UNFCCCs background information on the Kyoto Protocol, 3, accessed 5 September 2007 at http://unfccc.int/kyoto_protocol/background/items/3145.php. For Decision 24/CP.7 on "Procedures and mechanisms relating to compliance under the Kyoto Protocol, see the *Report of the Conference of the Parties on Its Seventh*

Session, Held at Marrakesh from 29 October to 10 November 2001, Addendum, Part Two: Action Taken by the Conference of the Parties, 64–77.

77. According to an analysis of preliminary data, "many parties to the Protocol may find that achieving their emissions reductions obligations will prove to be difficult or impossible within the commitment period." Susan R. Fletcher and Larry Parker, *Climate Change: The Kyoto Protocol and International Actions*, CRS Report for Congress RL 33836 (Washington, D.C.: Congressional Research Service, Updated 8 June 2007), 7.

78. From the UNFCCCs Background information on the Kyoto Protocol, 2.

79. On accounting tricks and other limitations of the Kyoto Protocol, see David G. Victor, *The Collapse of the Kyoto Protocol and the Struggle to Slow Global Warming* (Princeton, NJ: Princeton University Press, 2004), including the Afterword. Additional critiques of the Kyoto Protocol include Gwyn Prins and Steve Rayner, "Commentary: Time to Ditch Kyoto," *Nature* 449 (25 October 2007), 973–975, and Bjorn Lomborg, *Cool It: The Skeptical Environmentalist's Guide to Global Warming* (New York, NY: Alfred A. Knopf, 2007).

80. Daniel Bodansky, "The Emerging Climate Change Regime," *Annu.Rev. Energy Environ.*20 (1995), 425–461, at 425, 438, and 429, respectively.

81. Sonja A. Boehmer-Christiansen, "Global Climate Protection Policy: The Limits of Scientific Advice," *Global Environ.Change* 4 (1994), Part II (September), 185–200, at 192.

82. Sheldon Ungar, "Social Scares and Global Warming: Beyond the Rio Convention," *Soc. Natl Res.* 8 (1995), 443–456, at 448.

83. Daniel Bodansky, "The Emerging Climate Change Regime," 426.

84. These percentages were reported by the UNFCCC in 2009, based on total greenhouse gas emissions including LULUCF (adjustments for Land Use, Land-Use Change, and Forestry), and accessed at http://unfccc.int/ghg_data/ghg_data_unfccc/time_series_annex_i/items/3842.php.

85. M. Granger Morgan, "Managing Carbon from the Bottom Up," *Science* 289 (29 September 2000), 2285. Note that "bottom up" here refers not to people on the ground in local communities but to initiatives by individual countries.

86. Quoted in Thomas L. Friedman, "Live Bad, Go Green," *New York Times* (8 July 2007), WK12.

87. Quoted in John H. Cushman, Jr., "Why the U.S. Fell Short of Ambitious Goals for Reducing Greenhouse Gases," *New York Times* (20 October 1997), A15.

88. Compare Harold D. Lasswell, *Psychopathology and Politics* (Chicago, IL: University of Chicago Press, 1930 and 1977), 189: "The competition among symbols to serve as foci of concentration for the aroused emotions of the community leads to the survival of a small number of master symbols. . . . Symbolization thus necessitates dichotomization."

89. For overviews, see Climate Action Network Europe, *Emissions Trading in the EU*, accessed 19 October 2007 at http://www.climnet.org/EUenergy/ET.html;

and Larry Parker, *Climate Change: The European Union's Emissions Trading System (EU-ETS)*, CRS Report for Congress RL-33581 (Washington, D.C..: Congressional Research Service, 31 July 2006).

90. Calculated from 2009 UNFCCC data on total greenhouse gas emissions including LULUCF (adjustments for Land Use, Land-Use Change, and Forestry) for the 1990 baseline year and 2006. The data were accessed at http://unfccc.int/ghg_data/ghg_data_unfccc/time_series_annex_i/items/3842.php.

91. Climate Action Network Europe, *Emissions Trading in the EU*, 2.

92. However, ETS includes provisions for a member state to opt-in coverage of additional greenhouse gases or sectors, and with European Commission approval to opt-out of coverage of certain industrial installations in the first trading period.

93. Climate Action Network Europe, *Emissions Trading in the EU*, 3.

94. Ibid.

95. *EU Emissions Trading Scheme Delivers First Verified Emissions Data for Installations* (Brussels, 15 May 2006), accessed 19 October 2007 at http://ec.europa.eu/environment/climat/pdf/citl_pr.pdf. The first page notes that no information had been received from Cyprus, Luxembourg, Malta, and Poland because "their emission allowance registries are not yet operational."

96. Parker, *Climate Change: The European Union's Emissions Trading System (EU-ETS)*, 5, 9. The overallocation would be larger if the annual average of 73.4 million allowances held in reserve for new installations or auctioning were included.

97. *EU Emissions Trading Scheme Delivers First Verified Emissions Data for Installations*, 1.

98. Aggregate and national figures for each year are reported by the European Commission in *Emissions Trading: 2007 Verified Emissions from EU ETC Businesses* (IP/08/787, Brusssels, 23 May 2008), accessed at http://europa.eu/rapid/pressReleasesAction.

99. Climate Action Network Europe, *National Allocation Plans 2005-7: Do They Deliver? Key Lessons to Member States for 2008-12* (April 2006), 5–6. Accessed 5 November 2007 at http://www.climnet.org/euenergy/ET/0506_NAP_report.pdf.

100. Quoted in Parker, *Climate Change: The European Union's Emissions Trading System (EU-ETS)*, 8.

101. PricewaterhouseCoopers, *Building Trust in Emissions Reporting: Global Trends in Emissions Trading Schemes* (February 2007), 9, 24. PricewaterhouseCoopers is a global network of legally independent consulting firms. The authors are Jeroen Kruijd, Arnoud Walrecht, Joris Laseur, Hans Schoolderman, and Richard Gledhill. See also the harmonization issue in Parker, *Climate Change: The European Union's Emissions Trading System (EU-ETS)*, 16–18.

102. Stephen Castle and James Kanter, "EU Looks at Big Changes in Emissions Trading," *International Herald Tribune* (6 June 2007), 5.

103. Leila Abboud, "Europe Struggles to Meet Carbon Commitments," *Globe and Mail (Canada)* (3 April 2008), B11; Mark Milner, "Carbon Prices Rise Amid Tighter

Rules," *The Guardian (London)* (3 April 2008), 24; Danny Fortson, "Industrial CO$_2$ Emissions to Fall as Tougher EU Curbs Come In," *The Independent (London)* (3 April 2008), 44.

104. European Commission, *Emissions Trading: 2007 Verified Emissions from EU ETC Businesses.*

105. A. Danny Ellerman and Paul L. Jaskow, *The European Union's Emissions Trading Systems in Perspective* (Prepared for the Pew Center on Global Climate Change, May 2008), iii.

106. Elizabeth Rosenthal, "Europeans Switching Back to Dirty Fuel; Ecologists Alarmed by Return to Coal," *International Herald Tribune* (23 April 2008), 1.

107. Parker, Climate *Change: The European Union's Emissions Trading System (EU-ETS),* 21.

108. Ibid., 17.

109. For additional details on CCAP, see Ronald D. Brunner and Roberta Klein, "Harvesting Experience: A Reappraisal of the U.S. Climate Change Action Plan," *Policy Sciences* 32 (1999), 133–161.

110. Government Accountability Office, *Climate Change: EPA and DOE Should Do More to Encourage Progress Under Two Voluntary Programs,* GAO-06-97 (April 2006), Appendix I. According to the administration at http://www.climatevision. gov/right_more.html, accessed 1 September 2006, "The Climate VISION program is a public-private partnership program established in February 2003 to support President Bush's goal [announced in February 2002] of reducing the greenhouse gas emissions intensity of the U.S. Economy by 18 percent from 2002 to 2012."

111. Helen Dewar, "Senate Advises Against Emission Treaty that Lets Developing Countries Pollute," *Washington Post* (26 July 1997), A11; and Helen Dewar and Kevin Sullivan, "Senate Republicans Call Kyoto Pact Dead; Some Democrats Suggest Clinton Delay Submission to Ratification Vote," *Washington Post* (11 December 1997), A37.

112. John R. Justus and Susan R. Fletcher, *Global Climate Change* (Washington, D.C.: Congressional Research Service, Updated 12 May 2006), 6–7.

113. Susan R. Fletcher and Larry Parker, *Climate Change: The Kyoto Protocol and International Actions,*12.

114. Ibid.

115. Thomas L. Friedman, "The Capitol Energy Crisis," *New York Times* (24 June 2007), 4:14.

116. Paul Singer, "Uneasy Money," *National Journal* (12 August 2006), 22–30, at 22–23.

117. Jane A. Leggett, *Climate Change: Science and Policy Implications,* CRS Report for Congress RL33849 (Washington, D.C.: Congressional Research Service, Updated 2 May 2007), 1.

118. George H. W. Bush, "Remarks at the Opening Session of the White House Conference on Science and Economics Research Related to Global Change," 17 April

1990, *Public Papers of the Presidents* (Washington, D.C.: Government Printing Office, 1991), 585–586, at 585.

119. Leggett, *Climate Change: Science and Policy Implications*, 3.

120. Both quotations are from George E. Brown, Jr., "Global Change and the New Definition of Progress," *Geotimes* (June 1992):19–21, at 21. This paragraph is adapted from Ronald D. Brunner, "A Paradigm for Practice," *Policy Sciences* 39:(2006), 135–167.

121. Brown, Jr., "The Objectivity Crisis," 781.

122. Office of Technology Assessment, *Preparing for an Uncertain Climate—Volume I*, OTA-O567 (Washington, D.C.: Government Printing Office, October 1993), 144.

123. In particular, USGCRP would "substantially increase support for programs that develop tools for assessing policies and options, especially in the area of integrated assessment methods." The budget for "Assessing Policies and Options" increased from 1.2% of the total budget in FY 1994 to 1.9% of the total budget request for FY 1995, or $37.7 million out of the total USGCRP FY 1995 budget request of $1,819.8 million. *Our Changing Planet: The FY 1995 U.S. Global Change Research Program*, A Report by the Committee on Environment and Natural Resources Research of the National Science and Technology Council, A Supplement to the President's Fiscal Year 1995 Budget (1994), 110 for the figures and 23 for the quotations.

124. The letter dated 11 December 2006 was addressed to Dr. William Brennan of the CCSP. The first of 24 members who signed it were Jay Inslee (Democrat from Washington) and Wayne Gilcrest (Republican from Maryland).

125. The material in italics was printed in boldface in the Executive Summary, 3–4, of *Evaluating Progress of the U.S. Climate Change Science Program: Methods and Preliminary Results*, Committee on Strategic Advice on the U.S. Climate Change Science Program (Washington, D.C.: National Research Council, 2007). The Executive Summary was accessed 15 September 2007 at http://www.nap.edu/catalog/11934.html.

126. *Our Changing Planet: The FY 1990 Research Plan*, 3. Emphasis in original.

127. The Web site is http://americasclimatechoices.org/index.shtml, accessed 20 November 2008.

128. Harold D. Lasswell and Abraham Kaplan, *Power and Society* (New Haven, CT: Yale University Press), 133. On the multiple bases of power as control, see ibid., 85: "[I]t is of crucial importance to recognize that power may rest on various bases, differing not only from culture to culture, but also within a culture from one power structure to another." Compare Herbert A. Simon, *The Science of the Artificial* 3rd ed. (Cambridge, MA: MI T Press, 1996), 185: "the real flesh-and-blood organization has many interpart relations other than the lines of formal authority."

129. These percentages were reported in 2009 by the UNFCCC, and based on total greenhouse gas emissions including LULUCF (adjustments for Land Use, Land-Use Change, and Forestry) in Gg CO_2 equivalent, accessed at http://unfccc.int/ghg_data/ghg_data_unfccc/time_series_annex_i/items/3842.php.

130. Regional Integrated Sciences and Assessments Program, Alaska Exploratory Workshop, held in Anchorage, AK, February 18–19, 2004, 6. This summary of the workshop was prepared by Susanne C. Moser. RISA's address on the Web is www.climate.noaa.gov/cpo_pa/risa/.

131. *New Directions for Climate Research and Technology* Initiatives, Hearings before the Committee on Science, House of Representatives, 107th Congress, 2nd Session, 17 April 2002 (Washington, D.C: Government Printing Office, 2002), 5. The hearing charter prefaced the comment on RISA with the observation that USGCRP "is designed by scientists to address the big scientific questions related to global change. It has never considered the needs of users on the ground who may have to make decisions about environmental resources that are affected by weather and climate."

132. Ibid., 18–19.

133. An interpretation of Byerly's testimony along these lines was suggested by Josh Foster, who also provided supplementary information for the interpretation of figures on RISA's budget in the next paragraph.

134. By this time the Bush Administration had moved RISA from the USGCRP to the new Climate Change Research Initiative (CCRI), and both USGCRP and CCRI were subsumed under the U.S. Climate Change Science Program. The budget data can be found in the Budget Tables inserted in *Our Changing Planet: The U.S. Climate Change Science Program for Fiscal Year 2006*, A Report by the Climate Change Science Program and the Subcommittee on Global Change Research, A Supplement to the President's Budget for Fiscal Year 2006 (Washington, D.C.: November 2005), 11.

135. Ibid., 131.

136. A. K. Snover, L. Whitely Binder, J. Lopez, E. Willmott, J. Kay, D. Howell, and J. Simmonds, *Preparing for Climate Change: A Guidebook for Local, Regional, and State Governments* (Oakland, CA: ICLEI—Local Governments for Sustainability, 2007), 1. The Guidebook can be accessed at www.iclei.org/index.php?id=7066.

137. Ibid., 7.

138. National Assessment Synthesis Team, *Climate Change Impacts on the United States: The Potential Consequences of Climate Variability and Change* (Washington, D.C.: U.S. Global Change Research Program, November 2000), 8.

139. Ibid, 120. This conclusion was also emphasized in a box in the opening summary, p. 9.

140. Ibid., 122.

141. From the Web site at http://earth.usgcrp.gov/usgcrp/nacc/. See also Darren Samuelsohn, "Free-Market Group Sues White House Over Global Warming National Assessment," *Greenwire* 10 (7 August 2003) and "CEI, White House Resolve Suit; New Disclaimer Over Data Quality Act," *Greenwire* 10 (7 November 2003). In the background were allegations that publication of the National Assessment was an "October surprise," part of Vice President Al Gore's 2000 election campaign, and that Bush White House officials may have invited lawsuits from the CEI. See, respectively, Sen. Infhofe's press release available from *Congressional Press Releases*

(5 October 2000) and a press release by Charles Dow from the office of Steven Rowe, Attorney General of Maine (11 August 2003).

142. Paul D. Thacker, "Blowing the Whistle on Climate Change: Interview with Rick Piltz," *Environmental Science & Technology Online* (22 June 2005) at http://pubs.acs.org/journal/esthag.

143. W. N. Adger et al., "Assessment of Adaptation Practices, Options, Constraints and Capacity," in IPCC, Working Group II, *Summary for Policymakers*, in *Climate Change 2007: Impacts, Adaptation and Vulnerability*, Ch. 17, 71–743. The following quotations can be found on pages 728, 728, 729, 729, 734, 735 and 735, respectively.

144. Ibid., 723.

145. Ibid., 722.

146. Quoted in a recent documentary, "Meltdown in Nepal," and reported in "Climate change: The long-range forecast" by David Cyranoski, News Feature, *Nature* 438, 275–276 (17 November 2005), DOI: 10.1038/438275a.

147. Lawrence C. Hamilton and Richard Haerdrich, "Ecological and Population Changes in Fishing Communities of the North Atlantic Arc," *Polar Res.* 18 (1999), 383–388; Lawrence Hamilton, Per Lyster, Oddmund Otterstad, "Social Change, Ecology and Climate in 20th Century Greenland," *Climatic Change* 47 (2000), 193–211; and Lawrence C. Hamilton, Benjamin C. Brown, and Rasmus Ole Rasmusssen, "West Greenland's Cod-to-Shrimp Transition: Local Dimensions of Climate Change," *Arctic* 56 (September 2003), 271–282.

148. Hamilton, Lyster, and Oddmund, "Social Change, Ecology and Climate in 20th Century Greenland," 199.

149. Ibid., 200–201. These long-term shifts in water and air temperatures are related to the North Atlantic Oscillation. See the section on "The NAO and West Greenland Climate" in Hamilton, Brown, and Rasmussen, "West Greenland's Cod-to-Shrimp Transition," 273–274.

150. Ibid., 209. See Fig. 3, 199, for data on the total cod catch off west Greenland from 1920 to 1998.

151. Ibid., 201.

152. Ibid., 203. For the details, see Fig. 5, 202, which graphs west Greenland's cod and shellfish (primarily shrimp) catch from 1950 to 1998.

153. Ibid., 204.

154. Ibid., 207.

155. Hamilton, Brown, and Rasmussen, "West Greenland's Cod-to-Shrimp Transition," 276.

156. Ibid., 277.

157. Ibid.

158. Ibid., 278. In this Sisimiut was helped by the Home Rule government, which gained limited sovereignty from Denmark in 1979. It required that 25% of the shrimp catch be processed at onshore plants to maintain employment, rather than packed and frozen whole on factory ships at sea.

159. Ibid., 278.

160. Hamilton, Lyster, and Oddmund, "Social Change, Ecology and Climate in 20th Century Greenland," 200.

161. Hamilton, Brown, and Rasmussen, "West Greenland's Cod-to-Shrimp Transition," 277.

162. Ibid., 278.

163. Ibid., 280.

164. Ibid.

165. Ibid.

166. Michael Glantz, Ed., *Once Burned, Twice Shy? Lessons Learned from the 1997–98 El Niño* (New York, NY: United Nations University, 2001), 4–5.

167. For example, James Ford et al., "Reducing Vulnerability to Climate Change in the Arctic: The Case of Nunavut, Canada, *Arctic* 60 (June 2007), 150–166. The Nunavut case is complementary to the Barrow case described in the next chapter: While it focused on reducing the vulnerability of the "traditional resource use sector" to climate change, the Barrow case focused on reducing the vulnerability of residents and modern infrastructure to big storms.

168. Michael P. Hamnett, Cheryl L. Anderson, and Charles P. Guard, *The Pacific ENSO Applications Center: Lessons Learned and Future Directions* (24 March 2000). We are grateful to Mike Hamnett for introducing us to PEAC, and to Eileen Shea and Cheryl Anderson for comments and corrections on an earlier version of our account.

169. Ibid., 1.

170. Ibid., 1–2. This documentation focuses on rainfall forecasts and drought impacts and responses but also considers wildfires, agriculture, health, ecological, and fisheries impacts and responses.

171. Ibid., 2.

172. Ibid., 3.

173. Ibid., 4–5.

174. Ibid., 4.

175. Ibid.

176. Mike Hamnett, *The Pacific ENSO Applications Center and the 1997–1998 El Niño*, 1, accessed 4 March 2007 at http://www.soest.hawaii.edu/MET/Enso/reports/97ENSO.html.

177. Hamnett, Anderson, and Guard, *The Pacific ENSO Applications Center*, 5.

178. Ibid., 7.

179. Ibid., 5.

180. Ibid., 6.

181. Early in 1998 the Republic of the Marshall Islands requested reverse osmosis units from the government of Japan, which in February supplied three units capable of producing 56,000 gallons of water per day. Thus there were other initiatives and other sources of external assistance.

182. In a comprehensive study, Matthew Collins and his collaborators have found that "[t]he potential for the mean climate of the tropical Pacific to shift to more El Niño-like conditions as a result of human induced climate change is subject to a considerable degree of uncertainty. The complexity of the feedback processes, the wide range of responses of different atmosphere–ocean global circulation models . . . and difficulties with model simulation of present day El Niño southern oscillation . . . all complicate the picture. . . . The most likely scenario . . . is for no trend towards either mean El Niño-like or La Niña-like conditions." Collins and the CMIP Modelling Groups, "El Niño- or La Niña-like Climate Change?" *Climate Dyn.* 24 (2005), 89–104, at 89.

183. Note that droughts in Australia tend to run through the end of the calendar year, as does an El Niño event. Hence, the "water year" is often used to calculate rainfall statistics. The "water year" is the 12-month period that runs either from April in one year to March the next, or more commonly from July to June, depending on the location in the country. In some cases, a water year lasts less than 12 months. In the United States, the water year is standardized by the U.S. Geological Survey to run from October to September. Because the definitions vary, for clarity we use only calendar-year statistics in this section.

184. T. H. F. Wong, "Water Sensitive Urban Design—The Journey thus Far," *Australian J. Water Resour.,* Special Issue on Water Sensitive Urban Design, 10 (2006), 213–222.

185. Rebekah Brown and Jodie Clarke, *Transition to Water Sensitive Urban Design: The Story of Melbourne Australia* (Report 07/1, Facility for Advancing Water Biofiltration, Monash University, June 2007; ISBN 978-0-9803428-0-2), 23. According to the Public Record Office of Victoria, http://www.access.prov.vic.gov.au/public/, accessed 31 July 2008, Melbourne Water was formed by the merger of the public utility Melbourne and Metropolitan Board of Works (MMBW) and a number of smaller urban water authorities in 1991–1992. The MMBW was formed in 1891 to manage Melbourne's piped water supply and reservoirs. After the Victorian Liberal Party led by Jeff Kennett gained office in 1992, disaggregation was part of a process of privatisation of state-owned services, including gas, electricitiy, and water, which continued until the Liberal Party left office in 1999.

186. Ibid., 43.

187. The impacts of the drought are summarized from M. Horridge, J. Madden, and G. Wittwer, "The Impact of the 2002–2003 Drought on Australia," *J. Policy Model.* 27(2005); World Economy & European Integration, 285–308, DOI: 10.1016/j.jpolmod.2005.01.008; and from the 2002 special drought issue of the Australian Crop Report, available from www.abare.gov.au, accessed 11 March 2008.

188. Despite the uncertainties in the global picture, there was already evidence in 2003 that climate change worsens a drought associated with El Niño through increased temperatures. For example, a headline on 16 January 2003 read "Bushfires Worsened by Climate Change—Scientists" (http://www.abc.net.au/science/news/stories/s763013.

htm). The scientists in the story were David Karoly and James Risbey, whose report titled "Global Warming Contributes to Australia's Worst Drought" was published by the World Wide Fund for Nature. They documented observations indicating that increased temperatures (explained by human-induced climate change) caused a substantial increase in evaporation rates from soil, watercourses, and vegetation, causing a more devastating drought than previous periods with similar rainfall deficits.

189. Water supply is not the only factor in the problem of urban water management in Melbourne. Changes in the patterns of water demand contribute to making the problem more urgent. Bob Birrell and colleagues have suggested that domestic water consumption is likely to increase due to a combination of population growth, urban consolidation policies, and the decreasing number of persons per household. Population growth is the primary driver, but as average household size decreases, the per capita water use increases. B. Birrell, V. Rapson, and T.F. Smith, *Impacts of Demographic Change and Urban Consolidation on Domestic Water Use* (Occasional Paper No. 15, 2005, Water Services Associated of Australia, 469 Latrobe Street Melbourne 3000 VIC Australia).

190. Cedric Sandford outlined this process and its origins in "Open Government: The Use of Green Papers," which appeared in the *British Tax Review* in 1980. He noted that Green Papers were invented by the British Labour Government in 1967, while White Papers had been a standard element of formally articulating government policy for some time before that. Quoting Sir Harold Wilson, Sandford reports that a Green Paper "represents the best that the government can propose on the given issue, but, remaining uncommitted, it is able without loss of face to leave its final decision open until it has been able to consider the public reaction to it" (351). Green papers are also known as "consultation documents." The process has been formalized in the Code of Practice on Written Consultation (published by the British Cabinet Office in 2000 and available from www.cabinet-office.gov.uk), and a similar process is also followed by the European Commission and the Republic of Ireland.

191. Accessed at http://www.watersmart.vic.gov.au/downloads/final_report_-_summary.pdf, 23 December 2007. Now available at http://www.ces.vic.gov.au.

192. Except where noted, all quotes from John Thwaites in this section are from an interview conducted on 19 December 2007. According to the Our Water Our Future Web site, there were in fact 40 separate stakeholder briefings and 21 public forums attended by more than 1,300 people, resulting in 670 submissions (http://www.dpi.vic.gov.au/DSE/wcmn202.nsf/LinkView/0F296881DE2C2FF6CA256FFE0 00927FB16E9B1815F549080CA256FFF000B04E4, accessed 1 April 2008).

193. Department of Sustainability and Environment, *Stormwater and Urban Water Conservation Fund*, http://www.dse.vic.gov.au/DSE/wcmn202.nsf/LinkView/ 92C773A9748FD0BECA256FE1001CBE34CFB32E3D98756185CA256FDD00136E16, accessed 15 November 2007.

194. As an example of the system, Yarra Valley Water, one of the water retailers, uses a three-tier block tariff to encourage customers to become more water efficient:

If you use less than 440 liters per day, you will pay 85.17 cents per kiloliter on that amount of water. If you use over 440 liters per day, you will move up into the next block of pricing and will pay 99.92 cents per kiloliter over 440 liters, until you reach 880 liters a day. When your water use increases beyond 880 liters per day, each liter of water in this block will be billed at 147.63 cents per kiloliter.

195. From documents shown to Amanda Lynch during the December interview with John Thwaites. Consider a recent example from daily water–use reports by week on the Melbourne Water Web site: During the week of 11–17 October in the spring of 2007, water usage averaged 1,011 million liters per day, a significant reduction from the 1,245 million liters used on an average spring day between 1992 and 1999. On a per capita basis, the reduction is from about 380 liters per person per day in 1996 to about 290 liters in 2007.

196. John Thwaites, follow-up discussion on 11 April 2008.

197. *Draft Central Region Sustainable Water Strategy*, page 1. Obtained from http:// www.dse.vic.gov.au/DSE/wcmn202.nsf/LinkView/7E335A5BD331FBC6CA25716800 2490DADEE554DEB21669DBCA256FFE00103BF8, accessed 28 March 2008.

198. From *Our Water Our Future: The Next Stage of the Government's Water Plan*, released in June 2007, it was documented that Melbourne storage inflow in 2006 was only 165 gigaliters. The average annual inflow for 1997–2006 was 387 gigaliters, and the long-term average annual inflow calculated for the period 1913–2005 was 588 gigaliters. It was noted in the report, "The total inflows to Melbourne's water storages from 1997 to 2006 were 35% lower than the pre-1997 average. Yet even with the continuation of this reduced inflow, the system was capable of absorbing the impact and recovering. . . . It is possible that Victoria is suffering a major long-term reduction in average rainfall—a step-change in water availability due to climate change" (20).

199. Press release from the office of the Premier on Wednesday, 19 September 2007; and Department of Sustainability and Environment, *Our Water Our Future: The Next Stage of the Government's Water Plan* (June 2007), accessed 15 November 2007 at http://www.ourwater.vic.gov.au/ourwater/about_our_water.

200. Australian Science Media Centre, Rapid Roundup, Tuesday, 19 June 2007 (Updated Tuesday, 26 June 2007), accessed at http://www.aussmc.org.au/ Desalination_plant_for_Victoria.php.

201. Ibid.

202. Quoted in "Row over Victoria's Desalination Plant," news.sbs.com.au (20 September 2007).

203. Quoted in *Catalyst*, ABC Television (April 2005), transcript accessed 15 November 2007 at http://www.abc.net.au/catalyst/stories/s1340461.htm.

204. From Amanda Lynch's notes of Waller's remarks, as edited by Waller on 18 August 2008 and used with his permission.

205. This is the figure featured prominently on the Web site of the project primarily responsible, accessed 6 September 2008 at http://www.goingcarbonneutral. co.uk/.

206. http://www.goingcarbonneutral.co.uk/Press_releases.html accessed 22 August 2008.

207. Garry Charnock and Roy Alexander, *A Practical Toolkit for Communities Aiming for Carbon Neutrality,* 6, accessed 20 August 2008 at http://rccn.communitycarbon. net/2007/08/31/our-footprint-our-journey-the-ashton-hayes-*Toolkit/.*

208. Charnock and Alexander, *A Practical Toolkit for Communities Aiming for Carbon Neutrality,* 7.

209. Ibid., 8.

210. The RSK Group sponsored Roy Alexander's chair at the University of Chester, according to the University's Web site. Shell Global Solutions is a client of TES, according to the TES Web site.

211. Ibid., 11.

212. Ibid., 13.

213. Marie Friend, quoted from a 16-minute film by the Ashton Hayes Parish Council, *Tomorrow's Climate, Today's Challenge: Going Carbon Neutral, Ashton Hayes,* a Tes/Fifth Pictures Production, a Steve Holland Film, produced by Garry Charnock and narrated by Marie Friend. The film elaborated her comment in print: "These savings have amounted to a total of 39 tonnes of CO_2 per year." The Holland quote also is from the film. Links to the film can be found at http://www.goingcarbonneutral.co.uk/.

214. Charnock and Alexander, *A Practical Toolkit for Communities Aiming for Carbon Neutrality,* 24.

215. Ibid., 27.

216. Ibid., 27–28.

217. Ibid., 28. The eight Big Rules are constitutive, and listed on p. 29.

218. Ibid., 30.

219. Ibid., 5.

220. Each is quoted from and identified in the film, Ashton Hayes Parish Council, *Tomorrow's Climate, Today's Challenge: Going Carbon Neutral, Ashton Hayes.*

221. Elizabeth Kolbert, "The Island in the Wind," *The New Yorker* (7 and 14 July 2008), 68–77, at 68. For more details and a fuller analysis, see *Samsø, A Renewable Energy Island: 10 Years of Development and Evaluation* (Samsø: PlanEnergi and Samsø Energy Academy, 2007), accessed at www.energiakademiet.dk. This document tracks Samsø's transition to a renewable energy island through narrative histories of multiple initiatives and energy statistics for the years 1997, 1999, 2001, 2003 (estimates), and 2005. This account is supplemented by Søren Hermansen's presentation at the University of Colorado, Boulder, on 3 November 2008; we are grateful to him for correcting an earlier version. See also Warren Hoge, "Samsø Journal: In This Energy Project, No Tilting at Windmills," *New York Times* (9 October 1999), and Robin McKie, "Isle of Plenty," *The Observer (London) Magazine* (21 September 2008), 30.

222. Kolbert, "The Island in the Wind," 70–71.

223. Quoted in ibid., 71.

224. Ibid., 71.

225. Hoge, "Samsø Journal: In This Energy Project, No Tilting at Windmills," *New York Times* (9 October 1999).

226. Kolbert, "The Island in the Wind," 71.

227. *Samsø, A Renewable Energy Island*, 21.

228. Kolbert, "The Island in the Wind," 71.

229. *Samsø, A Renewable Energy Island*, 11.

230. Kolbert, "The Island in the Wind," 72.

231. Ibid., 73.

232. These rough figures in USD are from notes on Hermansen's presentation in November 2008. For earlier figures in DKK and Euros, see *Samsø, A Renewable Energy Island*, 48.

233. Ibid.

234. Ibid., 49–50. Emphasized in original.

235. Kolbert, "The Island in the Wind," 73.

236. Ibid., 70. For more on Samsø as a model, see McKie, "Isle of Plenty."

237. *Samsø, A Renewable Energy Island*, 7.

238. W. Henry Lambright, Stanley A. Chagnon, and L. D. Danny Harvey, "Urban Reactions to the Global Warming Issue: Agenda Setting in Toronto and Chicago," *Climatic Change* 34 (1996), 463–478, at 469. See also L. D. Danny Harvey, "Tackling Urban CO_2 Emissions in Toronto," *Environment* 35 (September 1993), 16–20, 38–43.

239. Peter Gorrie, "Blueprint Green; David Miller Says This Will Be North America's Greenest City," *Toronto Star* (17 February 2007), A1.

240. John Barber, "A Small City We May Be, But Our Impact Is Big," *Globe and Mail* (14 May 2007), A11.

241. ICLEI has an initiative on adaptation, Resilient Communities & Cities, but a survey of members' Local Agenda 21 plans in 2005 revealed that "only a few of these plans deal with disaster risk management." ICLEI staff suggested that few tools developed for disaster risk management "actually address the needs of local government decision-makers or staff. Local government representatives further pointed out that a local government-friendly tool will be very helpful in incorporating risk reduction in local policies and processes. This, however, would require engaging local decision-makers in tools application, planning and systems development." From ICLEI's Web site, http://www.iclei.org/index.php?id=805, accessed 15 October 2007.

242. L. D. Danny Harvey, "Tackling Urban CO_2 Emissions in Toronto."

243. *ICLEI International Progress Report: Cities for Climate Protection*, 15, assessed 15 October 2007 at http://www.iclei.org/documents/USA/documents/CCP/ICLEI-CCP International_Report-2006.pdf/.

244. Ibid. Emphasis in original.

245. *About ICLEI*, accessed 15 October 2007 at http://www.iclei.org/index.php?id=global-about-iclei.

246. *ICLEI USA 2006 Year in Review*, 1 and 2, accessed 15 October 2007 at http://www.iclei.org/documents/USA/documents/yearinreview/2006-USA-YearInReview.pdf.

247. *ICLEI International Progress Report: Cities for Climate Protection*, 4.

248. John Bailey, *Lessons from the Pioneers: Tackling Global Warming at the Local Level* (Minneapolis, MN: Institute for Local Self–Reliance, 2007), 3.

249. Nicholas D. Kristof, "Another Small Step for Earth," *New York Times* (30 July 2006), sec. 4, 13.

250. Harriet Bulkeley and Michele M. Betsill, *Cities and Climate Change: Urban Sustainability and Global Environmental Governance* (London, U.K.: Routledge, 2003), 171.

251. Ibid., 184–185.

252. Ibid., 187.

253. Jonathan L. Ramseur, *Climate Change: Action by States to Address Greenhouse Gas Emissions*, CRS Report for Congress RL33812 (Washington, D.C.: Congressional Research Service, 27 April 2007), Summary.

254. "Separately . . . nine Midwestern and governors and the premier of Manitoba signed an agreement to reduce carbon emissions and set up a trading system to meet the reduction targets. The Midwestern accord is modeled on similar . . . similar arrangements among Northeastern, Southwestern and West Coast states," John M. Broder, "Governors Join in Creating Regional Pacts on Climate Change," *New York Times* (15 November 2007), A16.

255. Felicity Barringer and Kate Galbraith, "States Aim to Cut Gases By Making Polluters Pay," *New York Times* (16 September 2008), A14.

256. Quoted in John M. Broder and Felicity Barringer, "E.P.A. Says 17 States Can't Set Greenhouse Gas Rules for Cars," *New York Times* (20 December 2007), A1, which further quoted EPA administrator Stephen L. Johnson: "Climate change affects everyone regardless of where greenhouse gases occur, so California is not exclusive. . . ." EPA granted the waiver at the end of June 2009. It allows uniform application of the California rule in the 17 states (with nearly half the U.S. car market) that adopted the California rule.

257. Ramseur, *Climate Change: Action by States to Address Greenhouse Gas Emissions*, 16.

258. From slide 20 of a lecture delivered by University of California, Berkeley, economist W. Michael Hanemann at Monash University, Australia, on 15 August 2008.

259. Ibid., Summary.

260. Broder, "Governors Join in Creating Regional Pacts on Climate Change."

261. Ramseur, *Climate Change: Action by States to Address Greenhouse Gas Emissions*, Summary.

262. The states as laboratories is a theme developed in Kirsten H. Engel and Marc L. Miller, "Sustainable Development in the States: Leadership on Climate Change: 2007," *Arizona Legal Studies* (Discussion Paper No. 07-37, December 2007).

263. Brunner and Klein, "Harvesting Experience," 133–161

264. Darcy Frey, "How Green Is BP? *New York Times Magazine* (8 December 2002), 99. See also John Browne, "Beyond Kyoto," *Foreign Affairs* 83 (July–August 2004), 20.

265. Andrew Martin, "In Eco-Friendly Factory, Low-Guilt Potato Chips," *New York Times* (15 November 2007), A1.

Chapter 3

1. Parts of this chapter are adapted from Ronald D. Brunner et al., "An Arctic Disaster and Its Policy Implications," *Arctic* 57 (December 2004), 336–346, and Amanda H. Lynch and Ronald D. Brunner, "Context and Climate Change: An Integrated Assessment for Barrow Alaska," *Climatic Change* 82 (2007), 93–111. Additional publications from our project are cited below.

2. Brower spoke as an official of the Inuit Circumpolar Conference to the annual conference of the Resource Development Council in Anchorage. This account including quotations is based on Margaret Bauman, "RDC Conference Attendees Get Firsthand Account of Arctic Warming," *Associated Press State & Local Wire* (29 November 2005).

3. Norman A. Chance, *The Iñupiat and Arctic Alaska: An Ethnography of Development* (Fort Worth, TX: Holt, Rinehart and Winston, 1990), 11. The larger problems of concern to Chance were "the unequal distribution of productive wealth among the world's peoples and . . . the deterioration of the environment" (11).

4. Chance, *The Iñupiat and Arctic Alaska*, 16. Our account of the evolution that followed the crossing is based primarily on this source and on Glenn W. Sheehan, *In the Belly of the Whale: Trade and War in Eskimo Society*, Alaska Anthropological Association Monograph Series—VI (Anchorage, AK: Aurora). Also consulted were Robert F. Spencer, *The North Alaskan Eskimo: A Study in Ecology and Society*, Smithsonian Institution, Bureau of American Ethnology, Bulletin 171 (Washington, D.C.: Government Printing Office, 1959) and Ernest S. Burch, Jr., *The Iñupiat Eskimo Nations of Northwest Alaska* (Fairbanks, AK: University of Alaska Press, 1998). We are grateful to Glenn Sheehan, executive director of the Barrow Arctic Science Consortium, and Anne Jensen, an archaeologist with UIC-Science in Barrow for comments and corrections on an earlier draft of this chapter.

5. Chance, *The Iñupiat and Arctic Alaska*, 17–18.

6. Ibid., 19.

7. Ibid.

8. The important sites are identified by Sheehan, *In the Belly of the Whale*, 6–7.

9. Peter Sloss, *Global Relief CD-ROM*, produced by the National Oceanic and Atmospheric Administration, National Geophysical Data Center, Digital media, Boulder, CO, June 2005.

10. Sheehan, *In the Belly of the Whale*, 6–7. On the whaling orientation, in addition to Sheehan, see Bill Hess, *Gift of the Whale: The Iñupiat Bowhead, A Sacred Tradition* (Seattle, WA: Sasquatch Books, 1999); Karen Brewster, Ed., *The Whales, They Give Themselves: Conversations with Harry Brower, Sr.* (Fairbanks, AK: University of Alaska Press, 2004); and parts of Charles Wohlforth, *The Whale and the Supercomputer: On the Northern Front of Climate Change* (New York, NY: North Point Press, 2004).

11. Spencer, *The North Alaskan Eskimo*, 19.

12. Sheehan, *In the Belly of the Whale*, 22.

13. Spencer, *The North Alaskan Eskimo*, 6.

14. Sheehan, *In the Belly of the Whale*, 20.

15. Ibid., 21.

16. Ibid., 22.

17. Ibid., 10.

18. Ibid., 14.

19. Ibid. Similarly, Burch in *The Iñupiat Eskimo Nations of Northwest Alaska*, 308, claims that the earlier sharp distinction between maritime and inland people "is worse than useless . . . obscuring the many subtle ways in which Iñupiat maximized their access to the various resources that Northwest Alaska provided."

20. Spencer, *The North Alaskan Eskimo*, 25–27.

21. Sheehan, *In the Belly of the Whale*, 1, 5.

22. Chance, *The Iñupiat and Arctic Alaska*, 21: Successful umialit "regularly won the right to lead through their personal qualities of hunting, trading, and human relations skills, energy and wisdom. These qualities were what gained them their following, and their following was what provided them their wealth." Spencer, *The North Alaskan Eskimo*, 443, elaborates: "The umealiq, whether inland or at sea, bribed his crew to keep them. Except for his own kin and his associated quasi-kin, he commanded little primary loyalty. Indeed, it was his competitive success which insured him a full crew."

23. Chance, *The Iñupiat and Arctic Alaska*, 24.

24. Sheehan, *In the Belly of the Whale*, 205.

25. Ibid., 167.

26. Ibid., 6.

27. Ibid., 205.

28. Ibid., 1.

29. Ibid.

30. Ibid., 6.

31. Chance, *The Iñupiat and Arctic Alaska*, 23.

32. Spencer, *The North Alaskan Eskimo*, 442

33. Chance, *The Iñupiat and Arctic Alaska*, 29.

34. Quoted in ibid., 31. A Russian expedition crossed the Bering Strait in 1741 and returned with valuable pelts. Aleuts to the south subsequently bore the brunt of subjugation and disease brought by Russian traders.

35. Sheehan, *In the Belly of the Whale*, 176–177.

36. For more on Brower, see his autobiography, *Fifty Years Below Zero* (New York, NY: Dodd, Mead, 1942), republished by the University of Alaska Press (Fairbanks, 1994) and The Long Riders' Guild Press. For a brief overview of the postcontact period, see Joe Stankiewicz, *North Slope Borough Local All Hazard Mitigation Plan* (Barrow: North Slope Risk Management Division, 2005), 6–7.

37. Chance, *The Iñupiat and Arctic Alaska*, 36.

38. Ibid., 37.

39. Sheehan, *In the Belly of the Whale*, 2, 178. Sheehan's figures on losses to disease are similar to Chance's. Sheehan, 178, quoting Charles Brower, adds that "[a]t Nuvuk in 1902, 'the village was nearly wiped out.'"

40. Chance, *The Iñupiat and Arctic Alaska*, 32.

41. Ibid., 40.

42. Ibid., 43.

43. Ibid., 41.

44. Ibid., 45. For the remaining quotations in this paragraph see pp. 46 and 48.

45. Ibid., 48.

46. Ibid., 50.

47. Ibid.

48. Ibid., 51. In 1898, the Bureau of Education officially sought "to prepare the natives to take up industries and modes of life established in the States by our white population, and by all means not try to continue the tribal life after the manner of the Indians in the western states and territories."

49. Ibid., 53.

50. See David W. Norton, Ed., *Fifty More Years Below Zero: Tributes and Meditations for the Naval Arctic Research Laboratory's First Half Century at Barrow, Alaska* (Fairbanks, AK: University of Alaska Press, 2001).

51. Kenneth Toovak was a regular contributor to our meetings in Barrow. Grace Redding and her husband were interviewed at their home in Sequim, WA, on 24 October 2003.

52. Bob Harcharek, project manager, *North Slope Borough: 2003 Economic Profile and Census Report*, Vol. IX (Barrow: North Slope Borough Department of Planning and Community Services, 2004), BRW-2.

53. The City Council's letter is reproduced under the headline, "Barrow Asks Assistance to Speed Townsite Survey," in the *Tundra Times* (4 March 1963), 5. For additional reporting on modernization at this time, see Thomas Snapp, "Barrow on Way to Modern City," *Tundra Times* (22 July 1963), 1; "BIA to Expend $400,000 for Barrow Streets," *Tundra Times* (19 August 1963), 5; "Many Improvements Coming to Barrow, But Eskimos Will Keep Independence," *Tundra Times* (2 December 1963), 7; and "Where's My House? Asked at Barrow," *Tundra Times* (2 March 1964), 6. In "Open Water Late in Coming to Farthest North Region," *Tundra Times* (22 June

1964), Guy Okakok reported that the realignment of houses according to the plan was completed in June 1964.

54. Chance, *The Iñupiat and Arctic Alaska*, 45.

55. Sheehan, *In the Belly of the Whale*, especially Ch. 5, "The Protohistorical Baseline for Social and Political Organization."

56. Spencer, *The North Alaskan Eskimo*, 2.

57. Quoted in Chance, *The Iñupiat and Arctic Alaska*, 56. Compare Eileen Panigeo-MacLean in 1988: "Sharing is what has made our culture strong. Only through sharing have we survived as a people in this land" (139).

58. Claude Levi-Strauss, *The Savage Mind* (Chicago, IL: University of Chicago Press, 1966), 92.

59. Pedro J. Schafer, "Computation of a Storm Surge at Barrow, Alaska," *Arch. Meteor. Geophys. Bioclimatol.*, Series A, 15 (1966), 372–393, at 374.

60. Guy Okakok, "Writer Reports on Arctic Storm Damage," *Tundra Times* (21 October 1963), 3. On the reported gusts, see J. D. Hume and M. Schalk, "Shoreline Processes near Barrow, Alaska: A Comparison of the Normal and the Catastrophic," *Arctic* 20 (1967), 86–103; and Howard Rock, "Northern Natives Have Learned to Live with Elements," *Tundra Times* (21 October 1963), 4.

61. Andrey Proshutinsky, personal communication, 22 September 2001.

62. Subjective estimates of wave heights are notoriously unreliable, but the combination of debris lines and photographs during the storm add some weight to this estimate. Z. Kowalik, "Storm Surges in the Beaufort and Chukchi Seas," *J. Geophys. Res.* 89 (1984), 10,570–10,578; Schafer, "Computation of a Storm Surge"; Hume and Schalk, "Shoreline Processes"; Rock, "Northern Natives"; and Anon., "Arctic Disaster: Storm Loss Estimated at $1 Million," *Tundra Times* (7 October 1963), 1.

63. Schafer, "Computation of a Storm Surge," 374.

64. Hume and Schalk, "Shoreline Processes," 86. See the same source for other details in this paragraph.

65. Leanne Lestak, William Manley, Cove Sturtevant, and James Maslanik, "Fifty-Four Years of Shoreline Change Along the Chukchi Sea Near Barrow, Alaska." Presented at NOAA CSC Geotools Workshop, Myrtle Beach, SC, 5–8 March 2007. Downloadable from http://nome.colorado.edu/HARC/seminars.html.

66. Quotations from Hume and Schalk, "Shoreline Processes," 96.

67. Data digitized by William Manley from the Barrow B4 USGS 15-min topographic map, Alaska Geospatial Data Clearinghouse (http://agdc.usgs.gov), Digital Raster Graphic (DRG), Barrow B-4, 1997. Photo inset by Leanne Lestak from the high-resolution (0.35 m) georectified 1964 grayscale aerial photographs, Barrow, northern Alaska.

68. Quotations from Rock, "Northern Natives." For other details in this paragraph see Anon., "Weather Bureau says storm was not expected," *Tundra Times* (7 October 1963), 7, and Anon., "Arctic Disaster."

69. Okakok, "Writer Reports."

70. Ibid.

71. Anon., "Guy Okakok, Tundra Times Writer, Loses His Home," *Tundra Times* (7 October 1963), 7.

72. Okakok, "Writer Reports."

73. Ibid.

74. Ibid.

75. Rock, "Northern Natives."

76. Hume and Schalk, "Shoreline Processes," 94–95.

77. Ibid., 96–97.

78. Ibid., 97.

79. Anon., "Lowell May Ask Disaster Fund from Lawmakers," *Tundra Times* (7 October 1963), 7.

80. Anon., "Barrow Getting New Houses with Aid of the Red Cross," *Tundra Times* (21 October 1963).

81. Rock, "Northern Natives." See also Anon., "Lake Contaminated by Sea Water, Fuel Oil, Sewage," *Tundra Times* (7 October 1963), 7.

82. Okakok, "Writer Reports."

83. Anon., "Arctic Disaster."

84. Ibid., and Anon., "Barrow Declared Eligible for SBA Loan Money," *Tundra Times* (21 October 1963), 3.

85. Anon., "Lake Contaminated by Sea Water, Fuel Oil, Sewage," *Tundra Times* (7 October 1963), 7; Anon., Barrow Electric Controversy Rages, *Tundra Times* (4 November 1963), 7; Anon., "Many Improvements Coming to Barrow; but Eskimos Will Keep Independence," *Tundra Times* (2 December 1963), 7; Anon., "Barrow to Get Natural Gas Soon; Negotiations Started," *Tundra Times* (21 October 1963), 5; and Anon., "Senator Says Barrow on Way toward Becoming Model City," *Tundra Times* (15 March 1963), 7. Arnold Brower, Sr., mentioned the decontamination process in informal remarks at a meeting in Barrow in March 2004.

86. Ron Brunner and Liz Cassono. Redding interview (24 October 2003); see note 51.

87. Rock, "Northern Natives Have Learned to Live with Elements," 4, is the source of all quotations in this paragraph, and includes stories illustrating these points. See also the editorial "Invaluable Knowledge," *Tundra Times* (4 February 1963), 2.

88. Hume and Schalk, "Shoreline Processes," 101.

89. Bill Hess, *Taking Control: The Story of Self-Determination in the Arctic* (Barrow, AK: North Slope Borough, 1993). For additional perspectives, see Chance, "Politics, Petroleum, and Profits," in *The Iñupiat and Arctic Alaska*, Part Three, 139–218; and Thomas R. Berger, *Village Journey: The Report of the Alaska Native Review Commission* (New York, NY: Hill and Wang, 1995). For non-Native perspectives more generally, see Clive S. Thomas, Ed., *Alaska Public Policy Issues: Background and Perspectives* (Juneau, AK: Denali Press, 1999).

90. Hess, *Taking Control*, 12. Similarly, consider Chance's assessment of the 20th Alaska Science Conference on the impact of oil in the future of Alaska, held at the University of Alaska at Fairbanks in August 1969. Chance in "Politics, Petroleum, and Profits," 158, reported he found at the conference "the implicit assumption that the land—tundra and forests, federal parks and wildlife refuges—belonged to the federal government and the state and by extension, its citizens. It was as if the claims of Alaska's Native to the land were nonexistent—or at least an inappropriate topic for discussion at a scientific meeting." He also reported that the Alaska Science Conference is a division of the American Association for the Advancement of Science (AAAS).

91. Hess, *Taking Control*, 11–12.

92. Chance, "Politics, Petroleum, and Profits," 143–145. See also Hess, *Taking Control*, 62.

93. Hess, *Taking Control*, 13; see also 39, 50.

94. Ibid., 13.

95. Chance, "Politics, Petroleum, and Profits," 147.

96. Ibid. 148.

97. Hess, *Taking Control*, 19–21; Chance, "Politics, Petroleum, and Profits," 149.

98. Chance, "Politics, Petroleum, and Profits," 149.

99. Hess, *Taking Control*, 21; Chance, "Politics, Petroleum, and Profits," 150.

100. Hess, *Taking Control*, 23.

101. Ibid., 24.

102. Ibid., 10.

103. Ibid., 25.

104. For this suggestion and the previous quotations in this paragraph, see Hess, *Taking Control*, 25.

105. Chance, "Politics, Petroleum, and Profits," 170.

106. Ibid., 166.

107. Ibid., 164.

108. Berger, *Village Journey*, vii.

109. Ibid., xii.

110. Ibid., xiv. See also Fae L. Korsmo, "Native Sovereignty: An Insoluble Problem?" Ch. 19, 263–274, in Thomas, *Alaska Public Policy Issues*.

111. Hess, *Taking Control*, 28–29.

112. Ibid., 32.

113. Ibid., 37.

114. Ibid., 62.

115. Ibid., 76.

116. Ibid., 78–80.

117. Ibid., 77–78.

118. *Comprehensive Annual Financial Report of the North Slope Borough, Alaska, Fiscal Year July 1, 2005–June 30, 2006*, ii. This report did not include in its Historical

Section the 10-year summaries of revenues and expenditures included in previous reports, the basis for Fig. 3.5.

119. Hess, *Taking Control*, 104.

120. From the *Comprehensive Annual Financial Report of the North Slope Borough, Alaska, Fiscal Year July 1, 1989–June 30, 1990*, xv: "The most significant decrease in expenditures relates to Debt Service. The decrease of 51.76% reflects normally scheduled reductions in debt service outstanding as principal debt matures, as well as the impact of Series H refunding issue which reallocated principal debt to fiscal years beyond 1989–90."

121. Hess, *Taking Control*, 103.

122. For more on the utilidor, see "The Utilidor: The End of Honeybuckets and Water Hauling," *Tundra Times* (24 February 1982), 12; and Army Corps of Engineers, *Barrow, AK, Section 905(b) (WRDA 86) Analysis, Shoreline Erosion, Flooding, and Navigation Channel* (Anchorage: U.S. Army Engineer District, Alaska, 24 May 2001), 3–4.

123. These population data are from Harcharek, *North Slope Borough: 2003 Economic Profile and Census Report*, Vol. IX, BRW-2. The informal estimates are from personal communications with Bob Harcharek in February 2008.

124. 1964 and 1997 planimetrics provided by the North Slope Borough GIS Department and Barrow Arctic Science Consortium. 1963 floodline digitized by Leanne Lestak from Hume and Schalk, "Shoreline Processes."

125. Harcharek, *North Slope Borough: 2003 Economic Profile and Census Report*, BRW 34–35.

126. Ibid., BRW-5 and BRW-34.

127. Consider Jacob Adams, who was elected to the board of directors of ASRC in 1971, served on the North Slope Borough Assembly, and was appointed Mayor of North Slope Borough after Eben Hopson died in 1980. Then he was elected to that office. One supporter observed, "Before all else . . . Jacob Adams is a whaling captain, a hunter with a keen commitment to the land. His appointment as mayor was strong recognition of this commitment." Quoted in Hess, *Taking Control*, 97.

128. For insights into the historical roots of these outcomes of the empowerment process, we are grateful to Glenn Sheehan and Anne Jensen. Also in Barrow, Bob Harcharek corroborated the increase in Natives returning home.

129. Hess, *Taking Control*, 121. On Eben Hopson and organization of the ICC, see pp. 83–84. Hess also reported that "Billy Neakok represented the North Slope Borough at the First International Conference of Indigenous Peoples, held October 27–31, 1975 in British Columbia" (77).

130. Hess, *Taking Control*, 130, from a section titled "The Fight for Whaling: A Culture Under Fire," 124–137.

131. Ibid., 130. Chance, "Politics, Petroleum, and Profits," 176, reported that "seventy whaling captains from the nine Arctic whaling villages" participated in the meeting in Barrow.

132. Hess, *Taking Control*, 128.

133. Ibid., 137.

134. R. S. Stone, "Variations in Western Arctic Temperatures in Response to Cloud Radiative and Synoptic-Scale Influences," *J. Geophys. Res.* 102 (1997) 21,769–21,776; J. Curtis, G. Wendler, R. Stone, and E. Dutton, "Precipitation Decrease in the Western Arctic, with Special Emphasis on Barrow and Barter Island, Alaska," *Int. J. Climatol.* 18, (1998), 1687–1707.

135. M. L. L'Heureux, M. E. Mann, B. I. Cook, B. E. Gleason, and R. S. Vose, "Atmospheric Circulation Influences on Seasonal Precipitation Patterns in Alaska During the Latter 20th Century," *J. Geophys. Res.* 109 (2004), DOI:10.1029/2003JD003845.

136. This change has been demonstrated by analyzing vorticity and surface pressure [J. E. Walsh, W. L. Chapman, and T. L. Shy, "Recent Decrease of Sea Level Pressure in the Central Arctic," *J. Climate* 9 (1996), 480–486] and cyclone frequency [G. J. McCabe, M. P. Clark, and M. C. Serreze, "Trends in Northern Hemisphere Surface Cyclone Frequency and Intensity," *J. Climate* 14 (2001), 2763–2768], and through a principal component analysis [D. W. J. Thompson and J. M. Wallace, "Regional Climate Impacts of the Northern Hemisphere Annular Mode," *Science* 293 (2001), 85–89].

137. This record of high-wind events is correlated significantly with the Northern Annular Mode (correlation coefficient of 0.3) and the Pacific North American Teleconnection (correlation coefficient of 0.45), even though cyclones in this region and *average* winds in Barrow are not correlated with either of these large-scale teleconnection patterns. The physical meaning of this teleconnection remains to be investigated.

138. W. L. Chapman and J. E. Walsh, "Recent Variations of Sea Ice and Air Temperature in High Latitudes," *Bull. Amer. Meteor. Soc.* 74 (1993), 2–16; R. S., Stone, "Variations in Western Arctic Temperatures in Response to Cloud Radiative and Synoptic-Scale Influences," *J. Geophys. Res.* 102 (1997) 21,769–21,776; M. C. Serreze et al., "Observational Evidence of Recent Change in the Northern High-Latitude Environment," *Climate Change* 46 (2000), 159–207; J. M. Stafford, G. Wendler, and J. Curtis, "Temperature and Precipitation of Alaska: 50 Year Trend Analysis," *Theor. Appl. Climatol.* 67 (2000), 33–44; A. H. Lynch, J. A. Curry, R. D. Brunner, and J. A. Maslanik, "Toward an Integrated Assessment of the Impacts of Extreme Wind Events on Barrow, AK," *Bull. Amer. Meteor. Soc.* 85 (2004), 209–221.

139. The National Snow and Ice Data Center defines multiyear ice as "ice that has survived at least one melt season; it is typically 2 to 4 meters (6.6 to 13.1 feet) thick and thickens as more ice grows on its underside." From the NSIDC Glossary, available at http://nsidc.org/cgi-bin/words/glossary, l, accessed 8 February 2008.

140. J. A. Maslanik, C. Fowler, J. Stroeve, S. Drobot, J. Zwally, D. Yi, and W. Emery, "A Younger, Thinner Arctic Ice Cover: Increased Potential for Rapid, Extensive Sea-Ice Loss," *Geophys. Res. Lett.* 34 (2007), L24501, DOI:10.1029/2007GL032043.

141. R. W. Lindsay and J. Zhang, "The Thinning of Arctic Sea Ice, 1988–2003: Have We Passed a Tipping Point?" *J. Climate* 18 (2005), 4879–4894.

142. Based on photogrammetric analysis of the last 50 years by Leanne Lestak, William Manley, and colleagues. See, e.g., L.R. Lestak, W. F. Manley, C. M. Sturtevant, and J. A. Maslanik, "Fifty-Four Years of Shoreline Change Along the Chukchi Sea near Barrow, Arctic Coastal Dynamics" (2006), Report of an International Workshop, Ber. Polarforsch. Meeresforsch.

143. J. H. Walker, "Bluff Erosion at Barrow and Wainwright, Arctic Alaska," *Z Geomorph. N.F.* 81 (1991), 53–61.

144. Analysis of aerial photographs demonstrates that the bluffs have retreated by only 0.35 m per year (on average, over the last 50 years), although at the point closest to Barrow, bluff-bottom retreat caused by erosion reaches a maximum of around 0.8 m per year. Thus, in general the erosion around Barrow is less than typical for the Arctic coasts, and hence might be considered a good site for a coastal village.

145. In this context, "normal transport" refers to the transport of sediment by episodic storms that are more moderate in strength than the 1963 storm, which occur regularly several times each autumn.

146. S. M. Solomon, D. L. Forbes, and B. Kierstead, "Coastal Impacts of Climate Change: Beaufort Sea Erosion Study," Atmospheric Environment Service, Canadian Climate Centre, Report 94-2 (1994), *Environment Canada*, 34 pp.

147. F. E. Are, "The Role of Coastal Retreat for Sedimentation in the Laptev Sea," in *Land-Ocean Systems in the Siberian Arctic*, H. Kassens et al., eds. (Berlin, Germany: Springer, 1999), 287–295.

148. This account is taken primarily from "Observations on Shorefast Ice Dynamics in Arctic Alaska and the Responses of the Iñupiat Hunting Community," *Arctic* (1 December 2004), by Karen Brewster, Hajo Eicken, Craig George, Richard Glenn, Henry Huntington, and Dave Norton, and also the account of the event by Karen Brewster as part of the University of Alaska at Fairbanks Oral History Program. Direct quotes of eyewitnesses come from the latter source.

149. This is known as the "inverse barometer effect." When atmospheric pressure is particularly low, the ocean surface beneath tends to be raised, and vice versa; 1 mb of atmospheric pressure decrease from the mean value corresponds to approximately 1 cm of increased sea level.

150. Winds of this minimum strength and duration account for almost 90% of the variance in flood forecasts. A. H. Lynch., L. R. Lestak, P. Uotila, E .N. Cassano, and L. Xie, "A factorial analysis of storm surge flooding in Barrow, Alaska," *Mon. Wea. Rev.* (2008), 898–912.

151. Anon., "Big Storms Follow Visit from Past Ancestor; A Doomed House Is Saved," *The Open Lead* (November 1986), 38–47.

152. Long-time Barrow resident and scientist Dan Enders, personal communication, 2001.

153. Near the NARL camp, a NOAA observing site owned by the NOAA Climate Monitoring and Diagnostics Laboratory is staffed by Dan Enders.

154. Anne Morkill, personal communication, 22 August 2000.

155. Army Corps of Engineers, *Barrow, AK, Section 905(b) (WRDA 86) Analysis, Shoreline Erosion, Flooding, and Navigation Channel.*

156. Amanda Lynch et al., "A Factorial Analysis."

157. Obtained from http://www.ak-prepared.com/community_services/barrow storm.htm.

158. George N. Ahmaogak, Sr., Letter to Colonel Steven T. Perrenot, Corps of Engineers, U.S. Army Engineers District, Anchorage, Alaska (Barrow: North Slope Borough Mayor's Office, 24 April 2001).

159. Dave Anderson, e-mail, Thursday, 7 February 2008.

160. North Slope Borough Assembly, Special Meeting and Public Hearing (22 September 1986), 1; Anon., "Emergency Action Taken," *The Open Lead* (November 1986), 7.

161. Frank Brown, *Barrow and Wainright Beach Nourishment Program: North Slope Borough Assembly Presentation* (November 1997), 4. CIPM was the division of the NSB government that managed the beach nourishment program.

162. BTS/LCMF, Ltd., *Mitigation Alternatives for Coastal Erosion at Wainwright and Barrow, Alaska: Active and Passive Considerations* (Anchorage, 1 April 1989), 48; see also I.

163. Ibid., 5. Compare this with the estimates of Leanne Lestak and colleagues using 50 years of aerial photography, radar and satellite images, and spatial analysis of 20 miles of coastline around Barrow. Lestak estimated a bluff top average (over 20 miles and 50 years) of 0.7 feet per year and a shoreline average of 0.16 feet per year, but the 20-mile averages varied between 3.3 feet erosion and 4.6 feet accretion from one year to the next. The spatial variation was even larger in time slices that included significant storms.

164. Ibid., 17. On the offshore material samples, see BTS/LCMF, Ltd., *Barrow Offshore Exploration: Core Samples and Field Information* (Barrow: North Slope Borough, September 1988). The press release included in the report acknowledged that BRS/LCMF was "charged with the responsibility of analyzing the samples to determine their compatibility and usefulness with future coastal nourishment." The report merely described 54 individual core samples—mostly silt, clay, sands with little or no gravel—without drawing conclusions on their suitability for beach nourishment.

165. BTS/LCMF, Ltd., *Mitigation Alternatives for Coastal Erosion at Wainwright and Barrow, Alaska*, I.

166. Ibid., 13–14. The utilidor-section seawall actually protected the landfill.

167. Brown, *Barrow and Wainright Beach Nourishment Program*, 5. Emphasis in original.

168. Ibid., 7.

169. Ogden Beeman & Associates, Inc., *Wainwright and Barrow Beach Nourishment Project and Plan Review: Final Report* (Portland, OR, September 1992), 1.

170. Brown, *Barrow and Wainright Beach Nourishment Program*, 7–8.

171. BTS/LCMF, Ltd., *Barrow and Wainwright Beach Nourishment Program: Budget and Cash Flow Update* (Barrow and Anchorage, November 1995), 7, Exhibit A.

172. Brown, *Barrow and Wainright Beach Nourishment Program*, 11.

173. Ibid.

174. BTS/LCMF, Ltd., *Mitigation Alternatives for Coastal Erosion at Wainwright and Barrow, Alaska*, 18.

175. Ibid., 25, 29–31. The 800,000 cubic yards of material allowed for a 42% loss during operations.

176. Steve Chronic, *1999 Barrow Beach Nourishment Post-Season Report* (Anchorage and Barrow, LCMF, Ltd., 2000), 3, adds that "[t]he unit price [80.00 per cubic yard] when applied to the number of feet of beach nourishment amount to $2,848 per foot of protection in 1999 dollars, discounting capital dollars, lay-up, maintenance, repairs, and modification that were done from 1996 through 1998."

177. BTS/LCMF, Ltd., *Mitigation Alternatives for Coastal Erosion at Wainwright and Barrow, Alaska*, II. The ambiguity in the Executive Summary is not resolved in the main text and tables so far as we can tell.

178. The lowest and highest estimates were verbal reports from different officials at CIPM who were informed about the program.

179. North Slope Borough Assembly, *Minutes, Regular Meeting and Public Hearing* (10 October 2000), 18.

180. Army Corps of Engineers, *Barrow, AK, Section 905(b) (WRDA 86) Analysis, Shoreline Erosion, Flooding, and Navigation Channel* (Anchorage: U.S. Army Engineer District, Alaska, 24 May 2001), 6, 5. The Corps believed that coarser material will stay on the beach longer.

181. Ogden Beeman & Associates, Inc., *Wainwright and Barrow Beach Nourishment Project and Plan Review*, 3.

182. Chronic, *1999 Barrow Beach Nourishment Post-Season Report*, 1.

183. Science Advisory Committee, North Slope Borough, *Review of the Barrow Beach Nourishment Project* (Fairbanks: University of Alaska Fairbanks, August 1994), 5. Among other things, the report (8) recommended, "The project should attempt to dredge material the same size or coarser than the present beach material," while noting that "[t]he fill texture will probably be finer than the normal beach."

184. George N. Amaogak, Sr., Letter to Colonel Steven T. Perrenot, Corps of Engineers, U.S. Army Engineer District, Anchorage, AK (Barrow: North Slope Borough Mayor's Office, April 24, 2001). On initiation by the borough and for an overview and updates on the project, go to www.poa.usace.army.mil, click on Civil Works and Planning, and select Barrow Coastal Storm Damage Reduction. Most of the public documents on the project cited below can be accessed there.

185. Army Corps of Engineers, *Barrow, AK: Section 905(b) (WRDA 86) Analysis, Shoreline Erosion, Flooding, and Navigation Channel*, 10. This analysis (2) reviewed Corps studies of beach erosion in Barrow in 1969, 1991, and 1999, but concluded that solutions were not economically justified under applicable legislation.

186. See the U.S. Army Corps of Engineers, *Project Management Plan: Coastal Storm Damage Reduction, Barrow Alaska* (2002), on the alternatives (4), on the schedule (35–37), on cost of the study (34), on the Project Delivery Team (8), and on technical requirements (10–11).

187. The alternatives evaluated are summarized in U.S. Army Corps of Engineers, *Feasibility Scoping Meeting/Progress Review Documentation: Barrow, Alaska, Coastal Storm and Flood Damage Reduction Feasibility Study* (Revised 8 September 2005), 55–57. See 50, 56, and 57 for the lines on Non-Structural Alternatives. What the Corps calls "non-structural alternatives" are approximately equivalent to what BTS/LCMF called "non-active" or "passive" alternatives.

188. Ibid., 41.

189. Ibid., 57. In the original "Completeness" is boldfaced and "complete solution" is underscored.

190. The most useful document is the Corps's summary of the meeting made from the borough's videotape, titled *Barrow Coastal Storm Damage Reduction Feasibility Study: Barrow Public Meeting—August 23, 2006.* The summary is not verbatim, but it is close when compared with an audio recording made by a member of the audience.

191. Ibid., 2. Although Brooks referred to our project's measurements of erosion in this connection, his interpretation differs from ours. Among other things, aggregation by time and place obscured the effects of the great storm and the 1968 ban on further mining of material from the beach.

192. Army Corps of Engineers, *Coastal Storm Damage Reduction Feasibility Study: Barrow Public Meeting—August 23, 2006*, 8.

193. Ibid., 9.

194. Ibid.; Army Corps of Engineers, *Feasibility Scoping Meeting/Progress Review Documentation: Barrow, Alaska, Coastal Storm and Flood Damage Reduction Feasibility Study*, 7.

195. Ibid., 3 and slide 10 of the Corps's presentation.

196. Ibid., 11. When told that the locations for the breaks were suggested by whaling captains who had met with the Corps, the audience pointed out that "the current locations are based on where the whaling captains live. This makes it easy for them to access the beach."

197. Ibid., 10.

198. Ibid., 1. These are round numbers slightly higher than those shown in slide 7 of the presentation,

199. See ibid., 3–6 for this example. The replies by the Corps's hydraulic engineer can be found on pp. 5 and 6, respectively. After the second reply, an audio recording of the meeting includes a skeptical and sarcastic question from an audience member: "Do they have *ivus* in New Hampshire?" At another point in the meeting, audience members pointed out that snow would accumulate around the dike, perhaps causing drainage, thermocarsting, and permafrost melting problems. Brooks replied,

"Thank you. That's why we come to the community and have public meetings so you can point out factors that we have not thought of. We don't live in the same area or climate as you do. This is the type of information we were hoping we would get by coming and talking with you" (11).

200. Ibid., 16. The note appended to the meeting summary elaborates: "Wave action directed against the Concertainer was able to wash the interior material out, causing partial to complete failure of the Concertainer seawall."

201. Phone conversation with Project Director Andrea Elconion, on 18 January 2008, supplemented by another phone conversation with Project Formulator Forest Brooks on 22 January 2008. The plan for the Independent Technical Review noted that "the purpose is to provide a technical review of all elements of the feasibility study, including models used, and to insure planning, analysis, and design conform to applicable USACE standards, policy, and guidance." The models used in the Barrow study did not have to be certified because they were in common use. Peer review external to the Corps was not required because "the scope and technical complexity of the feasibility study and report/EIS are not expected to be novel, controversial, or precedent setting." Eugene M. Ban, *Memorandum for Commander, Alaska Engineer District, Attn: CEPOA-PM-C*, Subject: Review Plan approval for the Storm Damage Reduction Measures, Barrow, Alaska, Study (26 July 2007).

202. National technical requirements may be understood as the institutionalization of cumulative lessons learned from mistakes in previous studies. To improve the adaptation of plans to local needs and opportunities, it would be constructive to field test selective exemptions from these requirements under specified circumstances.

203. Operations extend from the bluffs northeast to the culvert on the road to NARL; beyond that, UIC takes responsibility for its own property.

204. Anon., "Emergency Action Taken," *The Open Lead* (November 1986), 7.

205. Interview in Barrow with Donovan Price, 22 August 2001. Everyone with a VHF receiver in a 30–50+- mile radius can now get weather bulletins at the same time as NSB officials.

206. Anne Jensen, personal communication, e-mail, 10 September 2003.

207. Elizabeth N. Cassano, Amanda H. Lynch, and M. R. Koslow, "Classification of Synoptic Patterns in the Western Arctic Associated with High Wind Events at Barrow, AK," *Climate Res.* 30 (2006), 83–97.

208. Amanda H. Lynch, Leanne R. Lestak, Petteri Uotila, Elizabeth N. Cassano, and Lian Xie, "A Factorial Analysis of Storm Surge Flooding in Barrow, Alaska," *Mon. Wea. Rev.* 136 (2008), 898–912.

209. Rob Elkins, personal communication, 4 February 2008.

210. Interview with David Anderson at the NWS station in Barrow, 7 February 2008. Anderson is working with a colleague, Aimee Fish, to develop near-shore forecasts (up to 2 miles out) to compensate for the limitations of standard marine forecasts in this area. He hopes to involve Iñupiat elders in this initiative.

211. Emergency Response Institute International, Inc., *Comprehensive Emergency Management Plan (CEMP) for North Slope Borough, Alaska* (Olympia, WA: June 2000).

212. That account was later published in slightly revised form as Ronald D. Brunner, Amanda H. Lynch, Jon C. Pardikes, Elizabeth N. Cassano, Leanne R. Lestak, and Jason M. Vogel, "An Arctic Disaster and Its Policy Implications," *Arctic* 57 (December 2004), 336–346.

213. Joe Stankiewicz, *North Slope Borough Local All Hazard Mitigation Plan* (Barrow: North Slope Risk Management Division, 2005), 67. The following quotation is from p. 4.

214. Interview with Rob Elkins in Barrow, 4 February 2008.

215. Quoted in Arlene Glenn, *Can the Past Be Saved? Symposium on the Erosion of Archaeological Sites on the North Slope*, Barrow, AK, 10–11 September, 1991, 5.

216. Ibid., 76, where the other two recommendations also can be found.

217. Arlene Glenn, *Elder's Interviews on Beach Nourishment and Flood History*, translated by Jana Harcharek (Barrow: Iñupiat Heritage and Cultural Center, 19 July 2000), 2.

218. See endnote 53 for documentation on this early relocation.

219. Mentioned by a member of the audience in Army Corps of Engineers, *Barrow Coastal Storm Damage Reduction Feasibility Study: Barrow Public Meeting—August 23, 2006*, 13. "In 1970 we tried all this right here. All this area *(Barrow beachfront identified on the map and the area of Browerville southeast of the road with the AC Commercial and the Eskimo gas station)* is restricted under BIA. We tried moving from one side of lagoon, but couldn't. There were 22 residences that would not relocate because of restricted lots. Not one individual accepted. *(The Borough tried to relocate individuals in 1970. There were 22 individuals along the bluff that would not move because of the restricted lot status.)*"

220. Army Corps of Engineers, *Barrow, AK: Section 905(b) (WRDA 86) Analysis, Shoreline Erosion, Flooding, and Navigation Channel*, 2. The Corps had concluded, "Federal participation in structural measures was not justified" in Barrow under Section 103 of the 1962 River and Harbor Act.

221. BTS/LCMF, Ltd., *Mitigation Alternatives for Coastal Erosion at Wainwright and Barrow, Alaska*, III. Emphasis added.

222. Tekmarine, Inc., *Bluff and Shoreline Protection Study for Barrow, Alaska* (Pasadena, CA: 1987), Appendix D, 8.

223. BTS/LCMF, Ltd., *Mitigation Alternatives for Coastal Erosion at Wainwright and Barrow, Alaska*, III, 50.

224. North Slope Borough Assembly, *Minutes, Special Meeting and Workshop* (16 May 1989), 12. Dave Fauske of Administration and Finance made similar points in the same meeting and in North Slope Borough Planning Commission, *Minutes, Planning Commission Special Meeting* (18 April 1989), 2.

225. Stankiewicz, *North Slope Borough Local All Hazard Mitigation Plan*, 69.

226. North Slope Borough Assembly, *Minutes, Continuation of Joint Assembly and Planning Commission Special Meeting and Public Hearing* (17 June 1992), 84.

227. Science Advisory Committee, North Slope Borough, *Review of the Barrow Beach Nourishment Project*, 15, 4.

228. Brown, *Barrow and Wainright Beach Nourishment Program*, 3.

229. Ibid.: "To date, the Planning/Zoning component . . . has not been implemented and additional development in threatened coastal areas continues."

230. See Fig. 3.8.

231. Stankiewicz, *North Slope Borough Local All Hazard Mitigation Plan*, 69.

232. North Slope Borough Assembly, *Minutes, Regular Meeting and Public Hearing* (1 September 1998), 8–9. The second part of Resolution No. 41-98 would have resolved "To vest the Land Management Administrator with the responsibility, authority, and means" to carry out the first part.

233. Stankiewicz, *North Slope Borough Local All Hazard Mitigation Plan*, 70.

234. Army Corps of Engineers, *Barrow, AK: Section 905(b) (WRDA 86) Analysis, Shoreline Erosion, Flooding, and Navigation Channel*, 2.

235. Ibid., 4.

236. Information on the performance of the Concertainer seawalls came from an interview with the project director for the NSB, Curt Thomas, in Barrow on 6 February 2008. For information on the utilidor cross-section seawall and the Concertainer seawalls, see LCMF, *Construction Report: Coastal Erosion Pilot Project, Barrow Alaska, Wainwright, Alaska*, prepared for the North Slope Capital Improvements Projects Management Department (Anchorage: LCMF, 13 October 2004). CIPM is now part of Public Works.

237. Interview with Fred Brower, the NSB's disaster coordinator, in Barrow on 5 February 2008.

238. Tim Russell, personal communication, 23 March 2004.

239. Curt Thomas interview, 6 February 2008. The site selected is Site D. An early planning document noted that a Site F was "located at approximately the 8-foot contour [above mean sea level] and is approximately 500 feet from the ocean. Its flood potential is high. . . . This site has moderate to high risk for damage from pack ice." S. Hattenburg, *Barrow Wastewater Treatment Plant PAR: Analysis of Site F* (LCMF, 4 December 2000), 3.

240. Tekmarine, Inc., *Bluff and Shoreline Protection Study for Barrow, Alaska*, ii. Similarly, the report of the Scientific Advisory Committee on the beach nourishment program urged monitoring of the movements of material on the beach. Because relevant information was lacking, "the estimates of beach replenishment rates and timing are very uncertain and underline the need for an effective post-nourishment monitoring program." Scientific Advisory Committee (1994), 7.

241. *Local All Hazard Mitigation Plan*, 70: "The North Slope Borough should continue the new beach armament efforts initiated during the summer of 2004." Then it

expresses some skepticism (emphasized in the original): "Although it is improbable that such efforts [armarment] will completely halt continued erosion, the technology in use may be valuable in protecting critical areas. . . ."

242. The interview was conducted by Michele Norris and Melissa Block for *All Things Considered* and broadcast on National Public Radio 11 September 2007. The transcript, available at www.npr.org, is titled "Natural Resources Key to Alaska Town's Future." In the transcript, Richard Glenn is identified as a geologist, cocaptain of a whaling crew in Barrow, president of the Barrow Arctic Science Consortium, as well as an employee [actually vice president] of ASRC, and a key figure in bringing locally produced natural gas to Barrow. Glenn emphasized their dependence on natural resources: "And we have no significant tourism, no agriculture, no commercial fishing, no other local industry, and so our future is tied with resource development."

243. Quoted in Jad Mouawad, "Tension at the Edge of Alaska," *New York Times* (4 December 2007), C1. Another coalition of plaintiffs that includes Point Hope and the Iñupiat Community of the Arctic Slope filed suit to block oil exploration in the Chukchi Sea, alleging that "[t]he current environmental assessment . . . fails to adequately analyze the impact of the lease sale in the context of a warming climate." Felicity Barringer, "Suit Seeks to Block Oil Search Off Alaska," *New York Times* (1 February 2008), A13.

244. Edward S. Itta, "Arctic Sun Rises on Old Issues, New Hopes," *Arctic Sounder* (31 January 2008), 4.

245. Matters of concern to Alaska Natives, including those mentioned here, can be inferred from States News Service, "Senator Stevens Addresses Alaska Federation of Natives 2007 Convention" (27 October 2007), the text of the Sen. Stevens's speech as prepared for delivery in Fairbanks.

246. Networking as we have defined it was perhaps the main strategy for self-empowerment of the Iñupiat on the North Slope, as we reconstructed the story in an earlier section of this chapter. This experience also is available to inform constitutive decisions elsewhere.

247. Tekmarine, Inc., *Bluff and Shoreline Protection Study for Barrow, Alaska.*

248. Interview with Fred Brower, 5 February 2008. On the Kaktovik disaster, see Mary Pemberton, "Broken Pipes and a Flooding Problem in Kaktovik," *Associated Press State & Local Wire* (13 January 2005).

249. General Accounting Office, *Alaska Native Villages: Most Are Affected by Flooding and Erosion, but Few Qualify for Federal Assistance*, GAO-04-142 (December 2003), 32. The other quotations and information in this paragraph can be found on pp. 32–34. Shortly after this report GAO was renamed the Government Accountability Office.

250. General Accounting Office, *Alaska Native Villages*, 34.

251. K. Johnson, S. Solomon, D. Berry, and P. Graham, "Erosion Progression and Adaptation Strategy in a Northern Coastal Community, *Permafrost*, 489–494.

252. Ibid., 490–491.

253. Ibid., 491.

254. S. Solomon, *Tuktoyaktuk Erosion Risk Assessment 2001* (Dartmouth, NS: Geological Survey of Canada, 2001), 8. The failure mechanisms include overtopping; undermining at the ends and the toe; sapping of riprap when water penetrates holes in the geotextile base and erodes the sand and pebble beach below; subsidence from thawing of ice beneath shore protection measures; and possibly ice that plucks shoreline materials and floats them out to deeper water.

255. Jackie Jacobson quoted in Anon., "Arctic Village Losing Battle with the Sea," *The Record (Kitchener-Waterloo, Ontario)* (2 July 2004), D14. In October 1994, UMA Engineering Ltd. issued a final report commissioned by the government of the Northwest Territories. It concluded, "The least cost option to cope with shoreline erosion on the west shore of the peninsula is to gradually relocate all the buildings, the cemetery and historic objects off the peninsula and allow natural erosion to claim the land." It also recognized that "[i]t is possible that the community and particularly the residents on the peninsula will object" and noted that "a public information and consultation process will be beneficial." UMA Engineering Ltd., Tuktoyaktuk Shoreline Protection Study, Phases II and III, Interim Report No. 2, Final Draft (Edmonton, Alberta: October 1994), 3.

256. The quotations are from the General Accounting Office, *Alaska Native Villages: Most Are Affected by Flooding and Erosion, but Few Qualify for Federal Assistance*, GAO-04-142 (December 2003), 3–4.

257. The quotation is from the Senate report for the Energy and Water Development Appropriations Act, 2006, P.L. 109–103, 41.

258. Beth Bragg, "Inaction Erodes Stevens' Patience," *Anchorage Daily News* (12 October 2007), B1.

259. Luci Eningowuk, prepared testimony, 3.

260. Ibid., 5.

261. Bragg, "Inaction Erodes Stevens' Patience."

262. Luci Eningowuk, prepared testimony, 2, 3.

263. Owen K. Mason, "Living with the Coast of Alaska Revisited: The Good, the Bad, and the Ugly," in Orson P. Smith, Ed., *Coastal Erosion Responses for Alaska* (Fairbanks: Alaska Sea Grant College Program, 2006), 3–17, at 11. These are the proceedings of a workshop 4 January 2006, in Anchorage.

264. Owen K. Mason, *Geological and Anthropological Considerations in Relocating Shishmaref, Alaska*, Report of Investigations 96-7 (Fairbanks: State of Alaska, Department of Natural Resources, Division of Geological & Geophysical Surveys, 1996), 6.

Chapter 4

1. James R. Schlesinger, "Systems Analysis and the Political Process," *J. Law Economics* 11 (October 1968), 281–298. Schlesinger later served in the cabinet of President Ford as Secretary of Defense and in the cabinet of President Carter as Secretary of Energy.

2. In pragmatism, the major American contribution to world philosophy, "inquiry" connotes knowing-in-action. See Abraham Kaplan, *The Conduct of Inquiry: Methodology for Behavioral Science* (San Francisco, CA: Chandler Publishing, 1964), 43, as quoted before Chapter 1 of this book. Consider, e.g., Donald A. Schön, "Knowing-in-Action: the New Scholarship Requires a New Epistemology," *Change* 27 (November/December 1995), 27–34, at 31: "In the domain of practice, we see what John Dewey called inquiry: thought intertwined with action—reflection in and on action—which proceeds from doubt to the resolution of doubt, to the generation of new doubt." The policy sciences are part of the pragmatic tradition. In this chapter and elsewhere we have used conceptual and theoretical tools from the policy sciences as heuristics for contextual and problem-oriented inquiry. The major works include Lasswell and Kaplan, *Power and Society*; Harold D. Lasswell, *A Pre-View of Policy Sciences* (New York, NY: Elsevier, 1971), which includes a brief introduction to the framework in Ch. 2; and the most comprehensive statement in two volumes, Harold D. Lasswell and Myres S. McDougal, *Jurisprudence for a Free Society: Studies in Law, Science, and Policy* (New Haven, CT, and Dordrecht, Netherlands: New Haven Press and Martinus Nijhoff Publishers, 1992).

3. From the Executive Summary of IPCC, Working Group III, chaired by Frederick M. Bernthal, *Climate Change: The IPCC Response Strategies* (Washington, D.C.: Island Press, 1991), xxv.

4. National Assessment Synthesis Team, *Climate Change Impacts on the United States: The Potential Consequences of Climate Variability and Change* (Washington, D.C.: U.S. Global Change Research Program, November 2000), 9, 120.

5. IPCC, *Summary for Policymakers*, in *Climate Change 2007: Impacts, Adaptation and Vulnerability, Contribution of Working Group II to the Fourth Assessment Report of the Intergovernmental Panel on Climate Change*, M. L. Parry, O. F. Canziani, J. P. Palutikof, P. J. van der Linder and C. E. Hanson, Eds., (Cambridge, U.K.: Cambridge University Press, 2007), 12..

6. Compare Tim Sherratt, "Human Elements," in Tim Sherratt, Tom Griffiths, and Libby Robin, Eds., *A Change in the Weather: Climate and Culture in Australia* (Canberra: National Museum of Australia Press, 2005), 7: "Like the suffering of an elderly person in Chicago, or an immigrant mother in North Melbourne, the lived experience of drought is not contained within the event itself."

7. More of the policy sciences framework is visible in Boxes 4.1, 4.2, and 4.3, where it is used to summarize normative considerations. The framework also can be used to bring empirical and practical policy considerations into the focus of attention.

8. Compare Stephen Jay Gould, "Evolution and the Triumph of Homology, or Why History Matters," *American Scientist* 74 (January–February 1986), 60–69, at 61. "History is the domain of narrative—unique, unrepeatable, unobservable, large-scale, singular events." We would add small-scale and subjective events to the list.

9. Projections for the IPCC Fourth Assessment Report suggest the observed trend of increasing cyclones will continue, with the largest trends in the Bering Sea region: Cyclones originating in the Aleutians are projected to follow a more northerly storm track into the west coast of Alaska with increasing frequency. W. L. Chapman and J. E. Walsh, "Simulations of Arctic Temperature and Pressure by Global Coupled Models," *J. Climate* 20 (2007), 609–632.

10. As concluded in, e.g., Roger Pielke, Jr., and Chris Landsea, "Normalized Hurricane Damages in the United States: 1925–95," *Wea. Forecasting* 13 (1998), 621–631.

11. Thus, the researchers cited below may reject other conclusions we have drawn, and we do not accept all of what they wrote. Nearly all scientific and scholarly conclusions have been challenged by someone sometime. Nevertheless, the convergence of independent sources carries some evidentiary weight in favor of our proposal for more intensive research.

12. Steve Rayner and Elizabeth L. Malone, "Ten Suggestions for Policymakers," in Steve Rayner and Elizabeth L. Malone, Eds., *Human Choice & Climate Change, Volume 4: What Have We Learned?* (Columbus, OH: Battelle Press, 1998), 134.

13. W. Neil Adger, "Social Vulnerability to Climate Change and Extremes in Coastal Vietnam," *World Development* 27 (1999), 249–269, at 266.

14. Sarah Burch and John Robinson, "A Framework for Explaining the Links between Capacity and Action in Response to Global Climate Change," *Climate Policy* 7 (2007), 306–316.

15. When politics are factored into adaptive management, the technical connotations of "management" become less descriptive of reality than "governance." All quotations in this paragraph are from C. S. Holling, "What Barriers? What Bridges," in Lance H. Gunderson, C. S. Holling, and Stephen S. Light, Eds., *Barriers and Bridges to the Renewal of Ecosystems and Institutions* (New York, NY: Columbia University Press, 1995), 3–34, at 13. For similar views on adaptive management, see Donald Ludwig, Ray Hilborn, and Carl Walters, "Uncertainty, Resource Exploitation, and Conservation: Lessons from History," *Science* 260 (2 April 1993), 17, 36.

16. Some advocates of adaptive management reject trial-and-error learning in favor of the assessment of competing hypotheses based on the experimental ideal. See, e.g., Lance Gunderson and Stephen S. Light, "Adaptive Management and Adaptive Governance in the Everglades Ecosystem," *Policy Sciences* 39 (2006), 323–334, at 326.

17. Paul C. Stern, "A Second Environmental Science: Human-Environment Interactions," *Science* 260 (25 June 1993), 1897–1899, at 1897.

18. Ibid.

19. Ibid., 1898. At the macro level Stern recognized the importance of interactions among the multiple forces driving environmentally destructive behavior—forces such as population growth, economic development, technological change, changes in human values, and institutions. "What has become clear is that the driving forces interact—that each is meaningful only in relation to the impacts of others . . ." (1897).

20. Nearly a decade and a half later, other environmental researchers reiterated a similar demand "to promote efforts that create a science of coupled systems with both biophysical and socioeconomic elements fully endogenized." Matthew J. Kotchen and Oran R. Young, "Meeting the Challenge of the Anthropocene: Towards a Science of Coupled Human-Biophysical Systems," *Global Environ. Change* 17 (2007), 149–151, at 151.

21. Proponents include Jared Diamond in "Epilogue: The Future of Human Society as a Science," in his *Guns, Germs, and Steel: The Fate of Human Societies* (New York, NY: W. W. Norton, 2005), 403–425. His views on the historical sciences, especially pp. 420–426, are quite similar to Gould's as described below. The critics of reductionism include Robert Frodeman in "Geological Reasoning: Geology as an Interpretive and Historical Science," *GSA Bulletin* 107 (August 1995), 960–968, who distinguishes epistemologies and illustrates the epistemology appropriate for geology as an interpretive and historical science.

22. Stephen Jay Gould, *Wonderful Life: The Burgess Shale and the Nature of History* (New York, NY: W. W. Norton, 1989), 277.

23. Ibid., 278.

24. Ibid.

25. Ibid., 283.

26. Ibid., 289, 290.

27. Edward N. Lorenz, "Deterministic Nonperiodic Flow," *J. Atmos. Sci.* 20 (March 1963), 130–141. For a review of the social science equivalents, see Malcolm Gladwell, *The Tipping Point: How Little Things Can Make a Big Difference* (New York, NY: Little, Brown, 2002).

28. Philip W. Anderson, "More Is Different," *Science* 177 (4 August 1972), 393–396, at 393 for this and the following quotes.

29. John Horgan in "From Complexity to Perplexity," *Scientific American* (June 1995), 104–109, at 108.

30. Ibid., 109.

31. John H. Holland, "Complex Adaptive Systems," *Daedalus* 121 (Winter 1992), 17–30, at 20. Compare Herbert A. Simon, *The Science of the Artificial*, 3rd ed. (Cambridge, MA: MIT Press, 1996), 48: "As in any dynamic system that has propensities for following diverging paths from almost identical starting points, equilibrium theories of an economy can tell us little about either its present or future state." For more on the limitations of the equilibrium assumptions in economics, see Paul

Ormerod, *Why Most Things Fail: Evolution, Extinction and Economics* (New York, NY: Pantheon Books, 2005).

32. Holland, "Complex Adaptive Systems," 28.

33. Because of the magnitude of such obstacles, constructing an agent-based model cannot be considered analogous to constructing a global climate model or defended as such. Nevertheless, at least one climate researcher is optimistic about agent-based models: "The human, non-human, and inanimate agents in a complex adaptive system can be represented as nodes in a network, and the totality of the system's dynamics represented by the interactions across the links of this network." See John Finnegan, "Earth System Science in the Early Anthropocene," *Global Change NewsLetter No. 55* (October 2003), 8–11, at 10.

34. See Horgan, "From Complexity to Perplexity," for more on critics as well as proponents of complexity as a newly emergent discipline. In Matthew Berman, C. Nicolson, G. Kofinas, J. Tetlichi, and S. Martin, "Adaptation and Sustainability in a Small Arctic Community: Results of an Agent-Based Simulation Model," *Arctic* 57 (2004), 401–414, at 401, the authors note, "The economic and demographic outcomes suggest implications for less quantifiable social and cultural changes. The model can serve as a discussion tool for a fuller exploration of community sustainability and adaptation issues."

35. John H. Holland, quoted in Horgan, "From Complexity to Perplexity," 105.

36. Robert W. Kates and Thomas J. Wilbanks, "Making the Global Local: Responding to Climate Change Concerns from the Ground Up," *Environment* 45 (April 2003), 12–23, at 14.

37. Milton Friedman, "The Methodology of Positive Economics," in *Essays in Positive Economics* (Chicago, IL: University of Chicago Press, 1953), Part III, 16–23.

38. Ibid., 4.

39. Ibid., 25.

40. Friedman, "The Methodology of Positive Economics," 42–43.

41. The classic appraisal is William Ascher, *Forecasting: An Appraisal for Policy-Makers and Planners* (Baltimore, MD: Johns Hopkins University Press, 1978). Upon being asked, Professor Ascher informed us that it would be impractical to update this appraisal based on post-hoc accuracy because the relevant forecasts are now qualified by contingencies not often realized. The emissions scenarios used by the IPCC are contingencies in this sense. See also William Ascher, "The Forecasting Potential of Complex Models, *Policy Sciences* 13 (1981), 247–267; William Ascher, Political Forecasting: The Missing Link, *J. Forecasting* 1 (1982), 227–239; and Philip E. Tetlock, *Expert Political Judgment: How Good Is It? How Can We Know It?* (Princeton, NJ: Princeton University Press, 2005).

42. Alice M. Rivlin, "A Public Policy Paradox," *J. Public Policy Manage.* 4 (1984), 17–22. Compare Ascher, "The Forecasting Potential of Complex Models," 247: "[W]hile the accuracy of complex models in forecasting trends in such fields as economics and energy is, and will remain, undistinguished, complex models' special

virtues of preserving counter-intuitive results and representing subsystem inter-dependence could be used to better advantage than current practice permits."

43. Brian Greene, "The Universe on a String," *New York Times* (20 October 2006), A23.

44. Henri Pirenne, "What Are Historians Trying to Do?" in S. A. Rice, Ed., *Methods in Social Science*, 435–45, at 443 (Chicago, IL: University of Chicago Press, 1931). See also Carl J. Becker, "Everyman His Own Historian," *Amer. Historical Rev.* XXXVII (January 1932), 221–236.

45. Herbert A. Simon, *Models of Man: Social and Rational* (New York, NY: John Wiley & Sons, 1957), 198. Emphasis in original. The equivalent in economics is formulated in Ormerod, *Why Most Things Fail*, e.g., on 35, 56, 117. The equivalent in the policy sciences is the maximization postulate. "Since the actors are living forms, they participate selectively in what they do. We describe this selective characteristic by the *maximization postulate*, which holds that living forms are predisposed to complete acts in ways that are perceived to leave the actor better off than if he had completed them differently. The postulate draws attention to the actor's own perception of alternative act completions open to him in a given situation." Lasswell, *A Pre-View of Policy Sciences*, 16. Emphasis in original. Compare Simon, *Models of Man*, 199, on the first consequence of the principle of bounded rationality: "[T]he intended rationality of an actor requires him to construct a simplified model of the real situation in order to deal with it. He behaves rationally with respect to this model, and such behavior is not even approximately optimal with respect to the real world."

46. A chunk is the smallest meaningful unit of information; these limits are surmounted by rechunking information into larger chunks. Thus, an unfamiliar long-distance telephone number in the United States consists of 10 chunks. But the number of chunks drops to six if the three-digit area code and the three-digit local exchange are recognized as one chunk each. This and other basic parameters are reviewed and qualified in Herbert A. Simon, *The Sciences of the Artificial*, 3rd ed. (Cambridge, MA: MIT Press, 1996), Ch. 3, "The Psychology of Thinking." The seminal article is George A. Miller, "The Magical Number Seven, Plus or Minus Two," *Psych. Rev.* 63 (1956), 81–97.

47. Herbert A. Simon, "Human Nature in Politics: The Dialogue of Psychology with Political Science," *Amer. Pol. Sci. Rev.* 79 (1985), 293–304, at 302.

48. Ibid., 302.

49. Ibid., 301. Similarly, Lasswell and Kaplan, *Power and Society*, 70, spelled out one methodological implication of the maximization postulate: "Whenever a person chooses a course of action, we are to look for the specific values and expectations that make it appear to the actor most economical to him"—where "economical" refers to the expected ratio of gains to losses.

50. Simon, "Human Nature in Politics," 301. On the limited mileage from general laws in the hard sciences, see C. Jakob, G. Tselioudis, and T. Hume, "The Radiative, Cloud and Thermodynamic Properties of the Major Tropical Western Pacific Cloud

Regimes," *J. Climate*, 18 (2005), 1203–1215. On the call for more empiricism in economics, see Wassily Leontief, "Letter on Academic Economics," *Science* (9 July 1982), 104–105; and more recently Barbara R. Bermann, "Needed: A New Empiricism," *Economists' Voice* (March 2007), 1–4, at 3: "A knowledge of business behavior based on observation rather than conjecture could revolutionize the formation of public policy and the managing of the business cycle." This article is accessible through www.bepress.com/ev.

51. Harold D. Lasswell, Daniel Lerner, and Ithiel deSola Poole, "Political Symbols," in *The Comparative Study of Symbols: An Introduction* (Stanford, CA: Stanford University Press, 1952), 11. Emphasis in original.

52. Stephen Jay Gould, *The Hedgehog, the Fox, and the Magister's Pox: Mending the Gap between Science and the Humanities* (New York, NY: Harmony Books, 2003), 256.

53. Donald T. Campbell, "'Degrees of Freedom' and the Case Study," *Comparative Political Studies* 8 (July 1975), 178–193, at 182.

54. For examples, see Ronald D. Brunner, "A Paradigm for Practice," *Policy Sciences* 39 (2006), 135–167, at 137–143.

55. Preface to J. T. Houghton et al., Eds., *Climate Change 2001: The Scientific Basis, Contribution of Working Group I to the Third Assessment Report of the Intergovernmental Panel on Climate Change* (Cambridge, U.K.: Cambridge University Press, 2001), at ix.

56. Quoted in Andrew Revkin, "Skeptics on Human Climate Impact Seize on Cold Spell," *New York Times* (2 March 2008), 14.

57. Ibid.

58. Susanne C. Moser and Lisa Dilling, "Making Climate Hot: Communicating the Urgency and Challenge of Global Climate Change," *Environment* (December 2004), 32–46, at 38.

59. Quoted in Lasswell and Kaplan, *Power and Society*, 106n.

60. Walter Lippmann, *Public Opinion* (New York, NY: Free Press, 1965), 28, first published in 1922.

61. For the quote, components, and cross-cutting processes, see, respectively, 1, 2, and 3 in the Executive Summary to Bo Lim and Erika Spanger-Siegfried (Eds.) and Ian Burton, Elizabeth Malone and Saleemul Huq, *Adaptation Policy Frameworks for Climate Change: Developing Strategies, Policies and Measures* (Cambridge, U.K.: Cambridge University Press, 2005).

62. Lasswell, *A Pre-View of Policy Sciences*, 85–86.

63. If so, then Hans-Martin Füssel, "Vulnerability: a Generally Applicable Conceptual Framework for Climate Change Research," *Global Environ. Change* 17 (2007), 155–167, is more constructive as a guide to translation across frames of reference than as an attempt to establish a consistent terminology. A decade ago Rayner and Malone, "Ten Suggestions," 134, observed, "No standard framework exists for identifying different sources of vulnerability, but clearly they are many

and complex." A standard framework is not likely but neither is it necessary from our standpoint.

64. Lee J. Cronbach, "Beyond the Two Disciplines of Scientific Psychology," *Amer. Psych.* (February 1975), 124–125. Emphasis in original. This was a Distinguished Scientific Contribution Award address to the American Psychological Association.

65. On analogues, see William B. Meyer, Karl W. Butzer, Thomas E. Downing, B. L. Turner II, George W. Wenzel, and James L. Westcoat, "Reasoning by Analogy," in Steve Rayner and Elizabeth L. Malone, Eds., *Human Choice & Climate Change, Volume 3: Tools for Policy Analysis* (Columbus, OH: Battelle Press, 1998), 218–289.

66. Of course, predictions of climate change and its impacts are contingent—e.g., on socio-economic scenarios—and therefore not falsifiable unless or until the contingencies are realized.

67. Kaplan, *The Conduct of Inquiry,* 43. Pragmatism is one of the intellectual origins of the policy sciences.

68. Jim Collins, quoted in Michael Shellenberger and Ted Nordhaus, *The Death of Environmentalism: Global Warming Politics in a Post-Environmental World* (2004), 33, accessed at http://www.thebreakthrough.org/images/Death_of_Environmentalism.pdf.

69. Gould, *Wonderful Life,* 282.

70. Thomas S. Kuhn, *The Structure of Scientific Revolutions,* 2nd ed., Enlarged (Chicago, IL: University of Chicago Press, 1970), 188–189, for this and the previous quotes.

71. Ronald D. Brunner, "Myths, Scientific and Political," 5–12, at 6, in *The Objectivity Crisis: Rethinking the Role of Science in Society,* Chairman's Report to the Committee on Science, Space, and Technology, House of Representatives, 103rd Congress 1st Session, Serial D (Washington, D.C.: U.S. Government Printing Office, June 1993).

72. Harold D. Lasswell, *Democratic Character* (Glencoe, IL: Free Press, 1951), 524. Emphasis in original.

73. Lasswell and Kaplan, *Power and Society,* xxii–xxiii.

74. Lasswell, *A Pre-View of Policy Sciences,* 39. Emphasis in the original.

75. In general, compare George E. Brown, Jr., "The Objectivity Crisis," *Amer. J. Physics* 60 (September 1992), 779–781, at 789: "It seems to me that science—as articulated by scientists and policy-makers alike—has promised much more than it can deliver. . . . At the same time, our faith in science, and the knowledge and technology that it creates, may be an explicit roadblock to social action."

76. Lasswell, *A Pre-View of Policy Sciences,* 40.

77. Ibid, 28–29.

78. Compare Simon 1996, 124–125: "There are two ways in which design processes are concerned with the allocation of resources. First, conservation of scarce resources may be one of the criteria for a satisfactory design. Second, the design process itself involves management of the resources of the designer, so that his efforts will not be dissipated unnecessarily in following lines of inquiry that prove fruitless."

79. Gould, *Wonderful Life*, 290.

80. Amanda H. Lynch, principal investigator, and coprincipal investigators Ronald D. Brunner, Judith A. Curry, James A. Maslanik, Linda Mearns, Anne Jensen, and Glenn Sheehan, *SGER: To Explore the Feasibility of Collaborative Resident-Scientist Climate Policy Research on the Alaskan North Slope Coastal Region* (proposal to the National Science Foundation, 1999). The proposal also mentions that Jim Maslanik helped prepare the way: "[T]he objectives in this proposal were discussed with Barrow researchers and townspeople during a community meeting in Barrow [set up] by co-PI J. Maslanik on April 5, 1999, and during subsequent individual meetings. The response to the proposed idea was uniformly positive, and was reflected in an extended discussion period. . . ." After the proposal, we dropped the term "stakeholder" in favor of "participants" in the project.

81. Amanda H. Lynch, principal investigator, and coprincipal investigators Ronald D. Brunner, Judith A. Curry, James A. Maslanik, and James P. Syvitski, *An Integrated Assessment of the Impacts of Climate Variability on the Alaskan North Slope Coastal Region* (proposal to the National Science Foundation, 2000), sec. 4. Other information in this paragraph can be found in the Project Summary.

82. The endorsements based on meetings in Barrow in August 2000 and attached to the proposal included Bart Ahsogeak of UIC Real Estate; Dr. Tom Albert, senior scientist at the North Slope Borough Department of Wildlife Management; C. Eugene Brower of the Barrow Whaling Captains Association; Jon Dunham, deputy director of the North Slope Borough Planning Department; Richard Glenn, president of the Barrow Arctic Science Consortium; Edward Itta, Barrow Commission of the Alaska Eskimo Whaling Commission; and Jim Vorderstrasse, mayor of the City of Barrow.

83. Lynch et al., *An Integrated Assessment of the Impacts of Climate Variability on the Alaskan North Slope Coastal Region*, sec. 4.

84. As reported in Chapter 1, the *Arctic Climate Impact Assessment*, despite excellent science and graphics, did not attract the attention of various people actively involved in storm damage reduction in Barrow. Even if the assessment had been brought to their attention earlier, we believe they would have perceived regional Arctic climate issues as beyond their local capabilities and interests.

85. Bob Harcharek, Ph.D. in sociology, in a signed review for the editor of *Arctic* of a manuscript published in slightly revised form as Ronald D. Brunner, Amanda H. Lynch, Jon C. Pardikes, Elizabeth N. Cassano, Leanne R. Lestak, and Jason M. Vogel, "An Arctic Disaster and Its Policy Implications," *Arctic* 57 (December 2004), 336–346. The manuscript was received by the journal in October 2002; publication was delayed by other articles included in the same special issue.

86. From an interview 28 May 2006 with Roger Ebert promoting the film *An Inconvenient Truth*. The interview was published in the *Chicago Sun-Times* and accessed at http://rogerebert.suntimes.com.

87. Rayner and Malone, "Ten Suggestions," 120. Emphasis in original.

88. Ibid., 134, 136. Emphasis in original.

89. Ibid., 111, 114. Emphasis in original.

90. Ibid., 113–114. Compare Richard J. T. Klein, E. Lisa F. Schipper, and Surage Dessai, "Integrating Mitigation and Adaptation into Climate and Development Policy: Three Research Questions," *Environ. Sci. Policy* 8 (2005), 579–588, at 581: "[A] daptation typically works best on the scale of an impacted system, which is regional at best, but mostly local."

91. Thomas L. Friedman, "The Power of Green," *New York Times Magazine* (15 April 2007).

92. Rayner and Malone, "Ten Suggestions," 130. However, the index at the end of the four volumes they edited has no entry for "politics," only one entry under "political actor," and three entries under "nation states" (as political actors).

93. Box 18.3 in the full report notes that "the term 'mainstreaming' has emerged to describe the integration of policies and measures that address climate change into development planning and ongoing sectoral decision-making." R. J. T. Klein, S. Huq, F. Denton, T. E. Downing, R. G. Richels, J. B. Robinson, and F. L. Toth, "Inter-relationships Between Adaptation and Mitigation," 768, in M. L. Parry, O. F. Canziani, J. P. Palutikof, P. J. van der Linden, and C. E. Hanson, Eds., *Climate Change 2007: Impacts, Adaptation and Vulnerability. Contribution of Working Group II to the Fourth Assessment Report of the Intergovernmental Panel on Climate Change,* (Cambridge, U.K.: Cambridge University Press, 2007), 745–777.

94. Rayner and Malone, "Ten Suggestions," 129. Emphasis in original.

95. Timothy O'Riordan, Chester L. Cooper, Andrew Jordan, Steve Rayner, Kenneth R. Richards, Paul Runci, and Shira Yoffe, "Institutional Frameworks for Political Action," in Steve Rayner and Elizabeth Malone, Eds. (1998), *Human Choice & Climate Change, Volume 1: The Societal Framework* (Columbus, OH: Battelle Press, 1998), 345–439, at 428.

96. David. W. Cash and Susanne C. Moser, "Linking Global and Local Scales: Designing Dynamic Assessment and Management Processes," *Global Environ. Change* 10 (2000), 109–120, at 117–118. "Experiment-based" in this context is better understood as action-based because intervention in an open, evolving system is neither controlled nor replicable in all of its aspects.

97. Felicity Barringer, "Paper Sets off A Debate on Environmentalism's Future," *New York Times* (6 February 2005), 1. We accessed Michael Shellenberger and Ted Nordhaus, *The Death of Environmentalism*, 37 pp., at http://www.thebreakthrough. org/images/Death_of_Environmentalism.pdf. Their argument is developed further in Ted Nordhaus and Michael Shellenberger, *Break Through: From the Death of Envirionmentalism to the Politics of Possibility* (New York, NY: Houghton Mifflin, 2007).

98. Shellenberger and Nordhaus, *The Death of Environmentalism*, 15.

99. Ibid., 9. Emphasis in original.

100. Ibid., 8.

101. Ibid., 10. A case in point is Bjorn Lomborg, *Cool It: The Skeptical Environmentalist's Guide to Global Warming* (New York, NY: Alfred A. Knopf, 2007), in which the author accepts climate change science as sound but rejects the policy implications supported in the epistemic community, including the Kyoto Protocol.

102. Ibid., 26. Shellenberger and Nordhaus used the term "special interest" in approximately the same sense as we used it in Ch. 1.

103. Ibid., 26.

104. Ibid.

105. Ibid., 5.

106. For example, Friedman's green ideology supports public and private investments in the development of cleaner and more efficient energy technologies, which are expected to reduce greenhouse gas emissions, create new jobs, and reduce dependence on oil imports that finance terrorists and authoritarianism abroad. See Friedman, "The Power of Green." See also Joshua W. Busby, "Climate Change and National Security: An Agenda for Action," Council on Foreign Relations Special Report, CRS No. 32 (November 2007).

107. Accessed 2 September 2008 at http://www.theclimategroup.org/index.php/about_us/.

108. Steve Rayner and Elizabeth L. Malone, "Social Science Insights into Climate Change," Steve Rayner and Elizabeth L. Malone, Eds., *Human Choice & Climate Change, Volume 4: What Have We Learned?* (Columbus, OH: Battelle Press, 1998), 71–107, at 73.

109. Our search in June 2008 turned up 934 articles on "climate change" or "global change" or "global warming" in *Nature* between 1910 and 2008, although more than one or two a year did not appear until 1985. *Nature* published 75 articles on these topics in 2007 alone. *Science* published 867 articles on these topics between 1968 and 2008, and 96 articles in 2007 alone.

110. Dennis Wenger, "Hazards and Disasters Research: How Would the Past 40 Years Rate?" *Natural Hazards Observer* XXXI (September 2006), 1–3, at 2.

111. Drawing on the results of their survey of assessment participants and their own knowledge, the team prepared discussion notes for a workshop it convened in Washington in April 2004. Workshop invitees included assessment participants, potential users of assessments, and domain or policy experts not involved in the assessment. See M. Granger Morgan et al., "Learning from the U.S. National Assessment of Climate Change Impacts," *Environ. Sci. Technol. 39* (2005), 9023–9032.

112. Ibid., 9031.

113. Ronald D. Brunner, "Policy and Global Change Research: A Modest Proposal," *Climatic Change 32* (February 1996), 121–147, sec. 2.

114. Morgan et al., "Learning from the U.S. National Assessment of Climate Change Impacts," 9028.

115. Rayner and Malone, "Ten Suggestions for Policymakers," 126.

116. Cash and Moser, "Linking Global and Local Scales: Designing Dynamic Assessment and Management Processes," 118.

117. See, respectively, Harold D. Lasswell, "Technique of Decision Seminars," *Midwest J. Political Sci.* IV (August 1960), 213–236, and Ronald D. Brunner, Christine H. Colburn, Christina M. Cromley, Roberta A. Klein, and Elizabeth A. Olson, *Finding Common Ground: Governance and Natural Resources in the American West* (New Haven, CT: Yale University Press, 2002).

118. For advice on communicating with mass publics on climate change mitigation see Moser and Dilling, "Making Climate Hot," which is based on a similar concept of communication. More generally, see Chip Heath and Dan Heath, *Made To Stick: Why Some Ideas Survive and Other Die* (New York, NY: Random House, 2007).

119. Herbert A. Simon, *The Sciences of the Artificial,* 3rd ed. (Cambridge, MA: MIT Press, 1996), 114–115.

120. Simon's concept of the "inner environment" includes constraints on cognitive capacity and excludes goals, but otherwise it is approximately equivalent to what we have called participants' "perspectives" on a situation and to their internalized "predispositions" to act in a situation. The other component of an act is the external "environment" or Simon's "outer environment."

121. Simon, *The Sciences of the Artificial,* 111.

122. Ibid., 114–115.

123. Ibid., xi.

124. Herbert A. Simon, *Reason in Human Affairs* (Stanford, CA: Stanford University Press, 1983), 22.

125. Simon, *The Sciences of the Artificial,* 81. However, p. 66 notes evidence that "information can be added in a second or two to locations (variable places) in images that are already present in an expert's long-term memory. Such images are called retrieval structures or templates." If so, the difference between a second or two and eight seconds rewards reinforcement of structures or templates in long-term memory and penalizes their expansion. Retrievals from long-term to short-term memory are measured in fractions of a second.

126. Simon, *Reason in Human Affairs,* 16. The Olympian model is also called "Subjective Expected Utility" or SEU. For a fuller description and critique, see 12–17.

127. Ibid., 20.

128. Simon, *The Sciences of the Artificial,* 25–27. Emphasis added.

129. Simon, *Reason in Human Affairs,* 23.

130. Ibid., 19–20.

131. Ibid., 19.

132. Ibid., 20–22.

133. Simon, *The Sciences of the Artificial,* 128.

134. Simon, *Reason in Human Affairs,* 78.

135. Simon, *The Sciences of the Artificial,* 30. Emphasis in original.

136. Ibid.

137. Ibid., 35.

138. Ibid., 147–148.

139. Ibid., 149.

140. Ibid.

141. Compare Lasswell, *A Pre-View of Policy Sciences*, 56: "By definition a problem is a perceived discrepancy between goals and an actual or anticipated state of affairs."

142. Simon, *The Sciences of the Artificial*, 129.

143. Ibid., 122. In connection with computer programs that solve problems, Simon described these associations as "productions" that are reminiscent of "if-then" rules in the internal models of agents in Holland's models of complex adaptive systems. "Each production is a process that consists of two parts—a set of tests or *conditions* and a set of *actions*. The actions contained in a production are executed whenever the conditions of that production are satisfied" (102). Emphasis in original.

144. Ibid., 123.

145. Ibid., 193–194.

146. Ibid., 8.

147. Ibid., 148.

148. Ibid., 196. Simon elaborated the point on p. 105: "Suppose a problem-solving system is able to solve a particular problem but does it inefficiently after a great deal of search. The path to a solution finally discovered, stripped of all extraneous branching in the search, could serve as a worked-out example" to inform subsequent searches.

149. Ibid., 124.

150. Simon, *Reason in Human Affairs*, 106.

151. Simon, *The Sciences of the Artificial*, 43–44. Compare Harold D. Lasswell, *Psychopathology and Politics* (Chicago, IL: University of Chicago Press, 1930 and 1977), 185–186: "Political acts are joint acts; they depend upon emotional bonds. . . . People who are emotionally bound together are not yet involved in a political movement. Politics begins when they achieve a symbolic definition of themselves in relation to demands upon the world."

152. Professor Douglas Holt, a marketing specialist at Oxford University, suggested something along these lines in a lecture at the University of Colorado, Boulder, on 26 September 2008. He estimated that local environmental activists are perhaps 10% of the American public. They are in a position to organize another 15% who are concerned about climate change but passive, and eventually another 40% who are vaguely concerned. To go beyond individual action as consumers, an incremental strategy to organize these publics requires, he said, "a collective identification around efficacious action" at the local level and "an aggregation pathway to the national level."

153. Simon, *The Sciences of the Artificial*, 144.

154. Martin Landau, "Redundancy, Rationality, and the Problem of Duplication and Overlap," *Public Administration Review* (July/August 1969), 346–358; all quotations is this paragraph except the last are from pp. 355–356.

155. Martin Landau and Russell Stout, Jr., "To Manage Is Not to Control: Or the Folly of Type II Errors," *Public Administration Review* (March/April 1979), 148–156, at 155.

156. D. T. Campbell, "Reforms as Experiments," in F. Caro, Ed., *Readings in Evaluation Research* (New York, NY: Russell Sage Foundation, 1969), Ch. 18, 233–261, at 234. An earlier version was published in *Amer. Psych.* 24 (April 1969), 409–429. A related critique of scientific management, Martin Landau, "On the Concept of a Self-Correcting Organization," *Public Administration Review* (November/December 1973), 533–542, argued that organizations aspiring to scientific management are not scientific enough; they do not provide adequately for error correction, or feedback.

157. Campbell, "Reforms as Experiments," 234–235.

158. Ibid., 236.

159. Ibid., 233.

160. Ibid., 255.

161. Lasswell, *Psychopathology and Politics*, 188.

162. Lasswell, *A Pre-View of Policy Sciences*, 62-63. Emphasis in original.

163. Rayner and Malone, "Ten Suggestions," 127.

164. Saul D. Alinsky, *Rules for Radicals: A Practical Primer for Realistic Radicals* (New York, NY: Random House, 1971), 105. Emphasis in original. Alinsky's politics are almost exclusively class based, but his rules are springboards for insight into promotional politics generally. His rules were used, e.g., to train community organizers for Barack Obama's election campaign beginning in 2007.

165. Ibid.

166. Ibid., 106.

167. Ibid., 81. Alinsky also added to what we have found in the climate-related literature. For example, pp. 91–92: "Another maxim in effective communication is that people have to make their own decisions." A community organizer "will not ever seem to tell the community what to do; instead, he will use loaded questions" in the spirit of Socratic method. For example, "His response to questions about what *he* [the organizer] thinks becomes a non-directive counterquestion, 'What do you think?' His job becomes one of weaning the group away from any dependency upon him. Then his job is done."

168. On the amount of power, consider Harold D. Lasswell and Abraham Kaplan, *Power and Society: A Framework for Political Inquiry* (New Haven, CT: Yale University Press, 1950), 77. "The weight of power is the degree of participation in the making of decisions; its scope consists of the values whose shaping and enjoyment are controlled; and the domain of power consists of the persons over whom power is exercised. All three enter into the notion of 'amount' of power."

169. Rayner and Malone, "Ten Suggestions," 132.

170. Steve Rayner and Elizabeth L. Malone, "Why Study Human Choice and Climate Change," in Steve Rayner and Elizabeth L. Malone, Eds., *Human Choice & Climate Change, Volume 4: What Have We Learned?* (Columbus, OHhio: Battelle Press, 1998), 1–31, at 30.

171. Rayner and Malone, "Ten Suggestions," 132. The chapter mentioned is Timothy O'Riordan et al., "Institutional Frameworks for Political Action," which observes, p. 427, "The FCCC provides an important symbolic framework expressive of worldwide concern about climate and about the persistent issues of global development that are inextricably bound up with it. However, the real business of responding to climate concerns may well be through smaller, often less formal, agreements among states, states and firms, and NGOs and communities."

172. Rayner and Malone, "Ten Suggestions," 120.

173. William C. Clark, "Environmental Globalization," in Joseph S. Nye Jr. and John D. Donahue, Eds., *Governance in a Globalizing World* (Washington, D.C.: Brookings Institution Press, 2000), Ch. 4, 86–108, at 101.

174. Ibid., 104. Compare Ian Burton, Livia Bizikova, Thea Dickinson, and Yvonne Howard, "Integrating Adaptation into Policy: Upscaling Evidence from Local to Global," *Climate Policy* 7 (2007), 371–376, at 373: "Much of the current thinking is still mired in the false structures of the Convention dating from 1992. Extending the life of the Kyoto Protocol is still considered the first priority."

175. Thomas Dietz, Elinor Ostrom, and Paul C. Stern, "The Struggle to Govern the Commons," *Science* 302 (12 December 2003), 1907–1912, at 1907. They refer to Garrett Hardin, "The Tragedy of the Commons," *Science* 162 (13 December 1968), 1243–1248.

176. Dietz, Ostrom, and Stern, "The Struggle to Govern the Commons," 1908.

177. Ibid., 1910.

178. Ibid. The two sentences preceding this one affirmed both field testing (experimentation) and the first part of the linear model (emphasis added): "Sustained research coupled to an explicit view of national and international policies can yield the scientific knowledge *necessary* to design appropriate adaptive institutions. Sound science is *necessary* for commons governance, but not sufficient."

179. Thomas J. Wilbanks, "Scale and Sustainability," *Climate Policy* 7 (2007), 278–287, at 282, including the points quoted below. The more theoretical parts of this article generally assume a climate change mitigation context. Section 6, 284–285, considers "Climate Change Adaptation as an Example."

180. Ibid., 282.

181. Ibid., 284. Consider also David. W. Cash, W. Neil Adger, Fikret Berkes, Po Garden, Louis Lebel, Per Olsson, Lowell Pritchard, and Oran Young, "Scale and Cross-Scale Dynamics: Governance and Information in a Multilevel World, *Ecology and Society* 11 (2006), 8–18, on three responses to existing problems of scale:

institutional interplay across levels and scales, comanagement, and boundary or bridging organizations.

182. Wilbanks, "Scale and Sustainability," 281. We believe the point is even more important for adaptation. Wilbanks, at 281, also envisions procedurally rational decision making: "If the results are not sufficient to address imbalances and associated impacts, the process iterates further."

183. Herman A. Karl, Lawrence E. Susskind, and Katherine H. Wallace, "A Dialogue, Not a Diatribe: Effective Integration of Science and Policy through Joint Fact Finding," *Environment* 49 (January/February 2007), 20–34, at 33.

184. Madeleine Heyward, "Equity and International Climate Change Negotiations: A Matter of Perspective," *Climate Policy* 7 (2007), 518–534, at 518.

185. Burton et al., "Integrating Adaptation into Policy: Upscaling Evidence from Local to Global," 375.

186. Simon, *The Sciences of the Artificial*, 183–184.

187. Ibid., 186.

188. Ibid., 198.

189. See, e.g., the need for more than "one-off" solutions to erosion and flooding problems in Native Alaska villages as reported in Ch. 3, and ICLEI's original aspiration and premise as reported in Ch. 2.

190. Clinton as quoted, inter alia, in two reviews of literature on scaling up: Jeffrey Bradach, "Going to Scale: The Challenge of Replicating Social Programs," *Stanford Innovation Review* (Spring 2003), 19–25, at 19; and Arntraud Hartmann and Johannes F. Linn (2008), *Scaling Up: A Framework and Lessons for Development Effectiveness from Literature and Practice*, Working Paper 5 (Washington, D.C.: Wolfensohn Center for Development, Brookings Institution, October 2008).

191. Bradach, "Going to Scale," 19. Emphasis in original.

192. Ibid., 23.

193. Hartmann and Linn, *Scaling Up*, 36.

194. Everett M. Rogers, "Centralized and Decentralized Diffusion Systems," in his *Diffusion of Innovations*, 4th ed. (New York, NY: Free Press, 1995), 364–369, at 364.

195. Ibid., 363–364.

196. Etienne C. Wenger and William M. Snyder, "Communities of Practice: The Organizational Frontier," *Harvard Business School* (January–February 2000), 139–145, at 139, 142. See also Etienne C. Wenger, *Communities of Practice: Learning, Meaning, and Identity* (Cambridge, U.K.: Cambridge University Press, 1998). Recall Simon on the significance of such group identifications and commitments.

197. Wenger and Snyder, "Communities of Practice: The Organizational Frontier," 141.

198. Ibid., 142.

199. Ibid., 143.

200. Ibid., 140.

201. Ibid., 145.

202. Anne-Marie Slaughter, "America's Edge: Power in the Networked Century," *Foreign Affairs* (January/February 2009), 94–113, at, respectively, 94, 99–100, 112, and 98.

203. Landau, "Redundancy, Rationality, and the Problem of Duplication and Overlap," 356.

204. For example, on the Los Angeles Green Building Ordinance enacted in 2008, see Ashley Lowe, Josh Foster, and Steve Winkelman, *Ask the Climate Question: Adapting to Climate Change Impacts in Urban Regions* (Washington, D.C.: Center for Clean Air Policy, June 2009), 17. Similarly, on the multiple functions of widespread urban forestry initiatives see ibid., 12–13.

205. Landau, "Redundancy, Rationality, and the Problem of Duplication and Overlap," 356.

206. Ibid., 348.

207. Ibid., 352.

208. Ibid., 351.

209. Ibid., 354.

210. Robert A. Dahl, *The New American Political (Dis)order* (Berkeley, CA: Institute of Governmental Studies, University of California, 1994), 5; see also pp. 1–2. For more recent diagnoses and proposed reforms by former Senators, see David Boren, *A Letter to America* (Norman, OK.: University of Oklahoma Press, 2008), and Ernest F. Holling and Kirk Victor, *Making Government Work* (Columbia, SC: University of South Carolina Press, 2008).

211. Landau, "Redundancy, Rationality, and the Problem of Duplication and Overlap," 352.

212. Ibid., 356, anticipated that "large-scale organizations function as self-organizing systems and tend to develop their own parallel circuits: not the least of which is the transformation of such 'residual' parts as 'informal groups' into constructive redundancies."

213. Thomas L. Friedman, "Flush with Energy," *New York Times* (10 August 2008).

214. This should be distinguished from the derivation of predictions or other conclusions about the particular case from general propositions. For critiques of derivation that apply here, see "A Note on Derivations," in Lasswell and McDougal, *Jurisprudence for a Free Society*, 759–786; "Methodology of Morals," in Kaplan, *American Ethics and Public Policy*, 90–101; and earlier in the text of this chapter, Simon's critique of the Olympian or SEU model of rationality, and Kuhn's observations on the function of Newton's second law in problem solving.

215. Lasswell, Lerner, and Pool, "Political Symbols," Ch. 1, 1–25, at 4. Emphasis in original. Of course these writers are not responsible for our selection and use of these propositions. On the nature and functions of political myth, see same. Compare Brunner, "A Paradigm for Practice," 144: "A paradigm is a myth specialized to

explaining and justifying scientific practices and claims of scientific knowledge, just as an ideology is a myth specialized to explaining and justifying political practices and demands."

216. Lasswell and Kaplan, *Power and Society*, 61–62, e.g. What might be dismissed as unnecessary jargon is better understood as functional terms defined in relation to other terms in a comprehensive framework designed for contextual inquiry. In any case, translations across different frames of reference are sometimes necessary and always possible.

217. Lasswell, Lerner, and Pool, "Political Symbols," 4. Emphasis in original.

218. On the need for empirical inquiry, consider the response of a lobbyist on Capitol Hill when asked "what sort of understanding of Washington is essential" for his work. "You got to understand what motivates the politician. Dummies, even in this town, think that politicians just want to be reelected." Then he spelled out his understanding of a member of Congress. Charles Walker, as quoted in Elizabeth Drew, "A Reporter at Large: Charlie," *The New Yorker* (9 January 1978), 32f.

219. See Mary Robinson, "Climate Change Is an Issue of Human Rights," *The Independent (London)* (10 December 2008). See also the International Council on Human Rights Policy, *Climate Change and Human Rights. A Rough Guide* (Geneva, 2008).

220. Here we apply a proposition from Lasswell and Kaplan, *Power and Society*, 113: "Propaganda in accord with predispositions strengthens them; propaganda counter to predispositions weakens them only if supported by factors other than propaganda."

221. Lasswell, Lerner, and Pool, "Political Symbols," 5. Emphasis in original.

222. Ibid., 5, 6. Emphasis in original.

223. Karl et al., "A Dialogue, Not a Diatribe," 24.

224. Compare Lasswell, Lerner, and Poole, "Political Symbols," 10. Emphasis in original. *"Any general symbol may be elaborated in support of any specific proposition; any general symbol may be applied or nonapplied in practice. . . . During short periods of time for specific groups, the linking of general with specific symbols and of symbols with overt actions may be stable."*

225. McDougal, Lasswell, and Reisman, *International Law Essays*, 201. The authors continue, "The core test of constitutive and public order decision at any level of interaction is its immediate and prospective contribution to the realization of human dignity in a world commonwealth, sufficiently strong to protect the common interest and sufficiently flexible to permit the widest range of diversity to flourish."

226. Ibid., 209.

227. Ibid., 211.

228. Robert A. Dahl, *After the Revolution?* (New Haven, CT: Yale University Press, 1970), 59.

229. Ibid., 93. The Chinese Box is a useful simile suggesting a hierarchy of communities up to the most inclusive global community. However, the implication that each smaller box is wholly contained within a larger one is misleading. On the

problems caused by eliminating overlaps, see Christopher Alexander, "A City Is Not A Tree," *Architectural Forum* 122 (April and May 1965), 58–62 and 58–61,

230. Ibid., 102.

231. From the assembly's Web page on subsidiarity, http://www.aer.eu/en/main-issues/subsidiarity.html. The assembly is an independent network of more than 270 regions in 33 countries and 13 interregional organizations in Europe.

Chapter 5

1. On expansion and contraction dynamics in symbolic politics, see Harold D. Lasswell and Abraham Kaplan, *Power and Society: A Framework for Political Inquiry* (New Haven, CT: Yale University Press), 1950, 104–107. See also E. E. Schattschneider, "The Contagiousness of Conflict," in *The Semisovereign People: A Realist's View of Democracy in America* (Hinsdale, IL: The Dryden Press, 1975), Ch. 1, 1–19.

2. As another example of issue expansion, consider the "Epilogue: The Fundamental Rationale," in *Our Changing Planet: The FY 1998 U.S. Global Change Research Program*, 76. Emphasis in original: "Almost a decade ago, the first edition of *Our Changing Planet* ended with the following insight: '*In the coming decades, global change may well represent the most significant societal, environmental, and economic challenges facing this nation and the world. The national goal of developing a predictive understanding of global change is, in its truest sense, science in the service of mankind.*' The fundamental rationale for the USGCRP articulated in 1989 is the same now as it was then."

3. Models for intensive inquiry in support of action can be found in most of these examples, but also include RISA, the U.S. National Assessment, and the Tsho Rolpa case.

4. The anonymous reviewer commented on the first two chapters of an earlier manuscript that became this book. We have argued that every community is unique when its many relevant factors are considered comprehensively. This diversity continues to frustrate national and international solutions that presume one size fits all, as recognized by the IPCC's Response Strategies Working Group in the First Assessment Report in 1990. As another example of issue contraction, consider the renaming of the Global Climate Change Research Program to the Climate Change Science Program in the Bush Administration.

5. Paul Hawken, *Blessed Unrest: How the Largest Social Movement in History Is Restoring Grace, Justice, and Beauty to the World* (New York, NY: Penguin Books, 2007), 18.

6. Victorian Farmers Federation's president Simon Ramsey said on 4 December 2008: "The Victorian Government should explore all water options before spending millions of dollars on projects which divide rural communities and may not solve Victoria's water crisis. . . . The Government has explored scores of options for saving

water, from small scale suggestions like the use of water efficient shower heads, to medium scale proposals like last week's $5.4 million advertising campaign to minimise personal water use, to massive undertakings such as the North South Pipeline and the Wonthaggi Desalination Plant—both rejected by the rural communities they will affect." (ABC News Opinion Page, accessible from http://www.abc.net.au/news/stories/2008/12/04/2437561.htm).

7. For example, to address problems of coordination and continuity in veterans care, the report of the President's Commission on Care for America's Returning Wounded Warriors in July 2007 "recommended creating a 'recovery plan' for seriously injured military personnel and assigning one coordinator for each patient and their family to help them navigate the process of recovering and returning to duty or retiring from active service." Jim Rutenberg and David S. Cloud, "Bush Panel Seeks Upgrade in Military Care," *New York Times* (26 July 2007). On direction centers, see Garry D. Brewer and J. S. Kakalik, *Handicapped Children: Strategies for Improving Services* (New York, NY: McGraw-Hill, 1979). A more recent example is Centrelink in Australia, "an ambitious project that draws together under one roof a variety of social services from eight different federal departments as well as from various state and territorial governments. The goal is to offer one stop shopping across a variety of services for citizens." Centrelink is one example of "Joined-Up Government" mentioned in Stephen Goldsmith and William D. Eggers, *Governing by Network: The New Shape of the Public Sector* (Washington, D.C.: Brookings Institution, 2004), 15–17.

8. See Victor F. Ridgeway, "Dysfunctional Consequences of Performance Measures," *Admin. Sci. Quart.* 1 (1956), 240–247; Nancy Cochran, "Grandma Moses and the 'Corruption of Data,'" *Evaluation Quart.* 2 (1978), 363–373; Nancy Cochran, "Society as Emergent and More than Rational: An Essay on the Inappropriateness of Program Evaluation," *Policy Sciences* 12 (1980), 113–129; Nancy Cochran, Andrew C. Gordon, and Merton S. Krause, "Proactive Records: Reflections on the Village Watchman," *Knowledge: Creation, Diffusion, Utilization* 2 (September 1980), 5–18; W. Edwards Deming, *Out of the Crisis* (Cambridge, MA: MIT Center for Advanced Engineering Study, 1986); and the Congressional Budget Office, *Using Performance Measures in the Federal Budget Process* (Washington, D.C.: Congressional Budget Office, 1993). For an introduction and supplements to this literature, see Ronald D. Brunner, "Context-Sensitive Monitoring and Evaluation for the World Bank," *Policy Sciences* 37 (2004), 103–136, especially 127–128. See also the work of Martin Landau cited in Ch. 4.

9. U.S. General Accounting Office, *Global Warming: Difficulties Assessing Countries' Progress Stabilizing Emissions of Greenhouse Gases*, GAO/RCED-96-188 (November 1996), 4, 5. A subsequent report in December 2003 suggested improvements in emission reporting based on four economically developed and four developing nations, but drew no general conclusions. U.S. General Accounting Office, *Climate Change: Selected Nations' Reports on Greenhouse Gas Emissions Varied in Their Adherence to Standards*, GAO-04-98 (December 2003).

10. U.S. Government Accountability Office, *Carbon Offsets: The U.S. Voluntary Market Is Growing but Quality Assurance Poses Challenges for Market Participants*, GAO-08-1048 (August 2008), Highlights.

11. PricewaterhouseCoopers, *Building Trust in Emissions Reporting: Global Trends in Emissions Trading Schemes* (February 2007), 5. PricewaterhouseCoopers is a global network of legally independent consulting firms. Evidently, the report promotes services for developing "a new Global Emissions Compliance Language." The authors are identified as Jeroen Kruijd, Arnoud Walrecht, Joris Laseur, Hans Schoolderman, and Richard Gledhill.

12. Susan Moran, "Carbon Detectives Are Tracking Gases in Colorado," *New York Times* (2 December 2008).

13. Quoted in Cochran et al., "Proactive Records," 6.

14. Congressional Budget Office, *Using Performance Measures in the Federal Budget Process* (Washington, D.C.: Congressional Budget Office, 1993).

15. The NOAA Coastal Services Center sponsored the workshop, "Toward a Community Resilience Index: Exploring the Conceptual Framework," held 7 July 2006 in Boulder, CO. These observations are from the first author's participation and notes at the workshop.

16. From a presentation by Josh Foster on the Urban Leaders Adaptation Initiative at the 27th Policy Sciences Annual Institute, Boulder, CO, 25 October 2008.

17. From conference Web site http://ec.europa.eu/environment/climat/emission/ets_compliance.htm, accessed 2 November 2008.

18. Christina M. Cromley, "Community-Based Forestry Goes to Washington," 221–267, at 235, in Ronald D. Brunner, Toddi A. Steelman, Lindy Coe-Juell, Christina M. Cromley, Christine M. Edwards, and Donna W. Tucker, *Adaptive Governance: Integrating Science, Policy, and Decision Making* (New York, NY: Columbia University Press, 2005).

19. Crockett Dumas, quoted in Toddi A. Steelman and Donna W. Tucker, "The Camino Real: To Care for the Land and Serve the People," 91–130, at 118, in Brunner et al., *Adaptive Governance*.

20. Bob Muth, quoted in Lindy Coe-Juell, "The 15-Mile Reach: Let the Fish Tell Us," 47–90, at 58, in Brunner et al., *Adaptive Governance*.

21. Kira Finkler, a staff member of the Senate Energy and Natural Resources Committee, quoted in Cromley, "Community-Based Forestry Goes to Washington," 237.

22. David Obey, ranking member of the House Appropriations Committee, as quoted in Brunner et al., *Adaptive Governance*, 293, from the *Congressional Record* (16 June 2004), H4259.

23. Paul Polak, *Out of Poverty: What Works When Traditional Approaches Fail* (San Francisco, CA: Berrett-Koehler, 2008), 40.

24. James C. Scott, *Seeing Like a State: How Certain Schemes to Improve the Human Condition Have Failed* (New Haven, CT: Yale University Press, 1998), 4, 3.

25. Ibid., 51.

26. Ibid., 12.

27. Ibid., 15. According to Scott, p. 14, scientific forestry "originally developed from about 1765 to 1800, largely in Prussia and Saxony. Eventually it would become the basis of forest management techniques in France, England, the United States, and throughout the Third World."

28. Ibid., 19.

29. Ibid., 23.

30. Ibid., 24.

31. Ibid., 47.

32. Ibid., 24.

33. Ibid., 51–52, from the Introduction by J. L. Heilbron to Tore Frangsmyr, J. L. Heilbron, and Robin E. Rider, Eds., *The Quantifying Spirit in the Eighteenth Century* (Berkeley, CA: University of California Press, 1991), 22–23.

34. Ibid., 4, 3.

35. Ibid., 5.

36. Ibid., 4.

37. Scott, *Seeing Like a State*, 4. As noted in Ch.1, this ideological faith survives in constructs such as "Managing Planet Earth" and "Spaceship Earth."

38. Ibid., 353–354.

39. Ibid., 343. Emphasis in original.

40. John Ralston Saul, *Voltaire's Bastards: The Dictatorship of Reason in the West* (New York, NY: Vintage, 1993), 26.

41. Ibid., 15.

42. Ibid., 111. Voltaire is the name François-Marie Arouet (21 November 1694–30 May 1778) adopted in 1718 both as a pen name and for daily use.

43. Ibid., 6.

44. Ibid., 7. Emphasis added. As Saul conceived it, on p. 112, "The heart of reason is logic, but Voltaire had imagined this logic well anchored in common sense." Common sense, on p. 54, was conceived as "careful emotion or prudence. . . ."

45. For example, consider Voltaire in *The Philosophical Dictionary*, from the edition published in French in 1843, and selected and translated by H. I. Woolf (New York, NY: Knopf, 1924). "The real vice of a civilized republic is in the Turkish fable of the dragon with many heads and the dragon with many tails. The many heads hurt each other, and the many tails obey a single head which wants to devour everything." Obtained from the Hanover Historical Texts Project, Hanover College Department of History.

46. Saul, 15. Compare George E. Brown, Jr., "Guest Comment: The Objectivity Crisis," *Amer. J. Phys.* 60 (September 1992), 779–781, at 779–780, who found it "troubling . . . that we have elevated science—or, more precisely, scientific knowledge—to a position of predominance over other types of cognition and experience; that we have, unconsciously and ironically, imbued science with more value than other types of understanding which are overtly and explicitly value based."

47. Saul, *Voltaire's Bastards*, 42.

48. Ibid., 47.

49. Ibid. Compare Herbert A. Simon, *Reason in Human Affairs* (Stanford, CA: Stanford University Press, 1983), 7–8: "[R]eason is wholly instrumental. It cannot tell us where to go; at best it can tell us how to get there. It is a gun for hire that can be employed in the service of whatever we have, good or bad."

50. As documented by William Church in his classic history, *Richelieu and Reason of State* (Princeton, NJ: Princeton University Press, 1972).

51. Saul, *Voltaire's Bastards*, 24; see also p. 48.

52. Ibid., 319.

53. Ibid., 25. See also p. 40: "Napoleon had ridden in on the back of reason, reorganized Europe in the name of reason and governed beneath the same principle. The subsequent effect was to bolster the rational approach, not to discourage it."

54. Ibid.

55. Ibid., 40.

56. Ibid., 20.

57. Ibid., 81.

58. Ibid., 82.

59. Ibid., 83–84.

60. ibid., 25. Saul warned, "The great danger, when looking at our society, is that what we see encourages us to become obsessed by individual personalities, thus mistaking the participants for the cause."

61. Ibid., 85.

62. Ibid., 27.

63. Ibid., 17.

64. Ibid., 21.

65. Ibid., 135.

66. Ibid.

67. Ibid., 36.

68. Ibid., 136.

69. Ibid., 137.

70. Frederick Winslow Taylor, *The Principles of Scientific Management* (New York, NY: Harper & Brothers, 1911), 130. Emphasis in original. Taylor, p. 115, provides a more elaborate summary that consolidates the second and third principles into one, and emphasizes more the subordination of workmen. At the end, p. 140, Taylor summarizes scientific management as a combination of dichotomous alternatives to what came before: "Science, not rule of thumb. Harmony, not discord. Cooperation, not individualism. Maximum output, in place of restricted output. The development of each man to his greatest efficiency and prosperity."

71. Ibid., 25.

72. Ibid., 36.

73. Ibid., 26.

74. Ibid., 10. Emphasis added.

75. Ibid., 83. Emphasis in orginal.

76. Scott, *Seeing Like a State,* 337.

77. Ibid., 348.

78. Ibid., 99.

79. Saul, *Voltaire's Bastards,* 120.

80. Ibid.

81. Judith A. Merkle, *Ideology and Management: The Legacy of the International Scientific Management Movement* (Berkeley, CA: University of California Press, 1980), 244.

82. Vaclav Havel, "The End of the Modern Era," *New York Times* (1 March 1992), E15.

83. Brunner et al., *Adaptive Governance,* and Ronald D. Brunner, Christine H. Colburn, Christina M. Cromley, Roberta A. Klein, and Elizabeth A. Olson, *Finding Common Ground: Governance and Natural Resources in the American West* (New Haven, CT: Yale University Press, 2002).

84. William Easterly, *The White Man's Burden: Why the West's Efforts to Aid the Rest Have Done So Much Ill and So Little Good* (New York, NY: Penguin Press, 2006), 11. See also, William Easterly, "Was Development Assistance a Mistake?" *Amer. Economic Rev.* 97 (May 2007), 328–332. Amartya Sen, "The Man Without A Plan," *Foreign Affairs* 85 (March/April 2006), 171–177, argues that Easterly's book, while oversimplified, "could serve as the basis for a reasoned critique of the formulaic thinking and policy triumphalism of some of the economic development literature. The wide-ranging and rich evidence—both anecdotal and statistical—that Easterly cites in his sharply presented arguments against grand designs of different kinds deserves serious consideration."

85. Easterly, *The White Man's Burden,* 5–6, specifies the ideal types more fully: "Planners announce good intentions but don't motivate anyone to carry them out; Searchers find things that work and get some reward. Planners raise expectations but take no responsibility for meeting them; Searchers accept responsibility for their actions. Planners determine what to supply; Searchers find out what is in demand. Planners apply global blueprints; Searchers adapt to local conditions. Planners at the top lack knowledge of the bottom; Searchers find out what the reality is at the bottom. Planners never hear whether the planned got what it needed; Searchers find out if the customer is satisfied. . . . A Planner believes outsiders know enough to impose solutions. A Searcher believes only insiders have enough knowledge to find solutions, and that most solutions must be homegrown."

86. Ibid., 14–15. See also Sanjay Reddy and Antoine Heuty, "Global Development Goals: The Folly of Technocratic Pretensions," *Develop. Policy Rev.* 26 (2008), 5–28.

87. Easterly, *The White Man's Burden,* 22.

88. See, e.g., Richard J. T. Klein, E. Lisa F. Schipper, and Surage Dessai, "Integrating Mitigation and Adaptation into Climate and Development Policy: Three

Research Questions," *Environ. Sci. Policy* 8 (2005), 579–588; and Sarah Bruch and John Robinson, "A Framework for Explaining the Links Between Capacity and Action in Response to Global Climate Change, *Climate Policy* 7 (2007), 304–316, in which response capacity "represents a broad pool of development-related resources that can be mobilized in the face of any risk" (304).

89. Paul L. Doughty, "Ending Serfdom in Peru: The Struggle for Land and Freedom in Vicos," in Dwight B. Heath, Ed., *Contemporary Cultures and Societies of Latin America: A Reader in the Social Anthropology of Middle and South America*, 3rd ed. (Prospect Heights, IL: Waveland Press, 2002), 225–243, at 227. The other major source for this account is Allan R. Holmberg, "The Role of Changing Values and Institutions of Vicos," in Henry F. Dobyns, Paul L. Doughty, and Harold D. Lasswell, Eds., *Peasants, Power, and Applied Social Change: Vicos as a Model* (Beverly Hills, CA: Sage Publications, 1971), 33–63. Holmberg added that "Records show that all protest movements of Vicos had been squelched by a coalition of landlords, the clergy, and the police. . . . The rule at Vicos was conformity to the status quo" (43).

90. Some of the implications of human dignity were specified by Harold D. Lasswell in *A Pre-View of Policy Sciences* (New York, NY: Elsevier, 1971), 42–43, using the eight value categories listed in Box 4.2 and the U.N. Universal Declaration of Human Rights.

91. Holmberg, "The Role of Power," 44–45.

92. Ibid., 44. The value categories listed in Box 4.2 were specified for the Cornell–Peru Project, with emphasis added: "The principal goals of this plan thus became the devolution of *power* to the community, the production and broad sharing of greater *wealth*, the introduction and diffusion of new and modern *skills*, the promotion of health and *well-being*, the enlargement of the status and role structure, and the formation of a modern system of *enlightenment* through schools and other media. It was hoped that . . . this focus would also have some modernizing effect on the . . . institutions specialized to *respect* [including status], *affection* (family and kinship), and *rectitude* (religion and ethics), sensitive areas of culture in which it is generally more hazardous to intervene directly."

93. Doughty, "Ending Serfdom in Peru," 228.

94. Holmberg, "The Role of Power," 45.

95. Ibid., 45–46.

96. Ibid., 55.

97. Ibid., 45.

98. Ibid. 46.

99. Ibid., 47.

100. Ibid., 48. Doughty, "Ending Serfdom in Peru," 230, concurred: "[B]y 1956, when the community began to operate the estate directly for itself, serfdom had been abolished and the *hacienda* lands became community lands, with work done under the emerging elected community council."

101. Holmberg, "The Role of Power," 48–49.

102. Ibid., 49.

103. The celebration in Vicos that day, along with Holmberg and the project, were the subjects of a half-hour documentary by CBS News, *So That Men Are Free*, broadcast in 1962 as part of the series *The Twentieth Century*. Unfortunately, the documentary showed Holmberg looking down on dancing Vicosinos and supplied a Cold War framing to the event.

104. Holmberg, "The Role of Power," 56.

105. Ibid., 57. Doughty, p. 234, reports, "In the summer of 1960, as two students watched with horror, a detachment of police passed through Vicos to enter the adjacent hacienda of Huapra where the *colonos* of that estate were trying to build a school like that of Vicos, over the objections of the landlords. The police confronted the serfs in a wheat field and shot the defenseless *colonos*, leaving three dead and five wounded. . . ."

106. Doughty, "Ending Serfdom in Peru," 231.

107. Dobyns, Doughty, and Lasswell, *Peasants, Power, and Applied Social Change*, 14, from the Introduction by the editors, who review the shifting politics at the national level as they affected the Vicos experience. On reactions to the Cornell–Peru project by anthropologists, see Paul J. Doughty, "Vicos: Success, Rejection, and Rediscovery of a Classic Program," in Elizabeth M. Eddy and William L. Partridge, Eds., *Applied Anthropology in America*, 2nd ed. (New York, NY: Columbia University Press, 1987), Ch. 19, 433–459.

108. Doughty, "Ending Serfdom in Peru," 242–243, systematically compared changes in Vicos by value category from the 1952 baseline to the project decade 1952–1962 and the postproject changes to 1997. This documented an impressive record of sustained development in all categories, beginning with power. The record was corroborated by the equivalent of a multiparty evaluation, the Vicosinos' reaction to Doughty's public slide show on life in Vicos during the project, p. 238: "For the older people, it was a chance to 'prove' to youthful skeptics how they lived in the old days; for everyone under 50 it was the first time they had actually seen what the hacienda had been like, and how much Vicos had changed. . . . From their reactions of delight and curiosity, it seemed clear enough that this community, like so many others, needs to treat the past as a living experience which might inform contemporary decisions."

109. Holmberg, "The Role of Power," 61–62.

110. Doughty, "Ending Serfdom in Peru," 236.

111. Holmberg, "The Role of Power," 61–62.

112. Harold D. Lasswell, "The Transferability of Vicos Strategy," in Dobyns, Doughty, and Lasswell, Eds., *Peasants, Power, and Applied Social Change*, 167–177, at 176. Lasswell (p. 167; emphasis in original) anticipated efforts to contract the significance of Vicos; "After all, the objection runs, only two thousand villagers were

involved, while the number of peasant hamlets and villages on the face of the earth is in the millions. Part of the reply is this: *The program as a whole is transferable with little change to many social and physical environments.*"

113. Patrick Breslin, "Thinking Outside Newton's Box: Metaphors for Grassroots Development," *Grassroots Development* 25/1 (2004), 2.

114. Ibid., 7.

115. Ibid., quoting complexity theorist Chris Langton.

116. Ibid., 8.

117. Ibid., 7.

118. Ibid., 3.

119. Ibid., 8.

120. Ibid., 9.

121. Ibid.

122. Easterly, *The White Man's Burden*, 382–383, which includes "guiding principles" in a six-part strategy.

123. Frank J. Penna, Monique Thormann, and J. Michael Finger, "The African Music Scheme," in J. Michael Finger and Philip Schuler, Eds., *Poor People's Knowledge: Promoting Intellectual Property in Developing Countries* (World Bank and Oxford University Press, 2004), 95–112, and other documents under Penna's name at www.WorldBank.org.

124. That their advocacy was unsuccessful is one reason why Reynolds told her story at the World Bank, at the request of President Karzai. Reynolds's story is retold and analyzed in Brunner, "Context-Sensitive Monitoring and Evaluation in the World Bank."

125. Polak, *Out of Poverty*, 40. The "only realistic path" becomes "one realistic path" when a broader base of experience is taken into account.

126. Ibid., 46. In Ch. 1, Polak presents "Twelve Steps to Practical Problem Solving." Doctors Without Borders and Engineers Without Borders have direct and sustained contacts with poor people on the ground and rely on them as participants. See also Tracy Kidder, *Mountains Beyond Mountains: The Quest of Dr. Paul Farmer, a Man Who Would Cure the World* (New York, NY: Random House, 2004), a bestseller that reports on Farmer's hospital to care for the rural poor in central Haiti; and Greg Mortenson and David Oliver Relin, *Three Cups of Tea: One Man's Mission to Promote Peace . . . One School at a Time* (New York, NY: Penguin Books, 2007), a bestseller on Mortenson's mission to build schools for girls in remote regions of Pakistan and Afghanistan.

127. Easterly, *The White Man's Burden*, 30, advises his readers not to expect "a Big Plan to reform foreign aid. The only Big Plan is to discontinue Big Plans. The only Big Answer is that there is no Big Answer."

128. Consider, e.g., Amitav Ghosh, "Death Comes Ashore," *New York Times* (10 May 2008), A19, on "Mauritius, a small Indian Ocean island in a zone that meteorologists call a 'cyclone factory.' The islanders have evolved a sophisticated system

of precautions, combining a network of cyclone shelters with education (including regular drills), a good early warning system and mandatory closings of businesses and schools when a storm threatens. It's been a remarkable success: Cyclone Gamede of 2007, a monster of a storm that set global meteorological records for rainfall, killed only two people on the island. . . . Mauritius is a country that has learned, through trial and experience, that early warnings are not enough—preparation also demands public education and political will. In an age when extreme weather events are clearly increasing in frequency, the world would do well to learn from it." Another partial exception is Peter Newman, Timothy Beatly, and Heather Boyer, *Resilient Cities: Responding to Peak Oil and Climate Change* (Washington, D.C.: Island Press, 2009).

129. See Ashley Lowe, Josh Foster, and Steve Winkelman, *Ask the Climate Question: Adapting to Climate Change Impacts in Urban Regions* (Washington, D.C.: Center for Clean Air Policy, June 2009). For example, it concluded in part (35), "Urban Leaders partners are proving that with the right leadership, organizational structure and information, local governments can make tangible progress in improving resiliency to the impacts of climate change. . . . While many 'Levers of Change' exist for local governments to advance adaptation efforts from the bottom up, the CCAP Urban Leaders partners and other local governments also need support from other players at the national and state level to achieve full resiliency."

130. Ibid., 2.

131. See the proposed American Recovery and Reinvestment Act announced on 15 January 2009 by Rep. David Obey, chairman of the House Committee on Appropriations.

132. Consider the call for both feedback and support from the bottom up in Barack Obama's speech announcing his candidacy in Springfield, IL, in February 2007: "But too many times, after the election is over, and the confetti is swept away, all those promises fade from memory, and the lobbyists and the special interests move in, and people turn away, disappointed as before, left to struggle on their own. That is why this campaign can't only be about me. It must be about us—it must be about what we can do together. This campaign must be the occasion, the vehicle, of your hopes, and your dreams. It will take your time, your energy, and your advice—to push us forward when we're doing right, and to let us know when we're not." "Illinois Sen. Barack Obama's Announcement Speech (As Prepared for Delivery)," *Associated Press* (10 February 2007).

133. On this possibility, see Noam Cohen, "The Wiki-Way to the Nomination," *New York Times* (8 June 2008); Lawrence Downes, "Obama's Call to Change: What Is Everyone Waiting For?" *New York Times* (10 November 2008); and Jim Rutenberg and Adam Nagourney, "Retooling a Grass-Roots Network to Serve a YouTube Presidency," *New York Times* (26 January 2009). On a related possibility, the Serve America Act, see Bruce Reed and John Bridgeland, "Volunteers to Save the Economy," *New York Times* (23 January 2009); and an editorial, "The Moment for National Service," *New York Times* (25 January 2009). Christopher Twarowski, "Localities Make a Pitch

for Energy-Efficient Projects," *Washington Post* (11 December 2008), B8, reported on lobbying for local public works projects that are environment friendly. "'We want the president-elect to know that local governments are uniquely positioned to put the president's vision and plan into action,' said Ken Brown, executive director of Climate Communities, a national coalition of local governments, a sponsor of the event." The event was cosponsored by ICLEI.

134. Hansen as quoted in Dana Milbank, "Burned Up about the Other Fossil Fuel," *Washington Post* (24 June 2008), A3.

135. A related article in the same volume by David Cash and Susanne Moser, "Linking Global and Local Scales: Designing Dynamic Assessment and Management Processes," *Global Environ. Change* 10 (2000), 109–120, was cited 59 times according to ISI. All of these numbers obtained from the ISI Web of Knowledge have uncertainties but should be compared to the thousands of citations of the "classic" climate policy articles.

136. David G. Victor, Joshua C. House, and Sarah Joy, "A Madisonian Approach to Climate Policy," *Science* 309 (16 September 2005), 1820–1821, plus a letter in response in *Science* 311 (20 January 2006), 335–336.

137. "Major world publications" includes hundreds of newspapers and magazines (e.g., *The New Yorker, Newsweek, The Economist*) worldwide as well as trade publications (e.g., *Waste News*) and BBC overseas broadcasts.

138. See "in the press" at www.energiakademiet.dk.

139. The raw counts from Lexis Nexis Academic were adjusted to avoid double counting of stories listed twice in the same publication. The search for stories mentioning Samsø was complicated by the system's acceptances of "Sam" for "Samsø or Samso." The results were reduced to a manageable number by adding "wind turbines" to the search terms, then checked one by one to delete those stories that mentioned "Sam" only.

140. Similarly, the U.S. Climate Change Action Plan, introduced by President Clinton and Vice-President Gore in October 1993, was relatively neglected in major U.S. newspapers. Only 14 of the stories in the *New York Times* and *Washington Post* that referred to "climate change," "global change," "greenhouse effect," or "global warming" also referred to the Action Plan from January 1992 through June 1997. The average for that period was about 20 stories per month, excluding the peak at more than 90 per month around the Earth Summit in Rio in June 1992. For more details see Ronald D. Brunner and Roberta Klein, "Harvesting Experience: A Reappraisal of the U.S. Climate Change Action Plan," *Policy Sciences* 32 (1999), 133–161, at 138–139 and Fig. 1. The Action Plan was formally established under the UNFCCC, but major parts of it predated the UNFCCC and operated effectively outside the regime's main agenda: The Action Plan relied on existing science and technology to assist people on the ground—in business, government, and the military—directly and voluntarily in reducing emissions and energy costs based on fossil fuel consumption.

141. Stephen Skowronek, *Building a New American State: The Expansion of National Administrative Capacities, 1877–1920* (Cambridge, U.K.: Cambridge University Press, 1982), 46.

142. Ibid., 166. Skowronek noted the attempt to rise above politics: "By transforming ideological conflicts into matters of expertise and efficiency, bureaucrats promised to reconcile the polity with the economy and to stem the tide of social disintegration." Schools of public administration and think tanks like the Brookings Institution came later, to institutionalize the search for improvements in this pattern of governance.

143. There is nothing unusual about this. Compare Harold D. Lasswell, *A Pre-View of Policy Sciences* (New York, NY: Elsevier, 1971), 65: "The tendency of every group is to narrow its frame of reference, chiefly because quick and easy individual payoffs so frequently come by adding minor and rather obvious amplifications to the field of common reference. Hence the focus of attention needs to be redirected to neglected areas."

144. David M. Herszenhorn, "After Verbal Fire, Senate Effectively Kills Climate Change Bill," *New York Times* (7 June 2008), 12.

145. Elizabeth Rosenthal, "Amid a Hopeful Mood, U.N. Talks Set Countries on Path Toward a Global Climate Treaty," *New York Times* (13 December 2008), A7.

146. John M. Broder and Andrew C. Revkin, "Hard Task for New Team on Energy and Climate," *New York Times* (16 December 2008), A24. Rosenthal, "Amid a Hopeful Mood," reported from Poznan that "Perhaps contributing most to the hopeful mood were signs from high-level United States officials that the incoming administration of President-elect Barack Obama would be ready to hit the ground running with a new United States climate policy."

147. Ibid.

148. Quoted in Charles J. Hanley, "U.N.: Financial Chills are Ill Wind for Climate," *Associated Press Financial Wire* (9 October 2008). See also Vanessa Gera, "U.N.: Financial Crisis a Burden on Climate Change," *Associated Press Financial Wire* (27 November 2008).

149. Hanley, "U.N.: Financial Chills are Ill Wind for Climate."

150. Andrew C. Revkin, "Years Later, Climatologist Renews His Call for Action," *New York Times* (23 June 2008), A18.

151. Thomas L. Friedman, "Win, Win, Win, Win, Win . . . ," *New York Times* (28 December 2008), WK8.

152. Bob Inglis and Arthur B. Laffer, "An Emissions Plan Conservatives Could Warm To," *New York Times* (28 December 2008), WK10.

153. John M. Broder, "In Obama's Team, 2 Camps on Climate," *New York Times* (3 January 2009), A10.

154. John M. Broder and Andrew C. Revkin, "Hard Task for New Team on Energy and Climate," A24.

155. Bjorn Lomborg, *Cool It: The Skeptical Environmentalist's Guide to Global Warming* (New York, NY: Alfred A. Knopf, 2007), 121.

156. Ibid., 117.

157. Andrew C. Revkin, "A Shift in the Debate Over Global Warming," *New York Times* (6 April 2008), WK3.

158. Quoted in ibid.

159. Ibid.

160. Quoted in ibid.

161. The passive house uses the amount of energy consumed by a hair dryer to meet all of its needs for heat and hot water, at a construction cost that is 5% to 7% more than conventional houses in Germany. For more details, see Elizabeth Rosenthal, "No Furnaces but Heat Aplenty in 'Passive Houses,'" *New York Times* (27 December 2008), A1.

162. Søren Hermansen quoted in Robin McKie, "Isle of Plenty," *The Observer (London) Magazine* (21 September 2008), 30. Hermansen continued: "For example, Shell [the international oil company] heard about what we were doing and asked to be involved—but only on condition they ended up owning the turbines. We told them to go away. We are a nation of farmers, of course. We believe in self-sufficiency."

163. Lomborg, *Cool It*, 122.

164. Gregg Marland, "Could We/Should We Engineer the Earth's Climate?" an editorial introducing the special issue of *Climatic Change* 33 (1996), 275–278, at 275.

165. See the Environment Pollution Panel, President's Science Advisory Committee, *Restoring the Quality of our Environment*, Report to President L. B. Johnson (November 1965); and the Panel on Policy Implications of Greenhouse Warming, Committee on Science, Engineering, and Public Policy, National Academy of Sciences, National Academy of Engineering and the Institute of Medicine, *Policy Implications of Greenhouse Warming: Mitigation, Adaptation, and the Science Base* (Washington, D.C.: National Academy Press, 1992). For more on the history, see James R. Fleming, "The Climate Engineers," *Wilson Quarterly* (Spring 2006), 46–60.

166. Stephen Schneider, "Geoengineering: Could—or Should—We Do It?" *Climatic Change* 33 (1996), 291–302, at 296.

167. More than half the articles in the peer-reviewed literature using the term "geoengineering," coined in the mid-70s by Cesare Marchetti, have been published since 2006.

168. Paul J. Crutzen, "Albedo Enhancement By Stratospheric Sulfur Injections: A Contribution To Resolve A Policy Dilemma? An Editorial Essay," *Climatic Change* 77 (2006), 211–219, at 212.

169. John P. Holdren, "Presidential Address: Science and Technology for Sustainable Well-Being," *Science* 319 (25 January 2008) 424–434, at 431.

170. From a commentary by Peter G. Brewer, "Evaluating a Technological Fix for Climate," *Proc. Natl. Acad. Sci.* 104 (12 June 2007), 9915–9916, at 9916.

171. Edward Teller, Lowell Wood, and Roderick Hyde, "Global Warming and Ice Ages: I. Prospects For Physics-Based Modulation Of Global Change," UCRL-JC-128715, preprint (15 August 1997), prepared for invited presentation at the 22nd International Seminar on Planetary Emergencies, Erice (Sicily), Italy, 20–23 August 1997, 16, with emphasis added.

172. James R. Fleming, "The Climate Engineers," 46–47. Fleming continued: "Wood advanced several ideas to 'fix' the earth's climate, including building up Arctic sea ice to make it function like a planetary air conditioner to 'suck heat in from the midlatitude heat bath.' A 'surprisingly practical' way of achieving this, he said, would be to use large artillery pieces to shoot as much as a million tons of highly reflective sulfate aerosols or specially engineered nanoparticles into the Arctic stratosphere to deflect the sun's rays. . . . Far-fetched as Wood's ideas may sound, his weren't the only Rube Goldberg proposals aired at the meeting."

173. Fleming as quoted in the transcript of an interview with presenter and producer Wendy Carlisle, "The Climate Engineers," on *Background Briefing* (6 April 2008), an ABC Radio National program broadcast in Australia. The transcript is available at http://www.abc.net.au/rn/backgroundbriefing/stories/2008/2204410.htm.

174. Ralph J. Cicerone, "An Editorial Comment: Geoengineering: Encouraging Research and Overseeing Implementation," *Climatic Change* 77 (August 2006), 221–226, at 221. In Carlisle, "The Climate Engineers," Wally Broecker mentioned that he had coauthored a paper on climate model experiments with one approach to geoengineering in the mid-1980s. Then "we sent it around to prominent people in the field, and they said, 'By all means don't publish this; the world is not ready for it.' So we just put it on the shelf."

175. Daniel Bodansky, "May We Engineer the Climate?" *Climatic Change* (1996), 308–321, at 310 and 320, respectively. See also Stephen Schneider, "Geoengineering: Could We or Should We Make It Work?" *Philos. Trans. Roy. Soc. London, A—Mathematical Physical and Engineering Sciences* 366 (November 2008), 3843–3862.

176. Alan Robock, "20 Reasons Why Geo-Engineering May Be a Bad Idea." *Bull. Atomic Sci.* 64 (May/June, 2008), 14–18, at 17. The reasons were first written down at NASA's conference "Managing Solar Radiation." Teller and colleagues in "Global Warming and Ice Ages," note 5, anticipated concerns like Robock's and attempted to counter them: "While it may be argued that uncertainties in the accuracy and fidelity of modeling tools are sufficiently great that they should not be relied upon to forecast reliably the effects of enhanced sunlight scattering on the climate . . . it is these same modeling tools—not the presently quite ambiguous observations of the actual climate—that are considered sufficiently robust to motivate the present level of concern about man-made climate change."

177. Keith as quoted in Carlisle, "The Climate Engineers." Ambivalence of a different kind came up in a poll of 80 leading international climate scientists at the end of 2008. About 54% "agreed that the failure to curb emissions of CO_2 . . . has created

the need for an emergency 'plan B' involving research, development and possible implementation of a worldwide geoengineering strategy." About 35% "disagreed, arguing that it would distract from the main objective of cutting CO_2 emissions, with the remaining 11 per cent saying that they did not know whether a geoengineering strategy is needed or not." For more details, see Steve Connor and Chris Green, "Climate Scientists: It's Time for 'Plan B,'" *The Independent (London)* (2 January 2009).

178. Al Gore, "The Climate for Change," *New York Times* (9 November 2008), WK10, from which we quote extensively below. At 1423 words, the op-ed is almost twice the normal length. The length still forced Gore to leave out many relevant details but also to select what was more important to express his perspective.

179. Amory B. Lovins, "Energy Strategy: The Road Not Taken?" *Foreign Affairs* 55 (October 1976), 65–96, at 65. Lovins developed his ideas in subsequent books beginning with *Soft Energy Paths: Toward a Durable Peace* (Cambridge, MA.: Ballinger Publishing, 1977).

180. For example, see John M. Broder, "Geography Is Dividing Democrats Over Energy," *New York Times* (27 January 2009), A1.

181. Lovins, "Energy Strategy," 74.

182. For example, we quoted the IPCC's *Second Assessment Synthesis*, secs. 5.2 and 5.3, in Ch.1: "Significant reductions in net greenhouse gas emissions are technically possible and can be economically feasible. . . . The degree to which [this] technical potential and cost-effectiveness are realized is dependent on initiatives to counter lack of information and overcome cultural, institutional, legal, financial and economic barriers which can hinder diffusion of technology or behavioral changes." See also Elizabeth Rosenthal, "No Furnaces but Heat Aplenty in 'Passive Houses,'" on institutional barriers to diffusion of passive house technology in the United States.

183. Lovins, "Energy Strategy," 80–81. ERDA was the Energy Research and Development Administration, soon to be folded into the new Department of Energy.

INDEX

Revkin, Andrew, 306
Reynolds, Samantha, 293
Richelieu, Cardinal, 280
Rigor, Ignatius, 204–205
Rivlin, Alice, 202
Robinson, John, 195
Robock, Alan, 309
Rock, Howard, 120, 123, 127
Rogers, Everett, 248
Ruckelshaus, William, 21

Sachs, Jeffrey, 306–307
Sampson, Rick, 170
Samsø (Denmark) case study, 92–96, 229,
    236, 274
Saul, John Ralston, 5, 279–283
Schafer, Pedro, 119–120
Schalk, Marshall, 120, 123, 125, 127–128, 148
Schneider, Stephen, 9, 56, 308
Science, 221, 300
science of the concrete, 118, 178
sciences, historical, 197–211
Scientific American, 13
scientific management, 5, 8–12
    compared to adaptive governance, 19
    in the European Union, 57–61
    Intergovernmental Panel on Climate
        Change (IPCC), 41–52
    international decisions, 52–56
    International Geosphere-Biosphere
        Programme, 32–36
    origins of, 276–286
    in the United States, 61–67
    U.S. Global Change Research Program,
        36–41
Scott, James, 5, 277–278, 284
Seeing Like a State, 5, 20, 277–278, 283, 313
Shea, Eileen, 78
Sheehan, Glenn, 108, 111–112, 117, 174
Shellenberger, Michael, 220–221, 253
Simmonds, Samuel, 129
Simon, Herbert, 202–203, 223–228, 231, 246
Slaughter, Anne-Marie, 249–250
Smythe, William, 113
Snyder, William, 248

social responsibility, 209–211
Spencer, Robert, 113, 117
stabilization level, 3–4
Stamp, Sir Josiah, 272
Steffen, Will, 36
Stern, Paul, 196, 243
Stevens, Ted, 181, 238, 240, 264
Structure of Scientific Revolutions, The
    (Kuhn), 208
Suess, Hans, 7
Summaries for Policymakers (IPCC), 42–43,
    44, 46–48, 52, 219, 259

Taking Control (Hess), 128, 130, 131, 133, 142
tax, carbon, 305–306
Taylor, Frederick Winslow, 5, 283–284
Taylorism, 250, 283–284
Teague, Peter, 221
Tekmarine, 156, 157, 161–162, 166, 170, 177,
    179
Teller, Edward, 308
Tennekes, Hendrik, 13, 28, 209, 300
test of action, 206
Thwaites, John, 81, 83, 84, 85, 87, 194
Time, 301
Todhunter, Tracy, 92
Toronto (Canada) case study, 96–98
Tsho Rolpa case study (Nepal), 73–74,
    212–213, 236–237, 266

Udall, Stewart, 129, 130
uncertainty absorption, 29, 262
Ungar, Sheldon, 56
United Nations Climate Conference
    (Copenhagen), *vii*
United Nations Framework Convention on
    Climate Change (UNFCCC)
    and the common interest, 23, 25
    and decentralization of decision-
        making, 236, 257
    international decisions under, 52–53
    and limitations of centralized decision-
        making, 242, 251
    objective of, 2, 24, 302
    role of, 12